ACKNOWLEDGEMENTS

Grateful thanks are due to:

1 E. M. Nicholson, CB, lately Director-General of the Nature Conservancy, who first brought the work of the BSCS to my attention;
2 Dr Bentley Glass, Chairman of the BSCS Steering Committee, and Dr Arnold Grobman, First Director of the BSCS, who introduced me to the work of this project;
3 The Nature Conservancy, the University of Keele and the Staffordshire Education Authority for financial assistance which enabled me to carry out the research and development work which was an important preliminary to the adaptation;
4 The University of Stirling for financial assistance which made the work of adaptation possible;
5 The Director and Staff of the Biological Sciences Curriculum Study, Colorado, and through them all the teachers and biologists whose co-operative efforts produced the original 'Green Version'. Their names appear in the publications of the BSCS;
6 Kenneth Pinnock, Howard Jay and Patricia Winter of John Murray (Publishers) Ltd, to whom the adaptation team offer most sincere thanks for their guidance and help;
7 Finally we are particularly grateful for the forbearance and hard work of Mary Campbell who translated the team's handwriting into legible typescript, and of Doris Beedie who checked our proofs.

E. PERROTT
Director—British Adaptation

Thanks are also due to Ronald Stewart, University of Stirling, for permission to reproduce the cover photograph.

BIOLOGY: AN ENVIRONMENTAL APPROACH

Adapted from *High School Biology: Teacher's Guide* (the 'Green Version') by permission of Biological Sciences Curriculum Study, Boulder, Colorado

Adaptation team **Elizabeth Perrott** MSc, PhD, DipEd, FLS, FIBiol

Professor of Education, Visiting Professor, Department of Educational Research, University of Lancaster
Director of British BSCS Adaptation

Ian D. Campbell BSc, DipEd, MIBiol

Principal Biology Teacher, Madras College, St Andrews
(Formerly Research Fellow, Department of Education, University of Stirling)

David Hughes-Evans BSc, FLS, MIBiol

Lecturer and Head of Biology Section, Farnborough Technical College

TEACHERS' GUIDE

BIOLOGICAL SCIENCES CURRICULUM STUDY

JOHN MURRAY ALBEMARLE STREET LONDON

HIGH SCHOOL BIOLOGY: TEACHER'S GUIDE

Printed in Great Britain by Cox and Wyman Ltd, London, Fakenham and Reading

0 7195 2841 0

CONTENTS

Part One

Part Two Specific suggestions for the use of the texts and for laboratory work

KEY TO BOOKS IN THE SERIES

The following abbreviations are used in references to the five pupil's books:

WL = *The World of Life: the Biosphere*
DLT = *Diversity Among Living Things*
PLW = *Patterns in the Living World*
LO = *Looking into Organisms*
MHE = *Man and his Environment*

REFERENCES TO PAGES AND ILLUSTRATIONS

References preceded by the letter T are to this Teachers' Guide; other references are to the pupils' book concerned.

Part One

THE BIOLOGICAL SCIENCES CURRICULUM STUDY

INTRODUCTION

The Biological Sciences Curriculum Study was set up in 1959 by the American Institute of Biological Sciences, with financial support from the National Science Foundation totalling more than four million dollars, 'to seek the improvement of education in the biological sciences' at school level.

As a result, during three successive summers, 1960, 1961 and 1962, three parallel sets of course materials for American High School Biology were prepared by teams of writers working at summer writing conferences. The publications, which included texts, laboratory manuals and teachers' guides, were first released through commercial publishers in the United States in the autumn of 1963 (2nd editions, 1968).

THE GENERAL PATTERN

These BSCS course materials are unlike conventional treatments of high school biology in several ways:

A. Basic Emphases

All three versions of the BSCS course materials have significant unifying threads. This unity arises from the agreement of the writers that certain basic emphases should be woven into and through each of the three versions. These are:

1. Change of living things through time—evolution.
2. Diversity of the type and unity of pattern of living things.
3. Genetic continuity of life.
4. Complementarity of structure and function.
5. Biological roots of behaviour.
6. Complementarity of organisms and environment.
7. Regulation and homeostasis: the maintenance of life in the face of change.
8. Intellectual history of biological concepts.
9. Science as inquiry.

B. The Product of Team Work

They are the products of a well organized large-scale project in which there had been close co-operation, on a work party basis, between two main groups of people—on the one hand research biologists, university teachers, high school biology teachers and school science supervisors, and on the other, educationists with varied interests in science education. For instance, by this latter group, the position of biological education in the schools was surveyed and the problems pinpointed before the first writing conference (Hurd, 1961); school exercises in logical thinking called 'Invitations to Inquiry' were devised to test pupils' understanding of biological principles and concepts by discussion methods (Schwab, 1963); and special multi-choice tests for these courses were devised by a team of biologists working closely with a specialist testing agency.*

C. Large-scale Testing

This curriculum study also differed from others previously undertaken in the extensive and

* Psychological Corporation, 304E 45th Street, New York, N.Y. 10017.

thorough testing it has undergone in the schools. It was used in both large and small schools, in rural and urban areas, in every section of the United States. Some 1000 teachers, who received special courses of training, tested the materials on 165 000 pupils. During the first two years of testing, teachers sent back weekly reports on their classroom experience with BSCS courses and made suggestions for improvements and modifications. The materials were also tested against traditional programmes taught in control groups of schools.

D. Science as a Process of Inquiry

BSCS courses also differ markedly from traditional courses in the teaching methods employed, for the emphasis is on reasoning rather than recall, and away from laboratory and field work concentrated on verification, towards laboratory and field work involving real investigation. The writers aimed to teach science as a method of seeking answers, and to do this they stressed underlying concepts. Work is centred in the laboratory and field where real problems are explored, open-ended experiments and other materials being used as the media for conveying an understanding of science. Through emphasis on basic concepts and the illustration of such concepts in a variety of ways, the pupil is given practice in drawing generalizations, in seeking relationships and in finding his own answers; in other words, he is given experiences which aim to provide an understanding of the methods of science.

E. New Content

In addition, the different emphasis given in terms of content reflects modern biology more adequately than do conventional approaches. For example, where traditional texts emphasize the organ and tissue level of biological organization, BSCS texts place greater emphasis on other levels, viz. the molecular, cellular, individual, population, community and world biome.

F. Variety of Treatment

The production of three separate versions of the course materials emphasizes that there is more than one approach to the teaching of biology, and that it is for the teacher to choose the approach which is best suited to his own background and interests and to those of his pupils. For example: the Yellow Version (Moore *et al.*, 1968) puts its emphasis on the cell as the most elemental, independent structural and functional unit of living organisms, while the Blue Version (Welch *et al.*, 1968) puts main emphasis on molecular biology as the fundamental area of biology upon which all other biological knowledge is based. In the Green Version, on the other hand (Kolb *et al.*, 1968) a course is presented which gives primary importance to the ecological and behavioural aspects.

PURPOSE

The writers' primary purpose was to produce biology courses suitable for wide use in the average secondary school with average classes. The aim of the courses was to give pupils a basic understanding of science and of scientific processes, and, in doing so, to build scientific literacy in order to aid in preparing the pupil for responsible citizenship. They were prepared for the 10th Grade (i.e. for pupils of 15–16 years attending the comprehensive-type American high school), because most high school pupils take biology courses during this year in American school systems (see table).

THE RELEVANCE OF THE BSCS TO BRITISH EDUCATION

At first sight it seems that a course of study devised for 15–16-year-old pupils to be taught in one year on the basis of one hour per day or 5 to 6 hours per week has little relevance to the British system of education. However, on closer analysis one finds that our situation does not in fact differ so markedly. The results of a survey of 10 per cent of the schools in England and Wales carried out by the author and her co-workers in 1962 on behalf of the Nature Conservancy's Study Group on Field Biology and Education (Perrott *et al.*, 1963) showed that 70 per cent of secondary school pupils are taught integrated science and not biology as a separate subject between 11+ and 13 years. It is not until 14–15 years of age that biology, taught as a separate subject, is taken by the majority (over 70 per cent) of pupils. Another survey made by the senior author and her co-workers in 1967–68 revealed a similar pattern in Scottish secondary schools (Perrott, Martin and Campbell, 1969).

ENGLISH AND AMERICAN SCHOOL SYSTEMS

American Public Education			Pupils' chronological age (in years)		*English State Education*	
			5–6		Year 1	Primary
Elementary	1st Grade		6–7		Year 2	Primary
Elementary	2nd Grade		7–8		Year 3	Primary
Elementary	3rd Grade		8–9		Year 4	Primary
Elementary	4th Grade		9–10		Year 5	Primary
Elementary	5th Grade		10–11		Year 6	Primary
Elementary	6th Grade		11–12	General science courses	Year 7	Secondary (compulsory)
*Junior high	7th Grade	General science courses	12–13	General science courses	Year 8	Secondary (compulsory)
*Junior high	8th Grade	General science courses	13–14	General science courses	Year 9	Secondary (compulsory)
*Junior high	†9th Grade	General science courses	14–15	General science courses	Year 10	Secondary (compulsory)
Senior high	10th Grade	Biology	15–16	O-level	Year 11	Secondary (compulsory)
Senior high	11th Grade	Chemistry	16–17	A-level	Year 12	Secondary (optional)
Senior high	12th Grade	Physics	17–18	A-level	Year 13	Secondary (optional)

* In some American school systems general science is still an elective subject for only one year.
† Increasing numbers of American school systems are offering biology in the 9th Grade for able students. (This selected group takes chemistry in 10th Grade, physics in 11th Grade with advanced courses in chemistry physics or biology in 12th Grade.)

The situation is therefore not very different from that in the United States where integrated science is also commonly taught in the junior high school (see the table). The main difference is that the majority of pupils in England who study biology below the sixth-form level do so for two years instead of one year. But the time allocation during this two-year period is only equivalent to that allocated to biology in American high schools during one year.

These surveys also showed that very little time is spent on the teaching of ecology to pupils below sixth-form level in British schools, the average time being approximately $2\frac{1}{2}$ hours per annum, and that experimental or problem-solving work of the type devised for the BSCS Green Version course was undertaken by less than 10 per cent of the schools (Perrott *et al.*, 1963).

The feasibility of teaching this type of course in English secondary schools was tested by the author and her co-workers when an experimental programme was carried out, with 500 pupils from Staffordshire secondary schools, at the University of Keele in 1964. Lessons, laboratory work and field investigations, including many which were adaptations of the BSCS Green Version, were arranged. The topics chosen were taken from the GCE and other syllabuses followed by the schools and included such subjects as the relation of plants and animals to their natural environment, the interdependence of plant and animal life, and the formation and composition of the soil.

The aim of the programme was critically to assess the part which ecology can play in the learning of biology under normal school conditions at the pre-sixth-form level, and to estimate whether ecological investigations which are suitable for secondary school pupils at this stage are able to serve both the investigatory as well as the illustrative aims of science teaching.

The results of testing showed that the ecological approach can be effectively used to teach topics included in the GCE O-level syllabus and other terminal courses, and that such teaching can be integrated into the normal school programme of biology teaching as in BSCS courses. It is to be hoped that the raising of the school-leaving age to 16 years will provide even more opportunities for this type of course.

Accordingly, it was the second edition (1968) of the Green Version which was selected for adaptation. This was published in 1972 as five books (to be used together or singly) under the general title of *Biology: An Environmental Approach* (John Murray).

REFERENCES

HURD, P. *Biological Education in American Secondary Schools 1890–1960*. American Institute of Biological Sciences, Washington DC, 1961.

KOLB, C. H. *et al. High School Biology*. 2nd edition. Chicago: Rand McNally (distributed by John Murray), 1968.

MOORE, J. A. *et al. Biological Science: An Inquiry into Life*. 2nd edition. New York: Harcourt Brace & World (distributed by Hart Davies), 1968.

PERROTT, E. *et al. Science Out of Doors*. Longmans, 1963.

PERROTT, E. 'Biological Sciences Curriculum Study: its relevance to British education', *Science Teacher*, 1965, 9, pp. 14–17.

PERROTT, E., E. MARTIN, I. D. CAMPBELL. *Biological Sciences in Scottish Secondary Schools*. University of Stirling, 1969.

KLINCKMANN, E. *et al. Biology Teacher's Handbook*. 2nd edition. Wiley, 1970.

WELCH, C. A. *et al. Biological Science: Molecules to Man*. 2nd edition. Arnold, 1968.

BIOLOGY: AN ENVIRONMENTAL APPROACH

'The word "ecology" was proposed by Ernst Haeckel in 1870 to cover what he called "outer physiology". It is the point of view in biology that takes the individual organism as the primary unit of study, and is concerned with how these individuals are organized into populations, species, and communities; with what organisms do and how they do it.

'This contrasts with "inner physiology", the study of how the individual is constructed and how the parts work. Obviously the inside and outside of the organism are completely interdependent, and one cannot be understood without constant reference to the other. The division is arbitrary, but so are all of the ways in which biological subject matter might be split. We stress the outside rather than the inside on the assumption that this is more familiar and more easily understood. We believe, too, that it is more important for the citizen, who must participate in decisions about urban development, flood control, public health, conservation—always as a voter and sometimes as a member of the town council.

'For disorders of inner physiology the citizen should consult his physician. But there is no specialist for outer physiology, for disorders of the human biological community. Here each citizen shares responsibility, and biological knowledge is greatly needed for some kinds of decisions.'

Thus wrote Dr. Marston Bates, the first supervisor of the BSCS team entrusted with the development of the high school biology course that was to become known as the *Green Version*, of which *Biology: An Environmental Approach* is an adaptation. Sociological, educational, and biological events of the years that followed the introduction of the Green Version have given the statement even greater import.

GENERAL AIMS

Biology: An Environmental Approach has been developed on the basis of the following facts: (*a*) A very large number of pupils will study no biology after leaving school. (*b*) Very few will become research biologists, and only a slightly larger proportion will enter the biological professions. (*c*) All are potential voting citizens.

These facts mean that the secondary school biology course should provide the pupil with a background in biology that is as advanced as he is able to assimilate; that subject matter should be selected to increase his effectiveness as a future citizen as well as to help him assert himself in his universe. If, in addition, the biology course encourages a scientific viewpoint it will serve the interests of all.

Clearly, the writers believe that secondary school science should be presented as an aspect of the humanities. If, in some cases, a secondary school course also arouses the pupil to pursue further biological studies, fine—but this should be an incidental rather than a primary aim. The school is not the place to begin the training of professional biologists.

These general aims, logically derived from three simple and undeniable facts, are the foundation of *Biology: An Environmental Approach*. They explain what is included and what is excluded. They explain the manner as well as the matter. Whoever undertakes to use *Biology: An Environmental Approach* in his teaching needs to understand this position.

SPECIFIC AIMS

The Level

The course was planned for the middle 60 per cent (in interest and ability) of 14–16-year-old pupils. No classroom course will completely satisfy the needs of the upper range of pupils. However, by wise use of the 'Problems' and 'Suggested Readings' appended to each chapter and of the suggestions 'For Further Investigation' following many of the investigations, the teacher of *Biology: An Environmental Approach* may carry such pupils very much further into biology.

Teachers who regard 14–16-year-old pupils as children may be appalled by the degree of mental sophistication demanded in this course, by the lack of clear-cut definition, by the emphasis on shades of grey rather than on the black-and-white thought patterns. On the other hand, teachers who ignore the difference between the adolescent and the adult may be scornful of the control of running vocabulary, the attention to development of ideas, the paucity of esoteric detail.

The Scope

Biology: An Environmental Approach is not intended to provide an encyclopaedic account of biology. Many topics are omitted; many are treated cursorily. But those topics that have seemed to the authors to best fit the aims discussed above have been developed in depth. For example, the ecological concept of infectious disease seems to be more important to thinking citizens than theories concerning the mechanisms of immunity. The ecosystem concept likewise seems more important than the electron microscopy of cells. So also, the concept of speciation against the decipherment of protein structure. In each case topics that are exciting to the biological pioneers are slighted in favour of meat for the supporting troops. Yet, room has been left for the imaginative and adept teacher.

Laboratory Work

Laboratory work is an indispensable condition in this course. Two years might be devoted entirely to laboratory work. But the result would be either a myopic view of a narrow segment of biology or an episodic view such as might be obtained by glancing out of a car window at a landscape every 50 miles. In *Biology: An Environmental Approach* the text is used to provide continuity and perspective. Laboratory investigations are placed at points in the course where first-hand experience is most pertinent, most feasible, and most efficient in the utilization of pupil time.

'Laboratory' has been interpreted broadly. The laboratory is where the work of the scientist is done; it need not be bounded by four walls. Moreover, some of the investigations involve much reasoning and little or no doing. These also may be legitimately regarded as laboratory work.

Continuity

The course intentionally avoids the designation of its books as 'units'; paradoxically, it does so in order to unify. In practice, the unit organization tends to compartmentalize, though such was certainly not the original aim.

The writers of *Biology: An Environmental Approach* intend to give the pupil no relief at any time from the healthy tension of learning. The course is designed to build up ideas from beginning to end. Everywhere an effort is made to relate what is immediately in front of the pupil with what has gone before. At the end, the course returns to the beginning and relates all aspects of the course to a biological world view.

THE PUPIL'S BOOKS: AN OVERALL VIEW

Before specific suggestions for teaching can be meaningful, the teacher must be acquainted with the organization of the course, with variations in organization that experience has shown to be feasible, and with the teaching aids supplied to teacher and pupil. The books in the series are intended to provide such general background.

Organization

The books can be taken separately as resource materials but this Guide is written for their use as a series. The pupil's books and chapters are of unequal lengths, but the number of pages devoted to a topic is not necessarily an indication of that topic's relative importance or of the amount of time to be spent upon it. In general, the first three books are discursive, with a rather low density of ideas. The last two are more compact, with an increasing load of ideas per page.

The World of Life: The Biosphere

It is whole individual organisms with which the pupil has had experience. He is one himself. Therefore the course begins with biology at the level of the individual and treats the ways in which such biological units interact.

Chapter 1, 'The Web of Life', is designed chiefly to lay some groundwork, to establish the direction of the course. The interdependence of organisms in the transfer of energy and in the cycling of matter, the interdependence of the living system and the physical environment—these things will never be stated again quite so explicitly, but they will persist in the background throughout the course. 'Scientific method' is not preached; instead, the laboratory work is relied upon to introduce such basic matters as observation, measurement, experimentation, instrumentation.

The units of ecological study form a series from the individual (the most concrete) to the ecosystem (the most abstract). Chapter 2, 'Individuals and Populations', deals with individuals and the various groupings of individuals that can be called populations. The concept of a population keeps turning up: species populations form the basis of classification; populations interact in communities; contemporary genetic and evolutionary theories turn largely on population studies.

The actual field study of a community—even the community in a crack in a city pavement—is essential to Chapter 3, 'Communities and Ecosystems'. Since communities available for study differ greatly among schools, a brief description of a community is provided as a basis for comparison. The kinds of ecological relationships in communities are considered and, finally, the concept of an ecosystem is introduced.

Diversity among Living Things

The pupil should have some idea of the diversity of organisms and of their classification before going further into the patterns of ecological organization. But there is no need for a 'type study' of living things. Emphasis is on the variety of forms in which life can occur and on those aspects of form and function that are relevant to a useful and meaningful organization of such diversity.

On the principle of starting with the familiar, Chapter 1, 'Animals', begins with mammals. This, of course, makes for difficulties, but at this point there is no basis for a meaningful discussion of phylogeny anyway. Instead, diversity of form within the animal kingdom is stressed. But this diversity is not endless; patterns are discernible. Evolution is offered as a possible explanation for the apparent order within the general diversity.

The concept of classification does not need repetition in Chapter 2, 'Plants', but another abstract idea is developed—that of nomenclature. Historically, nomenclature and classification have developed together, but this is not sufficient pedagogical reason for presenting them to the pupil together. Experience has suggested that these concepts are better understood when they are presented to the pupil separately rather than as a single large block of abstract material.

In Chapter 3, 'Protists', a third kingdom is introduced. The teacher may not agree with the system of kingdoms that has been used in the texts or with the way particular groups have been assigned to these kingdoms. But there is no classification on which all biologists agree. Pointing out the reasons for disagreement ought to give pupils an idea of the nature of the problems of classification.

Patterns in the Living World

There are three bases on which patterns of biotic distribution within the biosphere may be constructed: ecological, historical, and biogeographical. Each of these is treated in this book.

A number of laboratory investigations involving micro-organisms have been started during the work on Chapter 3, *DLT*. Therefore, to facilitate continuity of laboratory procedures, Chapter 1, 'Patterns of Life in the Microscopic World', deals with ecological groupings of micro-organisms. Two such groupings, both of great importance to man's existence, are treated in some detail: soil organisms and the micro-organisms involved in disease.

The theme in Chapter 2, 'Patterns of Life on Land', is the distribution of macroscopic terrestrial organisms. The relation of physiological tolerances to the global distribution of abiotic environmental factors leads to the description of biomes. But ecological conditions do not wholly explain distributions. Explanation is then sought in evidence of past distribution and of artificial distribution by man.

Chapter 3, 'Patterns of Life in the Water', extends the principles of ecological distribution to aquatic environments. Ponds are probably more easily visualized as ecological systems than any other part of the biosphere. Bodies of running water are treated briefly. Finally it seems desirable that the pupil should obtain some understanding of marine life, which is certain to become increasingly important as a resource for man.

In Chapter 4, 'Patterns of Life in the Past', attention is given chiefly to the nature of the evidence in paleontology, to the kinds of work that paleontologists do, and to some principles of paleontological reasoning. Emphasis on ecosystems is maintained, and the temporal continuity of the biosphere is thereby stressed. The principle of evolution is not discussed here but simply assumed as the most reasonable basis for interpreting the evidence.

Looking into Organisms

Having spent nearly half its duration on the supra-individual levels of biological organization, the course now turns to the infra-individual levels. Some acquaintance with 'inner physiology' is essential, not only for appreciation of some rapidly developing areas of modern biology but also as background for topics, such as genetics and evolution, that are important parts of the biological understanding required by the functioning citizen.

The objective of Chapter 1, 'The Cell', is to provide the pupil with sufficient understanding of cellular structure, of some cell physiology, and of cell duplication to enable him to interpret subsequent chapters. Only those cell structures that have relevance to later discussion are treated, and the physiology deals principally with the relations of a cell to its environment. Two important topics, differentiation and ageing, are presented primarily as problems undergoing current investigation.

Energy-flow in living systems has been a fundamental idea from the beginning of the course. In Chapter 2, 'Bioenergetics', attention is focused on energy-storage and energy-release in cells. Here the pupil is exposed to some of the biochemical aspects of modern biology.

Chapter 3, 'The Functioning Plant', chiefly concerns the structure and function of those plants with which the pupil comes in contact most frequently—the vascular plants.

The theme of Chapter 4, 'The Functioning Animal', is the variety of ways in which the necessary functions of an animal body are carried out in different animal groups. In each case, the comparative physiology uses man as a principal example.

In Chapter 5, 'Behaviour', the reactions of organisms to external environment are considered as stemming from internal mechanisms. From the vast field of behavioural biology, topics have been chosen that are related to other parts of the course and that have proved stimulating and fairly comprehensible to 15–16-year-old pupils: learning, periodicity, territoriality, and social behaviour.

Man and His Environment

This may be considered the heart of the course. Perhaps the most fundamental thing to be said about life is that it goes on. In this book much of the matter from the preceding four is directed towards an understanding of this basic idea.

Reproduction is considered as a life process unimportant to an individual's continuity but, because of individual death, essential to the continuity of populations and all higher levels of biological organization. Chapter 1, 'Reproduction', continues the comparative method employed in the two previous chapters and again, as in Chapter 4 of *LO*, uses man as a principal example.

It is difficult to overestimate the importance of genetics in contemporary biology, but balance in the biology course demands that genetics not be allowed to get out of hand. In Chapter 2, 'Heredity', the topic is developed historically, and from this development some ideas concerning the logic of evidence are derived. Mathematics is not shunned; it can be either minimized or emphasized by the teacher.

The entire course—indeed, any modern biology course—can be regarded as a summary of the evidence for evolution. The main objective of Chapter 3, 'Evolution', then, is not to give the *evidence* for evolution; that has already been done in several ways, both implicit and explicit. Instead, the chief aim is to give the pupil some idea of the *mechanism* of evolution. Darwin is presented as one who provided an explanation of how evolution operates, not as the originator of evolution as a concept.

In Chapter 4, 'The Human Animal', some ways in which man differs anatomically and physiologically from fellow organisms are discussed. Since much of man's distinctiveness is behavioural rather than physiological or anatomical, the chapter inevitably becomes involved in borderline areas between anthropology and biology. Then the paleontological evidence for the origin of man is examined. Finally, racial variation within the human species is viewed biologically.

The authors hope they have woven the whole course together in Chapter 5, 'Man in the Web of Life'. The pupil is confronted with topics that will concern him in the future as a citizen, topics for which biological information has some relevance. These topics, of course, all go beyond biology; the primary aim is to provoke the pupil into continuing to think about them.

ORGANIZATIONAL ALTERNATIVES

The pupil's books have been organized so that vocabulary and concepts are sequential and cumulative. Full advantage can be taken of this only if the books are studied consecutively. Further, if this is done, the teacher is freed from problems of extemporaneous course design, which is expensive of time. More time can then be put into the tasks of actual teaching, especially into the time-consuming organization and supervision of laboratory work. Nevertheless, it is recognized that every wise teacher will look for adaptations to enhance the instructional opportunities available in his own situation or inherent in his own resources.

One of the most pressing problems in biology teaching is the seasonal availability of materials. Use of greenhouses, aquaria and refrigerators can do much to circumvent this problem but cannot entirely eliminate it. Local environment, weather, class timetables and other factors may create the need for a change in chapter sequence.

For various reasons it may seem desirable to omit or to treat lightly certain portions of the course. If the teacher uses a BSCS Laboratory Block (published by D. C. Heath & Co.), some omission will certainly be necessary. (A full list of Laboratory Blocks available in Britain is given at the end of this Chapter.) In this case the portion to be omitted will depend upon the nature of the Block that is used. For example, if the Block *Microbes: Their Growth, Nutrition, and Interaction* is used, Chapter 3, *DLT* and Chapter 1, *PLW* might be omitted or used as collateral reading.

If time is simply too short for study of the entire course, care must be taken in making omissions to ensure that sequences of thought are not disrupted. For example, it is difficult to make omissions from Chapter 3, *MHE*. In schools where much human anatomy is taught in connection with health courses in earlier years, much of Chapter 4, *LO*, might be omitted.

It is very easy to allow a course to fray out and expire merely by reason of the arrival of the last day of term. *This must never happen* with the *Environmental Approach*. No matter where the class may be in the course, no matter what the route through the course may have been, the teacher *must* reserve a few days at the end of the school year for the last chapter. Here an attempt is made to bring the whole course into focus. Here are discussed those biological problems man must face if he is to continue his existence on this planet. So that the teacher may put his laboratory in order, take stock, and close the laboratory accounts while classes continue to meet, no laboratory activities are proposed for the last chapter.

Conspectus of Teaching Resources

Usually 14–16-year-old pupils do not automatically recognize the uses of textual apparatus. Therefore the wise teacher will devote a little time to acquainting them with the books. But first, of course, he himself must investigate their resources. The following paragraphs discuss the intentions of the authors. No doubt teachers will find many other ways to use the resources of the pupil's books.

Book introductions. The several paragraphs introducing each of the five books are not to supply facts. Rather, they set the stage, present a viewpoint, relate the book to the whole of the course. The cover photographs bear no captions, but each has relevance to the theme of the book. Pupil speculation on this relevance would not be inappropriate as a terminal activity.

Italics. Within paragraphs, italics are used for either of two purposes: (*a*) they indicate the first occurrence of a technical term, or (*b*) they indicate an emphasis. They are used for the second purpose quite sparingly.

Vocabulary. In any science textbook, vocabulary is of two kinds: technical and running. Technical vocabulary is part of the course content because ideas cannot be divorced from terms. But the learning of terminology is not a proper *aim* of science teaching; it is a means only.

Secondary school biology has been particularly subject to criticism for its high density of technical terms. In the *Environmental Approach* considerable effort has been made to keep technical vocabulary within reasonable limits. This effort has inevitably resulted in the loss of some terms that may be favourites with teachers. There is, of course, no reason why a teacher cannot increase the load if he wishes. But the authors trust that he will approach each addition as they have done—with a penetrating *Why*?

Though each technical term has been carefully considered before being admitted to the text, the load remains great. Several levels of importance can be distinguished. Some terms are of pervasive importance throughout the course—*photosynthesis*, *environment*, *evolution*, for example. They must be applied again and again, examined in many contexts, approached from many viewpoints. Others are of less pervasive importance, but fundamental to some major section of biology—*meiosis*, *predation*, *natural selection*. Still others are not important in themselves but are essential as steps to larger ideas—*natality*, *crossing-over*, *tropism*. At the lowest level of technical vocabulary are names of things—*centromere*, *host*, *ATP*. Certainly it would be a pedagogical error to treat all technical terms in the same way.

The running vocabulary carries the narrative. In the first part of the text, it is rather simple, though no attempt is made to 'write down' to the pupils. In the latter part of the book, the authors have made less effort to find simple synonyms, and the running vocabulary probably approaches the recognition level of the average 15–16-year-old—with occasional excursions beyond.

Illustrations. In any textbook that stresses the biology of whole individual organisms, constant reference to kinds of organisms is necessary to provide examples of the general principles discussed. But 14–16-year-olds vary greatly in their previous experience with living things. Therefore pictures of many organisms are provided in many cases simply to strengthen general discussion with visual images.

Other illustrations are teaching materials of an importance coordinate to that of the text. No illustration is printed for decoration alone, though an effort was made to obtain pictures that are as attractive as possible. Captions tie in with the rest of the text and often extend it. The questions that frequently occur in the captions are by no means all rhetorical; they are worthy of class discussion.

Wherever the practice might be helpful, an illustration is accompanied by indications of the size of the pictured subject. In all such instances, the ratio of the picture size to life-size is given ($\times\frac{1}{4}$ = reduced to one-quarter life-size; $\times 1$ = life-size; $\times 2$ = enlarged to twice life-size).

Laboratory investigations. The first function of a teaching laboratory is to present from nature the evidence for the basic biological concepts. This *illustrative* function was probably the principal one in Thomas Henry Huxley's mind when he introduced laboratory work into science education. His insight was a simple one: Seeing is believing. In teaching science one must appeal not to the authority of a teacher or a book; one must look squarely at the facts, at the infinitely varied phenomena of nature. Unfortunately, the illustrative function came to be so heavily emphasized that pupils spent most of the laboratory time watching demonstrations, looking through microscopes, dissecting animals or plants, learning names, labelling drawings—but rarely doing an experiment in the sense of really investigating a problem, the answer to which is unknown.

Today something more must be expected from school laboratory work. Active participation of the learner in some scientific investigation is needed if he is ever to glimpse the true nature and meaning of science, ever to appreciate the forces that motivate and activate the scientist. This *investigative* function of laboratory work requires a different approach and a revision of goals. It requires different and (often) more expensive kinds of materials and many more of them. It makes desirable more extensive laboratory facilities.

The investigative function of laboratory work does not displace the illustrative function; it complements it. For at least thirty years biology teachers with quick mental insight have striven to elevate this twentieth-century function to a place beside the honoured nineteenth-century function. Both functions are represented in the investigations in the pupil's books.

The investigations are the heart of the course. At least 50 per cent of the pupil's class time should be centred on them: planning, performing, observing, recording data, interpreting data, drawing conclusions, and relating the work to other sources of information.

Investigations have been placed within chapters at points where they fit the development of ideas. Sometimes they are initial; sometimes they are terminal; sometimes two or more succeed each other without intervening text. At times two or more investigations must be in progress simultaneously. And usually pupils should be discussing their reading of text material, reference assignments and problems as work on the investigations proceeds. Nevertheless, to facilitate the management of such a complexity of activities in the classroom–laboratory and in the field, the authors have endeavoured to place investigations in the pupil's books in such a manner that the textual material is not unduly fragmented.

Because of the importance of the investigations in the course, a special section is devoted to them (pp. T12–15).

Summaries. Each chapter ends with a summary (set on a coloured background). Summaries are not outlines, and the pupil will soon discover that they are no substitute for studying the chapters. However, summaries do bring together major ideas in their chapters, sometimes in new relationships with each other; they should prove useful to pupils as consolidating devices and to teachers as bases for launching discussions.

Guide questions. These are based directly upon the chapter materials—both text and illustrations. They require recall rather than reasoning, though there is some departure from this generalization in later chapters. They are placed all together at the end of a chapter, and the sequence is exactly that of the ideas in the chapter. Thus, a teacher may break up a chapter for assignment in any way that he wishes, and the corresponding guide questions can readily be located.

It is intended that these questions be used by the pupil for his own guidance during his study. With some classes they may be used for checking pupil understanding of text materials. If confined to the guide questions, however, class discussion will proceed on a very low level and will result in the neglect of much material. Therefore, even with very slow classes the teacher needs to supplement the guide questions with some of his own, based on his understanding of the interests and attitudes of the pupils in each class. With average classes such supplementary questions should call for some reasoning. With above average classes the guide questions should receive little attention in class discussion.

Problems. Unlike the guide questions, the problems require reasoning, computation or research—sometimes all three. They are not intended as guides for the pupil while he studies the chapter, but as extensions beyond the chapter. Their sequence has no relationship to the sequence of ideas in the text. Although in some cases the problems may serve as material for class discussion, they should not be assigned *en masse*.

No attempt has been made to control either the vocabulary or the sentence structure of the problems. New terms are sometimes used without explanation. It is assumed that pupils who are sufficiently advanced to use the problems will also know how to use dictionaries and other reference books. It is not possible to rank the problems according to difficulty. Assignment should be made only after consideration of the interests as well as the abilities of individual pupils.

For the teacher, the principal value of the problems should be to provide suggestions for the invention of problems of his own. Problems with a local flavour or that bear upon biological topics currently receiving notice in news media are particularly valuable. Problems that lead the pupils into experimentation will, of course, be indistinguishable from the items entitled 'For Further Investigation'. Better pupils may be encouraged to develop their own problems.

Suggested reading. An effort has been made to limit the number of book titles. The intention is to encourage the formation of *classroom* libraries where books may be constantly available to pupils, leaving the school library as a place for more extensive research. There many of the books listed in Part Two of this Guide should be accommodated. Since it is desirable to encourage further reading on the part of all pupils—not only those who read well—some easy references have been selected as well as some that will challenge the most advanced.

Periodical references have, with few exceptions, been confined to articles in *Scientific American.* This sparse selection from periodical literature does not imply that worthwhile material is lacking elsewhere. But librarians usually keep back-numbers of this periodical and many of the *Scientific American* articles are available as reprints, while desired copies of many other periodicals are often hard to find.

The Appendix of Diversity among Living Things. In this appendix the scheme of classification outlined in *DLT* is presented in conspectus, enabling the pupil to see the levels of organization in close relationship to each other. The language has been kept as non-technical as possible, but not all terms are necessarily comprehensible at the time the pupil is studying *DLT.* Illustrations are juxtaposed with their descriptions. This material is, of course, strictly for reference purposes. It might be desirable for the teacher to present an outline of one or two other schemes of classification for comparison with this.

Indexes. By now the reader of this Guide must realize that memorizing definitions plays a very minor role in the *Environmental Approach.* The discussion of word meanings in context, the usage of words—these *are* the important matters.

But they are not served by a list of pat definitions. Therefore, the pupil's books contain no glossary. However, the indexes of the textbooks are quite comprehensive, so that the pupil has ready access to all the material in his book, including definitions in context whenever they occur (see, for example, 'Amino acids' p. 45, *LO*; 'Biosphere' p. 31, *WL*; and 'Conifers' p. 62, *DLT*).

Investigations: special considerations

Good secondary school biology work has been accomplished in spacious and fully equipped laboratories; it has also been accomplished in small classrooms with primitive and improvised equipment. Obviously, in widely different physical environments procedures must also vary widely. Therefore no prescriptions can be written for successful laboratory teaching. The following paragraphs are intended simply to alert the teacher to some sensitive areas of planning.

Nature of the investigations. The pupil's books contain 60 full-scale investigations and numerous suggestions 'For Further Investigation'. All of the 60 are an integral part of the course. They frequently present concepts or terms, an understanding of which is assumed in later development of ideas. None can be omitted without the same kind of checking and thoughtful balancing of time and objectives that should accompany the omission of any other material in the pupil's books. In fact, keeping the fundamental aims of the course in mind, omission of investigations should perhaps involve more deliberation than omission of textual material.

A few investigations are recommended as demonstrations. For the most part these are ones in which the replication of the procedure by many pupils seems to serve no useful purpose or involves an inordinate amount of equipment. The number of demonstrated investigations ought to be kept to a minimum, even though many not recommended for demonstration *could* be presented in this manner. It might be argued that a demonstration would gain in effectiveness if the superior skill of the teacher is employed. But from a pedagogical viewpoint, demonstrations that are performed by small groups of pupils are usually superior to those done by the teacher.

Organization of the investigations. Because individual investigations contribute in different ways to the advancement of the pupil's biological knowledge and scientific understanding, no set pattern of internal organization is followed. In the secondary school laboratory there should be no rigid scheme of scientific investigation; there is none in the scientist's laboratory. However, every investigation has a general 'Procedure'.

But every effort must be made to create in each pupil an awareness of the purpose of his activity. He may see the purpose and still flounder about. But if he sees the purpose, he is the less likely to flounder and the more likely to grasp and carry out the procedure.

Almost every investigation has a 'Materials and Equipment' list, which is mainly for the teacher and his laboratory technician. But such lists have also been found of value to pupils in setting about their work, so they have been retained in the pupil's books.

Most investigations have some kind of follow-up to the procedure. This is often a section called 'Studying the Data', in which directions are given for arranging data into tabular or graph form, and the meaning of data is elicited by suitable questions. It must be understood that the forms given in the pupil's books present only the scheme for recording data—size and amount of space required in his lab-book must be determined by the pupil.

Sometimes there is a 'Summary'. If an experimental situation justifies the use of the term, 'Conclusions' are called for.

Sometimes investigations include 'Background Information', material needed for understanding the procedure. Still other subheadings are used occasionally.

Suggestions 'For Further Investigation' are plentiful. These are materials that can be explored by individual pupils who have more than average energy and drive—and, hopefully, above-average ability. Some call for fairly simple extensions of the procedures in the main investigation; some entail original thought and attention to design. In most cases, specific directions are lacking; the pupil must work out his own procedures.

Questions are inserted wherever they seem appropriate, even in an introduction. Most, of course, occur in sections such as 'Studying the Data', 'Discussion', or 'Conclusions'. In some investigations questions are woven into the procedure (Investigations 1.3 and 1.4 of *WL* for example), and in others the whole procedure advances by means of questions (see Investigations 3.2 and 4.1 of *MHE*). These methods make it desirable to place an identifying letter (in parentheses) *before* a question. This placement of letters has been used consistently throughout all the investigations.

Laboratory books. Experience has shown that the most convenient way to handle the recording of data is by means of a bound notebook.

Each pupil should be encouraged to regard his lab-book as a place of primary record. As such, it must meet the hazards of constant use at the laboratory bench and will receive records hurriedly made. Under such circumstances a lab-book is not likely to be a thing of beauty. Slovenly work cannot be tolerated in science; but many pupils, particularly the 'better' ones, tend to equate slovenliness with mere lack of neatness. They must be taught to associate slovenliness with inaccuracy.

Although a lab-book is a personal record, not a report to the teacher, it should be checked occasionally to ensure that the pupil is using the most efficient methods of recording and organizing data. Checking should be accomplished on the spot, not by removing the book from the hands of the pupil, and should be dissociated from grading. Checking should be frequent at first, less so as the course progresses.

Laboratory reports. Communication is a most important part of science. Discoveries become a part of science only when they are made known to others—when they are published. In publishing scientific work, the writer must express himself so clearly that another person can repeat his procedures exactly. The reader must know what material was used (in biology, this includes the kind of organism) and be able to comprehend every detail of the work. Scientists must be free to communicate, but they can use this freedom only if they know how to communicate. Scientific reports are usually written in a rather standard form, somewhat along the following lines:

1. Title.
2. Introduction: Usually states how the problem arose and often gives a summary of past work.
3. Materials and equipment.
4. Procedure (or Method): Complete and exact account of what was done in gathering the data.
5. Results: Data obtained from the procedure, often in the form of tables and graphs.
6. Discussion: Relates the data to the purpose of the work.
7. Conclusion: Summary of the meaning of the results; often suggests further work that might be done.
8. References: Published scientific reports that have been specifically mentioned.

If pupils undertake work on an independent problem, their report should follow this form. But for the usual work in this course they do not have to be so elaborate. They are communicating with their fellow pupils and the teacher, who already know a great deal about the work.

The basis for the pupils' reports is the data from their lab-books. They should be reminded that in writing a report they are not writing for themselves but are trying to communicate ideas to others. Therefore stress the need for care about neatness, spelling and sentence structure, all of which contribute to clearness of expression. In short, the task of writing a report is very different from recording data. Both, however, are part of the scientist's work.

Modifications. In almost every investigation many variations are possible. Some of these are noted in Part Two of this Guide. Others will be dictated by necessity. In the early part of the year, it is best to stick as closely to the printed form of the investigations as the local situation will permit. At this time connection between variations in procedure and variations in results must be established. It is therefore desirable to train the pupil to follow procedures carefully; and the fewer changes he has to cope with, the better. If many changes must be made, it would perhaps be best for the teacher to re-write the investigation entirely.

Later, strict adherence to the printed procedures is not necessary, or even desirable. With more able classes, merely throwing out hints concerning possible changes may suffice to introduce valuable variations that may increase the teaching possibilities in the results. But beware of breaks in the reasoning process. Variations in procedure can degenerate into mere tinkering if links between procedure, data and conclusions are not strongly forged.

Finally, the investigations in the pupil's books should in no way inhibit the teacher's efforts to devise new ones suited to his own facilities and situation. Rather, they should encourage him to do so.

Looking ahead. Careful planning is a hallmark of good teaching. It is especially important in science teaching. In biological science it is crucial.

BSCS Biology, with its emphasis upon experimental procedures involving living materials, calls for the utmost skill and ingenuity in planning. Simultaneously the teacher must often consider the disposal of materials from a completed investigation, the care of materials in one or more current investigations, the provision of materials for work to be accomplished in the next few days, and the procurement of materials for investigations that are two, four, six or more weeks in the future.

Among the layers of planning responsibility, long-range foresight is the one most likely to be neglected. Yet it is essential. For example, in most parts of the country an investigation requiring young tomato plants is impossible to carry out in December unless the seeds have been planted in October. And the seeds may be difficult to obtain in October unless they have been bought in the spring or early summer.

Initiating the work. If the importance of laboratory work is to be established in the minds of pupils, there must be no delay in getting into it. Because very little procedure is involved, it is possible to set a class to work on Investigation 1.1, *WL*, at its second meeting. But before Investigation 1.2, *WL*, is attempted, some ground rules of laboratory work must be laid down. Among the matters to be considered are:

1. Need for thorough familiarity with the purpose (in the case of an experiment, the hypothesis) and the procedures to be followed.
2. Location of work stations and regulation of pupil mobility during laboratory work.
3. A scheme for distributing and collecting materials.
4. Principles of teamwork: leadership, acceptance of responsibility, and coordination of efforts.
5. Relationship between lab-book and the completed exercises.
6. Methods of evaluating pupil laboratory work.

Every pupil must assume responsibility for understanding the procedure of each investigation and, especially, his own part in it. The teacher has responsibility for relating the work to the accompanying textual materials and for adapting (when necessary) the procedure to his own classroom situation. The teacher is wholly responsible for the provision of the materials and equipment, though he may be able to delegate such responsibility to assistants—pupils or laboratory technicians. All required materials and equipment must be on hand when work begins. Nothing is more detrimental to good classroom laboratory work than lack of an essential item at a critical moment.

Finding time. Having understood the basic importance of laboratory work and having leafed through the investigations in the pupil's books, the teacher immediately asks, 'How can I find time to prepare all the material?'

No complete answer can be given to this question. Teachers *have* found the time, in one way or another. But every device suggested by teachers for providing more time seems merely to open up new opportunities for expanding laboratory activities. Thus teachers in the 'best' situations are often as busy as those in the worst.

When head teachers have become convinced that extensive, truly investigative laboratory work is educationally important, various kinds of administrative actions have been helpful. In some schools, science teachers are timetabled for one less class per day than are other teachers. The period freed from class instruction is available for preparation of laboratory and demonstration materials. In other schools, full-time laboratory technicians are employed.

Even without overt administrative backing, the individual teacher can improve his ability to provide worthwhile laboratory work for his pupils. Something may be achieved by increased care in planning. In schools having more than one biology teacher, the sharing of preparatory tasks is time and labour saving, particularly when special skills are utilized. And much time can be saved by attention to organization of materials, e.g. a stockroom or preparation room where all items have assigned places.

Most important, however, is the enlisting of voluntary aid from the pupils. In almost every class some pupils deem it a privilege to be permitted to help in laboratory preparations. Not only do such pupils make a real contribution towards freeing the teacher from routine chores, but they also come to realize that a certain amount of 'dishwashing' is an essential part of laboratory work.

There is, of course, nothing new in this practice, but some suggestions may be useful. Pupils should be selected on an informal basis and never to the exclusion of others who may later become interested in helping. Although their assistance may be greatest in routine work, they should be given instruction in some laboratory skills—partly in return for their aid and in encouragement of their continued effort, partly in hopes of discovering a few who may be especially deserving of guidance into careers in science or in laboratory technology.

Sometimes pupils who have already taken biology are willing to become laboratory assistants during their free time. They can be assigned a somewhat more formal status than current pupils, since they are not members of the classes in which they assist. Among the criteria to be considered for the selection of such assistants might be (*a*) the pupils' interest in biology, (*b*) their scholarship in biology and related subjects, (*c*) their general scholarship, (*d*) their available free time, (*e*) their willingness to help other pupils, (*f*) their ability to get along with others, and (*g*) their general reliability and dependability. (Caution: No matter how much help he may have, paid or voluntary, the teacher retains sole responsibility for laboratory safety.)

Checking the work. It is possible for a pupil to become so preoccupied with preparations for future laboratory activity, so enmeshed in the mechanics of present laboratory activity, and so fatigued by the chores of cleaning up past laboratory activity that little time is left for the rounding out of his experience. Since there is little virtue in mere activity, this should never be allowed to happen. Every investigation should be followed by class discussion.

The kind of class discussion may vary with the kind of investigation. When an investigation is observational only, the teacher must be sure that pupils relate observations to the purpose of the investigation. When the investigation is experimental, the teacher must be sure that the course of the reasoning, from hypothesis through experimental design and data to conclusions, is understood. In any case, class discussion following laboratory work is the milieu in which an understanding of the true nature of science is best developed. No other means is as effective in placing before the pupil the rationale of science, the difficulties of research, the uncertainties of knowledge.

For many investigations a class discussion constitutes a satisfactory termination. Written reports for all investigations are neither necessary nor feasible. The teacher may easily be lured into a never-ending race with paper work. The results are a poor use of pupils' time—in excessive writing—and a poor use of teacher time—in excessive reading. Which investigations to select for written report is a matter of personal choice. But it seems more reasonable to require such a report for an experimental investigation (such as 1.2, *WL*) than for an observational one (such as 1.3, *WL*).

A report that requires the pupil to copy lists of materials and long passages of procedure from his book is a sure way to undermine the purpose of the laboratory. In general, a written report might consist of (*a*) a title, (*b*) relevant data worked out from the lab-book, and (*c*) answers to the questions in the investigation. If complete statements are required in answering questions, grading is easier, and, more important, vague thinking is discouraged. At the beginning of the course, questions may be discussed in class before a written report is required.

Written reports should be submitted as soon as possible after completion of an investigation. Prompt evaluation and return of the report to the pupil make a follow-up discussion possible.

TESTS

No matter what the stated aims of a course may be, no matter how diligently the teacher may bend his efforts towards them, all is in vain unless the tests that are used to measure the pupil's progress reflect these aims. Whether much or little is made of marks in a school, pupils remain realists; they will work towards tests. Therefore, it is of utmost importance that tests be firmly based on aims.

If pupil memorization of arbitrary, clear-cut definitions is the teacher's aim, then he should not be using *Biology: An Environmental Approach*, which is certain to engender frustration in both himself and his pupils. But if this is *not* his aim, then he must be sure that his tests do not merely require rote memorization of terms. If the teacher really believes that laboratory work is as important as digestion of textbook information, then he must base his tests at least equally on laboratory work and textbook work.

All of this is probably self-evident. But granting the principle, the constructing of tests that reflect the aims of the *Biology: An Environmental Approach* remains difficult. For some of the aims, no good group-testing procedures are yet known. And for all of the aims, good testing procedures involve the investment of much time and effort.

During the years in which BSCS materials were being tried out in classrooms, a committee of teachers, biologists, psychologists and psychometricians worked to devise suitable tests to accompany the materials.

Frequent tests are needed to allow the teacher to determine his successes and failures, and to indicate the areas in which intensified review is required. There is no substitute for teacher-made tests. With the crowded timetables already existing for science teachers and the increased time requirements of laboratory methods, it is to be expected that, in his construction of test items, the teacher will not always be able to match his performance to his ideals. But something more than a casual nod to such ideals will be necessary if the pupil is not to be demoralized by his first encounter with the periodic tests. The teacher must have previously provided him with suitable test experience.

To assist the teacher in this task, the BSCS has provided two aids. In the *Biology Teacher's Handbook* (2nd edition. E. Klinckmann, Supervisor. Wiley, 1970), Chapter 8 is devoted to the construction of classroom tests. An experimental booklet of specimen test items, arranged by chapters, is published for *Biology: An Environmental Approach* by John Murray.

BSCS Laboratory Blocks and Teacher's Supplements

Six-week laboratory investigations that may be used individually or in combination to complement biology course materials or as an independent laboratory course in biology. Distributed in Britain by D. C. Heath Ltd., 1 Westmead, Farnborough, Hants.

Animal Behavior
Animal Growth and Development
Evolution
Field Ecology
Genetic Continuity
Life in the Soil
Microbes: Their Growth, Nutrition and Interaction
Physiological Adaptation
Plant Growth and Development
Regulation in Plants by Hormones–A Study in Experimental Design
The Complementarity of Structure and Function
The Molecular Basis of Metabolism

Part Two Specific suggestions for the use of the texts and for laboratory work

THE WORLD OF LIFE: THE BIOSPHERE

Whole living organisms are the centre of attention throughout this book. The classroom should be plentifully supplied with them. Whether the classroom is large or small, whether provided with the latest biological equipment or not, it can be made a home for a variety of living things.

Bad weather is always a bug-bear of the British climate, thus it is important for the teacher to retain a certain amount of flexibility in his planning to allow the field Investigation 3.1 to be undertaken at a suitable time. Nevertheless it is important that the logical order of the individual–population–community sequence is retained and made clear.

Throughout the book, the following ideas should be constantly kept before the pupils:

1. Individuals as biological units.
2. Verifiable observations as the foundation of biological concepts.
3. The flow of energy as the core of ecosystem function.

1

THE WEB OF LIFE

MAJOR IDEAS

The ideas listed below are not to be imposed upon the pupil. Rather they are intended to aid the teacher as he plans class discussions of the chapter.

1. An understanding of the spirit of science—particularly of the processes of scientific work—is basic to any worthwhile study of science.
2. Organisms tend to maintain a steady state, both internally and externally, in the face of environmental change.
3. Energy flows from the sun through the living system and back into the abiotic environment from which it is irrecoverable.
4. Having been introduced into the living system in the form of light, energy is passed from organism to organism in chemical form.
5. The matter of living things is the same matter that is found in non-living things.
6. In contrast to energy, matter moves cyclically between the living system and the non-living world.
7. Earth's living system together with its supporting abiotic environment may be conveniently termed the 'biosphere'.
8. Man—an integral part of the biosphere, yet possessed of extraordinary powers—is faced with enormous problems in maintaining a steady state in the biosphere. His own continued existence and that of all other living things will depend upon his biological understanding.

PLANNING AHEAD

Ideally, planning and ordering should have been done the year before and the materials for the first few investigations assembled or delivered a week or two before the first day of term.

If this has not been done, then:

First: procure and make labels for living organisms for Investigation 1.1. A photosynthetic flagellate may be the most difficult to collect locally. If so, telephone a supplier and ask for express delivery.

Second: obtain seeds for Investigation 1.2.

Third: prepare strips of small newspaper print and magazine photographs for Investigation 1.3.

Fourth: obtain micro-organisms for Investigation 1.4.

Fifth: collect or order snails and *Elodea* for Investigation 1.5.

Now, with the most immediate needs attended to, check the lists of materials and equipment in Appendix B against the supplies in your laboratory. Order what is missing or consider suitable alternatives.

Before work on Chapter 1 has proceeded very far, begin preparations for succeeding chapters.

1. Either order or prepare a stencil to run off copies of five-cycle semi-log graph paper for Investigation 2.1.
2. Check the number of test-tubes available for Investigation 2.2. The number will dictate the number of groups you can have. If your laboratory technician has little experience in preparing and sterilizing media, do these jobs well in advance to allow time for mistakes.

GUIDELINES

If materials are at hand, begin the year's work with Investigation 1.1. No previous reading assignment is necessary.

Investigations 1.1 to 1.4 are grouped together but they need not all be carried out before proceeding with text materials. As work in the laboratory proceeds, study assignments can be interspersed, e.g. reading of pp. 1–6 (guide questions 1–5); pp. 18–21 (guide questions 6–9), pp. 21–25 (guide questions 10–14). Short study assignments are preferable while you learn the reading capabilities of your pupils.

It is recommended that Investigations 1.1–1.4 be done with as little modification as possible. This will lessen confusion, emphasize the importance of a careful reading of procedures, and build up the pupils' confidence. Later, of course, you will often want to modify procedures to take advantage of special local circumstances.

Special attention must be given to proper use of the lab-book (p. T13). Most pupils will have gained sufficient physical and chemical knowledge in previous years to provide a sufficient background for Chapter 1. It might be an advantage, though, to see if the following basic terms are understood: element, compound, symbol, formula, chemical change.

The greatest danger the teacher faces in Chapter 1 is that of becoming enmeshed in the text. *Remember that this chapter is introductory.* If all the books in this series are used together then these ideas—the dependence of producers on solar energy, the dependence of consumers on producers, the interdependence of the living system and the physical environment—will always be in the background. Because they will be continually reappearing, the depth of understanding achieved at this point need not be great. Therefore, beware of becoming bogged down in a mass of detail at this point.

TEACHING NOTES

Rabbits and Blackberries (pp. 1–3)

p. 2. The terms 'producer' and 'consumer' are approximately equivalent to 'autotroph' and 'heterotroph'. In some classes it may be desirable to use the latter pair of terms, but the former have a firm place in ecological literature.

p. 2, Fig. 1.2. Energy passes from sunlight to producer (acacia) to first-order consumer (giraffe) to second-order consumer (lion). It is desirable to use a variety of organisms from a variety of places to illustrate food chains. Supplement the text with examples from the pupils' own environment, such as plankton–minnow–trout–otter, and grass–grasshopper–shrew–fox.

The Scientist's Viewpoint (pp. 3–6)

pp. 3–6. This section does not lay down a 'scientific method', it merely describes the ways in which scientists work. Observation and verification are regarded as the fundamental processes without which there can be no science. The text material requires no discussion. The four investigations are the correct elaboration.

pp. 4–5. p. 4, *upper*: field observation; *lower*: experimentation and use of instruments; p. 5, *upper*: measurement; *lower*: assembling a report and graphing data.

INVESTIGATION 1.1

OBSERVING LIVING THINGS (pp. 6–7)

The teaching objective of this investigation is the *process* of observation.

Setting up the Laboratory

In early autumn a sufficient variety of organisms is probably obtainable in most school situations. Although it is desirable to start this Investigation promptly, there are some advantages in enlisting the aid of pupils in gathering the materials. Pupil contributions are likely to run to the more conspicuous animal forms, so the teacher will have to help. Aim for a balance among plants, animals and protists.

The following are organisms that have been found useful in this exercise:

For observing under compound microscopes–*Euglena*, *Volvox*, *Spirogyra*, yeasts, rotifers, nematodes.

For observation under stereo-microscopes (or hand lenses)—*Rhizopus*, liverworts, mosses, lichens, *Planaria*, small annelids, small insects.

For observation by the naked eye—seaweeds, mushrooms, ferns, *Pelargonium* (flowering), *Begonia* (flowering), sensitive plants (e.g. mimosa, sundew, Venus flytrap), *Lemna*, sponges, earthworms, snails, crabs, centipedes, millepedes, spiders, flies, beetles, butterflies and moths, goldfish, frogs, toads, snakes, lizards, canaries, budgerigars or other birds, rabbits, rats, mice, gerbils, hamsters.

Euglena, *Volvox* or similar green motile organisms are important in giving the pupil an opportunity to see the difficulty of grouping all organisms as either plants or animals.

Set up microscopes so that illumination and focus are optimal. Provide cultures that are rich enough for no pupil manipulation of microscopes to be necessary. Of course microscopes will arouse much pupil interest, but at this time they should concentrate on what is to be observed. Do not attempt too many things at once.

Directing the Work

The number of specimens to be observed must be gauged by the time available in the period. Observation time per specimen will vary with the age and ability of the group. It is useful to allow somewhat longer for the first few specimens.

Go over the directions with your pupils. The purpose must be made clear. Assign each pupil (or pair of pupils) the number of the specimen with which he will begin and announce the total number of specimens. Agree upon a signal for changing from one specimen to the next.

During the observation period the teacher merely gives the signals for the change of station and makes sure that the operation proceeds smoothly. Pupils often feel insecure and may try to ask innumerable questions about what to look for and how to record their observations. Answer *no* questions at this stage, but encourage pupils to write in their data books any questions that arise while they are observing.

Directing Discussion

By the day following observation, pupils should have written up their data ('Studying the Data'). During class discussion the teacher's primary job is to stress accuracy of observation. For example, does a pupil say there are roots on a geranium plant because he *sees* roots or because he thinks that a plant in soil must have roots? Here the distinction is between observation and inference. Secondly, the teacher should try to arouse a *critical attitude* towards points that pupils propose as characteristics of plants and animals, using the pupils' own knowledge of organisms to the greatest extent possible. Obviously no definite conclusion concerning the problem posed in the 'Purpose' can be drawn from such limited data. This needs to be pointed out and the pupils' natural uneasiness with the tentative and unresolved must be pacified.

The *process* of observation rather than the *product* is the teaching objective of this investigation.

INVESTIGATION 1.2

THE GERMINATION OF SEEDS: AN EXPERIMENT (pp. 7–10)

This investigation is concerned primarily with the concepts encompassed by the terms 'hypothesis', 'variable', 'control', 'data' and 'conclusions'. The pupil should gain some insight into these concepts and should note the role of numbers and use graphic records in recording observations. He should also see the value of the team approach and the possibility of designing a procedure for simultaneously gathering data on two related hypotheses.

The information concerning seed germination is incidental, but not without value for chapters where the influence of environmental factors on populations is discussed.

Materials

Garden seeds are generally available in shops only during the spring and early summer. They may be purchased in advance and stored in insect-proof containers in a cool, dry place. However, seeds used for pet food or for attracting wild birds may be obtained throughout the year. The following are satisfactory: wheat, oats, barley, turnip, radish, carrot, sunflower, marigold, parsnip (slow), parsley (very slow). Note that bean seeds are subject to mould even when treated with fungicide.

Many seed fungicides are sold commercially. They may be used according to the directions on the packages. Several substances generally available in the laboratory are also satisfactory. They are:

ethanol or propan-2-ol (ethyl or isopropyl alcohol)–70 per cent (soak seeds for one minute);

methanal (formaldehyde) (dilute commercial formalin to 1:500 and soak seeds for twenty minutes);

sodium chlorate(I) (hypochlorite) (dilute commercial bleach to 1:4 and soak seeds for fifteen minutes).

Baby-food jars make satisfactory containers for soaking seeds. If there is a shortage of petri dishes, the soaked seeds can be planted in sand in half-pint milk cartons. Punch holes in the bottoms of the cartons for drainage. The plastic bags are recommended because classrooms are usually rather dry and the dishes dry out quickly. If they are not available, add water as needed with a medicine dropper to avoid disturbing the seeds.

Procedure

Group sizes depend upon availability of materials; 4 or 5 pupils per group is good. To test the second hypothesis each group should use a different kind of seed. Kinds that show a wide range of response should be selected. Radish represents one extreme, parsley the other.

Treatment of seeds with fungicide reduces—but does not eliminate—loss of germination due to injury caused by fungi.

All seeds must be 'planted' at the same time. Therefore, a schedule must be drawn up to ensure that all groups of seeds are soaked for the proper length of time.

If necessary, the soaking time for the last set of seeds may vary from two to four hours, depending upon the time at which the class meets on planting day. It is convenient to start the soaking schedule on a Friday and either have some pupils take jars home, where soaking can begin on Saturday and Sunday, or arrange for this to be done. In this way a week of observation is available before another weekend occurs.

Completing the Investigation

The actual work occupies little time except on the first day and the planting day, but the procedure for this exercise extends for many days. During this time it is easy to lose sight of the purpose. Particular attention must be placed on rounding off the experiment after the procedure is completed.

Pupils may need help in drawing bar graphs. Check the tables of data to make sure that data are accumulated from day to day. Make sure a table for the recording of data from different kinds of seeds is provided on the blackboard or elsewhere. Check to see that groups record on this table only the results from seeds soaked for 0 hours.

These are small matters, but attention to such details is especially needed at the outset. The reward for such attention will be a clearer discussion at the end of the Investigation, and *better work later*.

Notes on Pupil Responses

Class discussion of the 'Conclusions' and of the lettered items in the 'Discussion' is essential. Note that the questions are concerned entirely with experimental method, not with seed germination *per se*.

p. 9, no. 2. It may well happen that the difference in germination time for soaked *v.* non-soaked seeds is small. Raise the question of how *big* a difference is necessary before a Yes or No conclusion is valid. In better classes introduce the *idea* of (but not the method of) statistical tests of differences.

p. 10, E. The comparison of germination in different kinds of seeds involves a non-quantitative variable. Therefore, there can be no 'set-up' that serves as a control except on the purely arbitrary basis of picking one kind against which others are judged. A 'control' in the technical sense is not a necessary part of an experiment; however, some basis for comparison is needed.

INVESTIGATION 1.3

USE OF THE MICROSCOPE: INTRODUCTION (pp. 10–14)

Most pupils have a great interest in the microscope and the world it opens up for them. This motivation can be of value in introducing some basic ideas about the role of instrumentation in biology. Among these are the following:

1. The microscope is a tool that enables the biologist to extend the range of his observations beyond that afforded by his unaided vision.
2. The way the tool is used determines the kind and amount of information that the scientist can obtain.
3. Information made available through instruments has given rise to problems that would never have been recognized if these tools had not existed. If biologists had never been able to see *Euglena*, *Paramecium*, or bacteria, the problem of deciding which organisms are plants and which are animals would be far simpler than it is.
4. The development of science has been closely tied to the development and improvement of instruments that made crucial information available. The microscope is an excellent example of such an instrument.

Materials

While not suitable for precise work at high magnification, No. 2 glass cover slips are satisfactory for pupil use at both low and high power. They do not break as easily as the thinner No. 1 cover slips.

Lens paper is usually supplied in booklets or

large sheets. Store it in a dustproof container. Each piece of lens paper should be used once then discarded. Be sure pupils understand that lens paper is to be used for cleaning *lenses* only—*not* slides. Slides may be cleaned with ordinary paper towels, but a softer grade of paper is preferable. Paper tissues are satisfactory.

The print used for financial reports and sports statistics in newspapers is small enough to fit into the low-power field of view. Try to find pieces of newspaper with printing on one side only. The strips should be only about 1 cm long.

Ordinary transparent plastic rulers (with a metric scale on one edge) may be cut into several pieces with a saw. Short pieces are easy to handle on the stage of the microscope.

Look out for pieces of magazine photographs without printing on the back.

Procedure

Description of the microscope is necessarily presented in general terms. Fig. 1.5 represents the type that was standard for most of the last half century and is still found in many schools. The best course, then, is to exhibit an example of the type to be used by the pupils, pointing out on it the names of the parts.

Previous pupil experience, the amount and kind of equipment available, the level of pupil ability—these and many more factors determine the rate of progress of any particular class. Most classes, however, require part of a period for the introduction of the microscope by the teacher and two full periods for pupil procedure. Undue haste at this stage may result in faulty techniques and the development of poor attitudes towards careful laboratory work. It is important that sufficient time is given, so that the pupil *can* proceed carefully and *can* have an opportunity to do work of high quality.

The questions in the procedure concern observations or simple calculations from observations. Pupils should briefly note answers in their data books *at the time of observation*. There is no point in having these answers written out and handed in. Check notes informally as the pupils work.

Laboratory book notes should serve as a basis for a short class discussion at the completion of the observation periods. Since the notes are based on simple observation only, disagreements between pupils reflect differences in interpretation of instructions. Do not impose by decree the 'right' observation, but attempt to discover the source of difficulty and then establish consensus. Pupils often have an inordinate desire for recourse to higher authority; only by iron self-discipline will you be able to resist such flattery. But you must.

Measurement with the Microscope

For most pupil microscopes the diameter of the low power field of view is 1.4 mm or 1400 μm. Expect considerable variation in the estimation of this figure. Point out that for future reference it is convenient to reach a class decision. And that this is not a matter of 'He is right, you are wrong.'

For most pupil microscopes (if 1400 μm is accepted as the diameter of the low power field) the high power field is 350 μm in diameter.

Be sure that clearing up at the end of the laboratory period is quick and complete. If this is neglected now, bad laboratory habits will plague you for the rest of the course.

INVESTIGATION 1.4

USE OF THE MICROSCOPE: BIOLOGICAL MATERIAL (pp. 15–17)

The pupil now uses his microscope to observe biological materials.

Materials

To prepare iodine–potassium iodide solution, dissolve 15 g of potassium iodide in 1000 cm^3 of water. Dissolve approximately 3 g of iodine in this solution. Small bottles with dropper tops are useful for dispensing the stain.

Potatoes should be cut into 0.3 cm cubes (approximately). If placed in a little water in a dish, the pieces can be kept for several periods.

To prepare a yeast culture, add approximately 1 g of dried yeast to about 5 cm^3 of water and mix to form a paste. In a glass jar of approximately 400 cm^3 capacity, place about 250 cm^3 of a 10 per cent solution of glucose (10 g glucose per 100 cm^3 water). Pour the paste into the jar and stir to disperse the yeast. Place the jar in a warm, dark place. The culture should be set up several hours before use. Dispense in small beakers.

Procedure

As in Investigation 1.3, pupils should jot down in their lab-books the answers, *in the laboratory and at the time of observation*.

A brief class discussion should follow the laboratory work, but there is no need to achieve unanimity on all the observations. Instead, emphasize the possible reasons for *differences* in observations.

p. 15, B. The detection of the laminated structure of starch grains depends on lighting. Many pupils will not see the structure. Does it exist, or are pupils who report it seeing something that is not there? This is an opportunity for a brief discussion of artifacts in microscopy.

p. 16, C. Differences depend on a gradient of stain concentration and may easily be missed as the stain spreads.

p. 17, A. If starch grains turn up in the yeast preparations, pupils have not used sufficient care in cleaning slides.

p. 17, B. Pupils may have difficulty recognizing that the presence of smaller organisms attached to larger ones indicates the occurrence of budding. An explanatory drawing may be needed. This could be recalled when the concept of an 'individual' is discussed in Chapter 2.

p. 17, C. It is unusual to see much structure in unstained yeast cells. Nuclei might be seen in stained organisms.

The Foundations of Life (pp. 18–25)

p. 19, ¶ 1. The consumer level of each kind of organism is not invariable, but may change according to what is eaten. Relatively few organisms, however, are as omnivorous as man.

p. 20, ¶ 1. It is a common opinion that sunshine is health giving. Point out that animals live in caves and in the deep sea in total darkness. If time permits, vitamin D synthesis may be mentioned. Indicate that no way is known by which the energy absorbed while a human organism is lying on a beach can be transformed into the energy required for muscle contraction.

p. 20, ¶ 3, lines 5–6. The reference is to chemosynthetic bacteria.

p. 20, ¶ 4. The idea is developed further in Fig. 1.10.

p. 20, ¶ 5. The term 'saprovore' is later used at times in the wider sense of anything that eats dead organic matter.

p. 24, ¶ 4. Make use of a periodic chart of the elements to show how few of the natural chemical elements are involved in the chemistry of organisms.

p. 24, Fig. 1.13. The absence of living things would have little effect on the water cycle but it would virtually eliminate the carbon cycle.

p. 25, Fig. 1.16. The biogeochemical cycles of all elements except hydrogen, oxygen, nitrogen and carbon are similar to that of calcium.

INVESTIGATION 1.5

INTERRELATIONSHIPS OF PRODUCERS AND CONSUMERS (pp. 26–28)

Unlike the other investigations in Chapter 1, this one is concerned primarily with the subject matter of the chapter rather than the methodology of science. It can easily be performed as a demonstration set up by a small group of pupils.

If the class has been taught little or no previous science, the teacher may demonstrate the properties of oxygen and carbon dioxide. Avoid becoming too deeply involved in the chemistry. If such a demonstration seems unnecessary, a brief, silent demonstration of the effect of exhaled air on bromothymol blue is useful.

Materials

Test-tubes 20 mm × 120 mm can be used if screw-cap culture tubes are not available. After the experimental materials are placed in them, they should be tightly closed with rubber bungs and sealed with paraffin wax.

If *Elodea* and small snails cannot be easily collected locally they may be obtained wherever aquarium supplies are sold and usefully cultured in the laboratory.

Bromothymol blue is used because of its narrow pH range. pH 6.0 (yellow) to pH 7.6 (blue). A 0.1 per cent stock solution may be prepared by dissolving 0.5 g of bromothymol blue powder in 500 cm^3 of distilled water. To the stock solution add, drop by drop, a very dilute solution of ammonium hydroxide until the solution turns blue. If the water in your area is alkaline, the addition of ammonium hydroxide may not be needed. (Solutions of bromothymol blue purchased from suppliers are usually made up with alcohol; these solutions kill organisms.)

The use of fluorescent light will avoid excessive heat.

Discussion

In light a green plant *appears* to play a role in the exchange of CO_2 and O_2 opposite to that of animals.

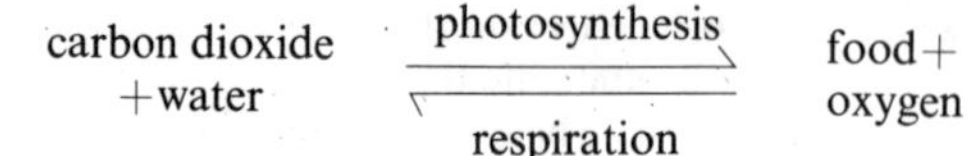

But respiration of course goes on continuously in both animals *and* green plants. In light, green plants re-use almost immediately in photosynthesis the CO_2 released from respiration, so it does not accumulate in the environment. In darkness, however, photosynthesis does not occur; in which case the carbon dioxide is not re-used. Therefore, bromothymol blue in a culture tube containing *Elodea* in darkness turns green, because carbon dioxide from respiration accumulates in the water, acidifying it. When photosynthesis begins on exposure to light, carbon dioxide is used up more rapidly than it is released. After a short time, the plant must begin to extract carbon dioxide from the water. As the amount of carbon dioxide in the water decreases, the bromothymol blue reverts to its blue colour.

So much for the theory of the experiment. In reality many strange things may happen, testing the ingenuity of teacher and pupils in finding an explanation. For example Tube A2 may be green or yellow at the top and blue at the bottom. (In this case, it is usually found that the snail is at the top of the tube and the *Elodea* at the bottom.)

Studying the Data (p. 27)

A. In most cases the snail in Tube 1 dies first, but it may live longer than the snail in Tube 2 in the *dark*—since, in darkness, both plant and animal are using O_2 from the water.

E. In Tubes 4 a change in colour of the indicator may be brought about by carbon dioxide produced by micro-organisms present in pond water.

Conclusions (p. 28)

B. Hypotheses about the carbon cycle—there are many that can be made, but focus attention on statements that *can be tested.* Examples: Both plants and animals give off CO_2 as a by-product of life processes. Light destroys CO_2 in the presence of plants.

The Web of Life (pp. 28–31)

p. 28, ¶ 3. Although frequently used in this sense, the definition of the term 'food' is certainly not restricted by biologists. The reason for the arbitrary restriction is simplification. Attention is focused on energy flow and much ambiguity is avoided.

p. 29, Fig. 1.16. Note that individual arrows do not go to each kind of saprovore, but rather to all the first-order saprovores, all the second-order saprovores, etc., and that many arrows could run from the various saprovores to things that eat them. Organic objects (a dead leaf, for example) may be chewed by millepedes, ants, springtails, snails, mites, etc. This action helps to break down the leaf. The partially decomposed leaf as well as the faeces from these animals, makes a good substrate for the growth of bacteria and fungi, which further reduce the substances. Finally the entire 'mess' goes through the intestines of earthworms or roundworms. Whenever a saprovore is working on substances derived from plant material, it is a first-order consumer.

p. 30, Fig. 1.17. This diagram introduces an important concept—continuity of biological processes. Can both Figs. 1.16 and 1.17 be read with equal confidence? Here is an opportunity to encourage the habit of evaluating sources of data.

Man and the Biosphere (pp. 31–33)

From the viewpoint of the aims expressed in the Introduction (pp. T5–6) to the Guide, this is the crucial and climactic section of Chapter 1 in *WL*. But excessive discussion at this stage will not increase its effectiveness.

p. 32, ¶ 1, line 1. Note the 'if'. In their reaction to anti-evolutionists, biologists sometimes become excessively dogmatic. We prefer to let the evidence speak for itself over the whole series of books.

p. 32, Fig. 1.19. The question is, of course, rhetorical.

p. 33, ¶ 2. Class discussion of these problems should be left until the study of this course of biology is completed.

GUIDE QUESTIONS (p. 34)

8. Note that this question requires that the caption of Fig. 1.10 be read. This makes the point that most of the illustrations have content and must not be ignored.

PROBLEMS (pp. 34–35)

1. Coal, oil and gas are 'fossil hydrocarbons' —remains of plants and animals that were only partially decomposed by saprovores before they were buried and that were later altered by heat and pressure within the Earth.

2. The balance of nature would remain but the point of equilibrium might be considerably shifted. Homeostatic adjustment of other populations would compensate for change in one population. How long is 'permanent'? If an algal population increases greatly, populations of consumers might also increase. Effects of an increase in a fish population would depend primarily upon the consumer order of the fish.

3. If the Earth is not itself considered to be derived from solar matter, then the exceptions are volcanic and hot-spring energy, energy of radioactivity, nuclear energy, chemical energy in compounds formed by processes within the Earth (but not chemical energy in substances, such as nitrogen compounds, formed by atmospheric processes).

4. It is virtually impossible to define life in a satisfactory way—and this is the point here. If a pupil works at it, he will be able to think of some good *characteristics* of life.

6. A decrease in oxygen in the atmosphere and subsequent changes in the thermal climate resulting from changes in energy absorption and transmission in the atmosphere.

8. The main point here is to design a low-weight but highly efficient system that *recycles* materials. Consider starting with the widest *variety* of small producers and consumers, putting them in the kind of space environment they will be subjected to, and then letting the *system* select the proper group of organisms that will maintain a steady state.

9. This is another approach to the problem of limits to scientific inquiry. What kind of *data* can be collected to prove something *absent*? This leads to a discussion of the nature of proof.

TEACHER'S REFERENCE BOOKS

ABERCROMBIE, M., C. J. HICKMAN, M. L. JOHNSON. *A Dictionary of Biology*. Penguin, 1971.

BOTTLE, R. T., H. V. WYATT—Editors. *The Use of Biological Literature*. Butterworth, 1966. (A useful and reliable reference work.)

JAEGER, E. C. *A source book of biological names and terms*. 3rd edition. Blackwell, 1959.

PHILLIPSON, J. *Ecological Energetics*. Arnold, 1970.

SCIENTIFIC AMERICAN. *The Biosphere*. Freeman, 1970. (This goes deeper than is necessary for teaching Chapter 1 but is still very useful.)

2
INDIVIDUALS AND POPULATIONS

MAJOR IDEAS

1. The individual organism, regardless of difficulties in definition, is a primary unit of biological study.
2. Individuals may be grouped in different ways. The indefinite but versatile term 'population' is useful for such groupings.
3. Quantitative study of populations involves the idea of density—the number of individuals per unit of space.
4. Populations continually change in size. Such changes are determined by the interaction of four rates—natality, mortality, immigration and emigration.
5. These factors are affected by biotic and abiotic environmental factors that are continually interacting.
6. Therefore, the study of natural populations is immensely complicated and the application to natural populations of results from experimental populations is fraught with uncertainties.
7. In general, however, natural populations appear to be maintained in a steady state by the continuous operation of homeostatic mechanisms.
8. Mathematics is an essential tool of biology.
9. The species population is an entity of great practical and theoretical importance in biology.

PLANNING AHEAD

Consider the season and the weather prospects in your locality with respect to the planning for Investigation 3.1. Arrange for an alternative lesson plan in case of bad weather.

Check the number of thermometers (Celsius) available for Investigation 3.2.

It is not too early to begin accumulating living animals for Chapter 1 of *DLT*. Obtain as great a variety of individual animals as possible. Collections should be made or orders should be placed for *Hydra*, planarians, earthworms, frogs, crayfish, and brine shrimp or *Daphnia* required for Investigation 1.3, *DLT*.

GUIDELINES

The introductory and more superficial treatment recommended for the teaching of Chapter 1 cannot be continued into Chapter 2. The concepts of 'individual' and 'population', first encountered here, form a foundation on which much of the remainder of the course in this and the four other books in the series rests. In Chapter 3, communities are regarded as interacting populations. In the book *Diversity Among Living Things* the species population is the unit of taxonomy. The 'patterns' discussed in the book *Patterns in the Living World* are composed of individuals grouped as species populations. Though these concepts sink into the background in the first part of the book *Looking into Organisms*, the individual emerges again as a unit of biological organization in the last chapter. Subsequently, in *Man and his Environment*, individuals and populations remain the focus of attention. So thorough study of Chapter 2 in this volume is essential.

Work on this chapter should begin with the textbook. Chapter 2 can best be assigned in three sections: pp. 36–42 (guide questions 1–6), pp. 45–52 (guide questions 7–10), pp. 56–68 (guide questions 11–16). Soon after the pupils have been exposed to the first assignment they can be introduced to Investigation 2.1. Later, after commencing Investigation 2.2, the mathematical ideas in the first assignment can be more extensively developed. The second and third assignments can be done during the time required for growing the yeast populations of Investigation 2.2.

The pupil encounters many difficulties in this chapter and the teacher must be prepared to assist him. The major difficulties are: abstract ideas; new vocabulary (sometimes disguised as familiar words); and a long and more demanding laboratory procedure in Investigation 2.2.

For teachers and pupils who regard textbooks merely as fodder for memory work, this series of books will prove to be disconcerting: they present issues and then fail to resolve them. On the other hand, when textbooks are used as a jumping-off point—a basis for discussion—this is a useful characteristic. Such unresolved, open-ended description recurs throughout the series, but specific examples can be pointed out in Chapter 2. On p. 37 (paragraph 5), opportunity exists for pursuing the matter of definition in science. On p. 38, the third paragraph raises a point that reappears as an important theme in Chapter 3 in *MHE*. On p. 58, paragraph 1, and on p. 65, paragraph 1, other unresolved problems appear. The teacher who disregards such opportunities cannot begin to achieve the objectives of this course.

The methods used for the study of population density and of the interaction of rates should be a matter for discussion with the mathematics staff. Examples must be worked through with the class and then pupils must practise with numerous problems. Some practice may be obtained from problem 1, p. 69 of the text and from the additional problems given in the Guide on p. T34. But teachers will no doubt prefer to devise problems of their own, since problems with a local flavour are, of course, the best. Such problems should be reviewed occasionally during the several months following 'completion' of work on Chapter 2.

Both at the beginning and at the end of Chapter 2, the difficulty of definition is discussed —first in connection with individuals, finally with regard to species. From this the pupil may have formed the idea that slipshod language is characteristic of science. The introduction of the concepts of density and rate, with their mathematical expression, helps to counteract such an impression.

TEACHING NOTES

Individuals (pp. 36–37)

The term 'individual' is appropriate. As a living unit of the biosphere, an individual can be thought of as indivisible. Where the ability to divide exists, new individuals result. Portions of an individual can be maintained, as in tissue cultures, but this is obviously artificial.

p. 36, Fig. 2.1. The question in the caption is rhetorical.

p. 37, ¶ 5. The use of pseudo-precise language where ideas are imprecise involves just as much falsity as does the use of imprecise language for ideas that are precise.

Populations (pp. 37–42)

p. 40, Fig. 2.4. The density is 0.0144 trees/m^2, or 144 trees/hectare.

p. 40, Fig. 2.5. The density of clover is 0.0147 plants/m^2 in the field as a whole and 0.0424 plants/m^2 in the north-west quarter. Densities of dandelion and buttercup can also be calculated and compared.

pp. 40–41. A rate of change in population size without consideration of the space factor can be calculated in the same way as a rate of change in density. For some pupils the size calculation may appear simpler; if so, the teacher may provide an example before considering the density case.

p. 41, Fig. 2.6. Relate the meaning of a change in slope to the interpretation of the line graph drawn in Investigation 2.1. The text example uses a decline because the investigation deals with an increase.

p. 42. Be sure the pupils actually consider each of the rhetorical questions.

INVESTIGATION 2.1

POPULATION GROWTH: A MODEL (pp. 43–45)

This investigation of a hypothetical population provides a basis for comparing the real populations encountered in Investigation 2.2 (a closed population with a fixed food supply and no provision for elimination of wastes), and Investigation 2.3 (an open, natural population). It also continues the task begun in Investigation 1.1—to introduce the pupil to scientific methodology. Therefore attention must be directed, at first, to p. 43, where the use of conceptual models and the role of assumptions are discussed.

Materials

Semi-log graph paper may be obtained in 1, 2, 3, 4, 5 or more cycles. Such paper is logarithmic on one axis and is regularly spaced on the other. The 1975 summer population is 31 250 birds, and 5-cycle paper is sufficient for this. Paper satisfactory for this exercise can, if necessary, be produced on school spirit duplicators from stencils prepared by the teacher, the finer subdivisions shown on commercial paper being omitted.

Procedure

The computations and the construction of the graph on the ordinary grid can be done at home. Some previous ground-work should have been done in the class. This should include some discussion of assumptions. In addition, many teachers find a review of line-graph construction advantageous. Often pupils attempt to solve the problem of choosing a suitable scale for the ordinary graph by arbitrarily changing the value of intervals on the vertical axis. Such pupils, of course, have no idea of the meaning of gradient. At this stage, however, it is only necessary to point out that a scale must have some mathematical consistency.

To some pupils the semi-log grid appears as arbitrary as their own. But it is not necessary to become deeply involved in mathematics. Briefly develop the series 10^1, 10^2, 10^3, 10^4, etc., note the correspondence of the indices to the number of zeros in the series 10, 100, 1000, 10 000, etc. Direct pupils to label the cycles on the semi-log graph paper: units, tens, hundreds, etc. Point out that each succeeding cycle represents numbers ten times greater than those represented by the preceding cycle. Link this idea with the fact that within each cycle the system of second-order subdivisions separates spaces of decreasing width. Then illustrate the plotting of points, using numbers different from those used in the investigation.

p. 44, C. The principal advantage of semi-log paper is that it permits the plotting of very large numbers in later generations while showing clearly the small increments in earlier generations. The straight line obtained on the semi-log paper indicates a constantly accelerating growth rate. If it is known that a rate is of this kind, the plotting of only two points will establish the gradient. Extrapolation is then easy. With most classes only the principal advantage need be stressed.

Studying the Data (p. 44)

A–B. Unfortunately many pupils with competency in making graphs have little knowledge of how to interpret them. Therefore a great deal of attention should be given to the concept of gradient. The relationship of gradient to rate is basic in the interpretation of graphs. It is discussed simply here, since all gradients in the graphs for this exercise are positive. But if there seems to be no danger of confusing the pupil, the significance of zero gradients and negative gradients can also be discussed, reducing the difficulties to be encountered in Investigation 2.2. The graphs in Chapter 2 should be helpful—especially Fig. 2.6.

D. Continuing to use the same set of assumptions for an indefinite number of years, the line on the graph paper would approach the vertical. On semi-log graph paper the line would continue on the same gradient.

E. The main point about the growth of this hypothetical population is that the rate is accelerating.

F. Whether any population might grow in this way should be left open to argument at this stage. The idea of limiting factors and the impossibility of an infinitely large population will come later.

Further Investigations

The last of these investigations invites the pupils to devise other problems. The usual difficulty encountered here is vagueness in stating the assumptions.

Populations and Environment (pp. 45–52)

p. 46, Fig. 2.8. This is a photograph of an experiment being carried out at the Pest Infestation Control Laboratory, Surbiton, by Drs. P. Crowcroft and F. P. Rowe.

p. 47, ¶ 3. This illustrates the importance of careful description of methods in a published report of a scientific experiment. Without it no verification is possible.

p. 47, ¶ 5. Uncooked plant foods are, of course, frequently alive when eaten. Pupils are often surprised at this idea.

INVESTIGATION 2.2

STUDY OF A YEAST POPULATON
(pp. 52–56)

Pupils must obtain some first-hand experience with population dynamics. This investigation forms the hard core of such experience in the laboratory and without it Investigations 2.1 and 2.3 are meaningless. Investigation 2.2 builds upon the pupil's experience of experimental methods (Investigation 1.2) and applies his developing ability to use the microscope (Investigations 1.3 and 1.4). Furthermore, it greatly extends his conception of teamwork in science and his appreciation of the value of replication. The teacher *must* overcome any temptation to retreat before the numerous and varied difficulties that are likely to be encountered. Remember, this investigation *has been done* over and over again in a wide variety of classroom situations!

Considerations of time and space severely limit the choice of organisms. On both counts microscopic organisms are ruled out. Algae reproduce somewhat too slowly; bacteria are too small for the direct observation needed to provide a sense of reality. Yeast seems to be the most suitable organism. It reproduces rapidly, its requirements are simple, it is easily visible with the high power lens of the microscope, it responds readily to a decrease in food concentration and an increase in toxic substances and it is an economically important organism already well known to pupils by name and appearance (Investigation 1.4).

Because of the time span it requires, this investigation should be set up early in the work on Chapter 2. Ideally, Investigation 2.1 should be completed before 2.2 is started, but this is not absolutely necessary. The purpose of Investigation 2.2 will be clear any time after 2.1 has begun.

From start to finish this exercise requires two to three weeks, but much of the class time during this span is utilized in other work.

Timing

1. Procedure I requires about a single 40-minute period.
2. On each subsequent day, inoculation requires not more than five minutes and involves only three pupils.
3. Procedure II requires a double period.
4. Discussion of the results, if it is properly done, will require at least one full period.

Materials

The medium can be prepared by a group of pupils. For this job the following will be required:

2000 cm^3 Erlenmeyer flask
Measuring cylinder
Large stirring rod
Source of heat
Test-tube basket
Autoclave or pressure cooker
Test-tubes of about 20 cm^3 capacity, 1 per pupil plus 1 per group of 10 pupils
Square of aluminium foil, large enough to form a cap over a test-tube, 1 per pupil
Balance
Spatula
Clean, soft cloth

Yeast extract, 2.5 g
Potassium dihydrogenphosphate(v) (dihydrogen orthophosphate), 2.0 g
Glucose, 40.0 g
Peptone, 5.0 g
Distilled water, 1000 cm^3

10 cm^3 of medium is needed for each pupil, plus 10 cm^3 for each group. (The quantities listed above will make a little more than 1000 cm^3 of medium.) Calculate the amount of each material needed. Weigh out the amounts of dry materials and add them to the required volume of water. Dissolve the materials by stirring continually over a low-heat source. When properly prepared, the medium is sparkling clear and slightly yellow in colour.

Pour 10 cm^3 of the medium into each tube (1 tube per pupil plus 1 tube per group). Shape a square of aluminium foil as a cap over the mouth of each test-tube. With the caps fitted lightly over the tubes, sterilize at 15 lb/in^2 (10^5 N/m^2) in an autoclave or pressure cooker for fifteen minutes. Tighten the foil caps and store the sterile, cooled test-tubes in a refrigerator.

Few schools have access to an autoclave. But a large pressure cooker of the kind used at home is an essential investment. Later investigations, involving micro-organisms, require it. For Investigation 2.2, however, it is *possible* to manage without a pressure cooker. With luck, a sufficient degree of sterilization may be obtained by boiling the medium and the tubes in water for ten minutes. If this is done, an uninoculated tube of medium should be boiled as a control.

Aluminium foil caps are easy to make. Use a square of foil a little more than three times as wide as the test-tube. Place the mouth of the test-tube in the centre of the square, fold the foil edges down by running the tube through a hole made by curving the forefinger around the base of the thumb. Bacteriological cotton plugs *can* be used instead of foil, but making them requires extra skill—and they are messy.

Brewer's yeast tablets can be substituted for yeast extract. Mix the tablets in about one-fifth the amount of water to be used in making medium sufficient for one class. Let it stand overnight, decant and then filter.

Sodium dihydrogenphosphate(V) or dipotassium hydrogen phosphate(V) can be substituted for potassium dihydrogenphosphate(V).

Sucrose may be substituted for glucose, but is less desirable than the substitutions mentioned above.

Beef bouillon cubes may be substituted for peptone. Two cubes are sufficient for 1000 cm^3 of hot water. To remove the fat, cool the mixture and filter.

If the mineral content of the local water is low, tap water may be used instead of distilled water. Or the water can be treated with a de-ionizer.

Though 10 cm^3 is a convenient and efficient amount of medium per pupil, as little as 8 cm^3 may be used—or as much as 15. The amount in each test-tube must be the same. If all the tubes have the same diameter, time can be saved in dispensing the medium by making a water 'blank'. Carefully measure out 10 cm^3 of water and pour into the 'blank' tube, then pour medium into each tube to the level of the water in the 'blank'.

When a pressure cooker is used, care must be taken to bring the contents back to room pressure slowly, to avoid the caps being blown off the tubes.

Procedure I

The investigation is written with a class of 30 pupils in mind—three groups of 10. Each pupil is responsible for inoculating a tube. Pupils are numbered according to tube numbers, each tube number corresponds to the number of days the tube is to be incubated before counting day. On counting day pupils work in pairs—No. 1 with No. 2, No. 3 with No. 4, etc.

Of course, few classes consist of exactly 30 pupils. If there are fewer than 30, the number of incubation days can be reduced to 8 (requiring groups of 8 pupils), but then the chances of obtaining a clear decline in populating are reduced. A better plan would be to have some pupils take responsibility for more than one tube. Such doubling up should be done with tubes having lower numbers, because they are less likely to require dilution on counting day.

For classes of more than 30 pupils, other kinds of adjustment are necessary. First, an additional pupil may be assigned to work with the tube inoculated on counting day, making the number of pupils in each group 11. Second, extra pupils may be assigned general duties connected with distribution of and accounting for materials, and they can always substitute for the inevitable absentees. Finally, a small separate group may be assigned the task of preparing the medium.

Each group should have a leader and a co-leader. They are responsible for seeing that the tubes are inoculated according to schedule by group members. Make a chart showing classes, groups and tube numbers, with the name of the assigned pupil opposite each of the numbers. Placed on the notice board, such a chart helps remind pupils of their tube numbers and may be used to check daily inoculations.

For ease in handling the sets of tubes, it is advisable to designate each group with an individual letter that is not repeated in other classes. Groups A, B, and C might be in one class. Groups D, E, and F might be in another, and so on. With this system, containers holding each group's set of tubes can be distinctively labelled. The containers may be bacteriological culture baskets, or beakers.

Pupils may experience some difficulty using the glass-marking crayon on test-tubes. Small pieces of masking or labelling tape may be used instead. Both group letter and tube number may then be written with ordinary pencil. Test-tubes with ground areas on which pencil may be used are also available. If desired, the eleventh tube may be placed with the group sets and labelled 0.

Experience has shown that fairly uniform initial populations can be obtained by the apparently haphazard method of inoculation. It *is* important to exercise some care in picking grains that are as nearly the same size as possible. It is not necessary that uniformity of grain size be obtained among groups—only within groups.

No matter which day of the week is chosen for beginning the inoculations, at least one weekend will be involved in the schedule. The teacher will have to decide how inoculations can best be made at weekends.

The cultures may be incubated at room temperature (normally about 22 °C). At this temperature the populations should be declining (the 'death phase' of the growth curve) during the ninth and tenth days. It is important to keep the cultures from draughts and sudden changes in temperature. The cultures may conveniently be kept in an incubator or a cupboard against an inside wall of the building. The development of the population may be accelerated by increasing the temperature of incubation, or conversely, it may be slowed down by lowering the temperature. If facilities for temperatures above and below room temperatures are available, some classes might be made to incubate their cultures at such temperatures. On the whole, however, this is not recommended, and for the following two reasons:

1. It introduces a variable that, while interesting, is irrelevant to the purpose of the experiment.
2. It reduces the number of replications available for 'smoothing out' the population curve.

Procedure II

In Investigation 1.4, pupils observed yeast organisms, but they have not attempted counts before and they are still novices in manipulating the microscope.

A day or two before counting day, which is the day after Tube 1 has been inoculated, pupils may be given opportunity to practise the counting technique. First, ask for counts of the 20 individuals in Fig. 2.14, p. 53. Then provide a fairly dense culture of yeasts and have counts made under the microscope. Just as they will on counting day, have the pupils work in pairs, checking each other's counts. Have all pupils practise diluting, even though not all will need to do this on counting day. The pipettes or droppers should be calibrated for number of drops per cm^3; not all deliver 20 drops/cm^3.

If counting cannot be completed in one day, the cultures must be stored in a refrigerator, so that the population of the first day is maintained with as little change as possible. There is no need for sterile techniques on counting day, because the cultures will not be used again.

One of the most common sources of error in making cell counts is an uneven distribution of cells in the medium when the sample is removed. The pupil should shake the tube vigorously and then make a quick transfer to the slide. Since maintenance of sterile conditions is not essential, adequate shaking should not be any problem. The pipette should be thoroughly rinsed after each sampling. Before counting is begun, the floating organisms should be allowed to come to rest.

Adequate instructions for making counts are included in the pupil's book. At this point the best advice for the teacher is to read instructions carefully and to anticipate (on the basis of his knowledge of pupils) as many difficulties as possible, taking steps to obviate them. He should insist that pupils follow directions. Link this insistence not to obedience but to the scientific necessity for uniformity of procedure.

Discussion

Check to see that all pupils have included the dilution factor ($\times 1$, if no dilution is made) in their calculations. The table shown in Fig. T-1 is a convenient one for gathering data of all groups.

A wide range of numbers is likely to appear when counts of different groups are gathered together. Furthermore, fluctuations in the population, as measured by any one group, are likely to be so great that to detect any pattern of growth may be difficult. When, however, the data from many groups are averaged, a fairly good growth curve usually results. Fig. T-1 shows data obtained by three classes of somewhat less than average ability, and Fig. T-2 is a graph based on the average of data from all nine groups.

To obtain a curve that is explicable in terms of population theory is, of course, desirable and very satisfying to pupils. But failure to obtain such a curve must not be interpreted as failure of the investigation.

An explanation of the similarities and differences in the graphs, on the basis of the headings (*a*) to (*d*), p. 56, is the pivot on which the investigation turns. No matter what the results, they will provide material for a fruitful discussion of sources of error in an experimental procedure and of the need for teamwork in some kinds of scientific work.

Conclusions (p. 56)

A–C. Responses depend, of course, upon the nature of the graph obtained from experimental results. Comparison of Fig. T-2 with the graph obtained in Investigation 2.1 shows that, initially, the growth rate of the yeast population accelerated in a manner similar to that of the hypothetical sparrow population. The yeast population, however, soon reached a peak and began to fluctuate. Superimposed upon the fluctuations is a decline in the population. To explain the similarities and differences in simplest

GROUP	DAYS										
	0	1	2	3	4	5	6	7	8	9	10
A	18	218	219	162	355	95	175	132	167	485	136
B	24	63	69	283	281	161	147	365	199	227	314
C	39	61	363	56	20	114	322	41	66	87	38
D	36	53	75	710	56	240	230	190	200	630	340
E		30	210	45	59	46	82	453	93	60	88
F	47	71	73	170	20		242	660	73	110	55
G	16	25	35	980	540	50	350	165	14	160	212
H	48	42	36	650	760	500	305	356	313	65	69
K	23	344	60	45	90	330	54	250	37	138	74
Total	251	907	1140	3101	2181	1536	1907	2612	1162	1962	1326
Average	31	101	127	345	242	192	212	290	129	218	147

Fig. T-1

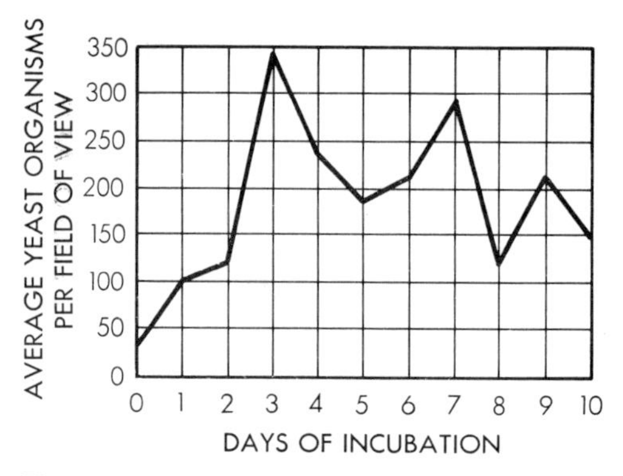

Fig. T-2

terms, you need only point out that both the sparrow and the yeast populations were initially small populations in a presumably favourable environment. Assumptions in Investigation 2.1 made no allowances for the 'facts of life'—the finite quantity of resources on the island. The yeast exploited a constantly declining food supply and encountered an increasing accumulation of waste products in the environment. So environmental resistance operated only in the yeast experiment.

Regardless of the nature of the results, the hypothesis set up at the beginning of the investigation must be considered at its end.

Kinds of Population Changes (pp. 56–61)

p. 57, ¶ 2. Note that this discussion depends for its meaning on the work done in Investigations 2.1 and 2.2.

p. 58, ¶ 1. The word 'cycle' should be given special attention because its sense here is different from its sense in Chapter 1—particularly when Figs. 1.12, 1.13 and 1.14 are compared with Fig. 2.18.

p. 59, Fig. 2.20. The graph is based on studies by J. Davidson, *Transactions of the Royal Society of South Australia*, 1938, **62,** 342–6. Since sheep are entirely dependent upon man in Australia, changes in population reflect man's activities.

INVESTIGATION 2.3

POPULATION CHANGES IN OPEN SYSTEMS (pp. 62–64)

The graphs for the investigation can be prepared at home—and at any time after Investigation 2.1 has been completed. Consideration of the questions, however, should be delayed until Investigation 2.2 has been discussed. The conclusions to Investigation 2.3 constitute a summary of all three of the investigations.

Procedure

Field mouse. p. 61, A–F. The July–October portion of the curve resembles the growth of the hypothetical sparrow population, while the October–April portion resembles the declining phase that may have been obtained in your yeast population. The mouse curve stabilized from April to June. This is a result of limiting factors, which were ignored in Investigation 2.1. There may be little apparent difference between the mouse and yeast curves. The mouse population is an example of an open population and, as such, illustrates seasonal increases and decreases around some steady-state level. Natality is highest in autumn and early winter. The effect of mortality is most apparent in late winter and early spring.

Grey squirrel. p. 62, A–C. The seasonal fluctuations of squirrels is due to the addition of young in autumn. This is a good example of a 'new' population that has not yet reached the carrying capacity of the habitat. Counts at some time after 1958 would undoubtedly have shown a decrease in the rate of population growth. It is not possible to predict from these data when this would have occurred.

Heath hens. pp. 62–63, A–D. The increase of heath hens in 1915 is due to preservation efforts. An excessive number of males not only reduces natality but also causes increased aggression. The population became extinct in 1932.

Conclusions (p. 63)

A–E. The characteristic curve of 'new' populations shows an exponential increase. The chief difference between the graph of the hypothetical population and those of the real populations is that it shows none of the variability inherent in real data. Also, given the assumptions, it can never show decrease. The mouse curve best illustrates steady state. It shows annual fluctuations around a steady-state level.

Species Populations (pp. 64–68)

p. 57, Plate III. Dimorphism related to sex and age is quite common in many organisms. Birds are used as examples because they are colourful and generally familiar. For other examples of dimorphism, see Fig. 3.7 and Plate II in *MHE*.

PROBLEMS (pp. 69–70)

1. Simply add the rates of natality and immigration (43/year) and subtract the total of mortality and emigration (38/year). This means there was an addition of 5/year, mostly owing to surplus natality. In 10 years the snail population would equal 65.

2. Ecologists are interested in *changes* in density. For this, estimates are quite useful if they are made by a consistent method and if the amplitude of population fluctuations is greater than the margin of error in the estimates.

3. The information needed can be obtained from p. 602 of the 1971 edition of *Whitaker's Almanack*. (*a*) The curve to date resembles that of the hypothetical population of Investigation 2.1. (*b*) The decline in rate of increase of population during the decade 1911–21 can be attributed to great numbers of men killed during the 1914–18 World War. (*c*) In the future the difference between immigration and emigration is likely to be negligible. The death rate will presumably continue its downward trend, and, though the birth rate may decline slightly, a continuing increase in population may be predicted. (*d*) To calculate densities, the area of the UK is required. (*e*) The plot of density parallels closely the total population curve.

4. If A breeds with B and B with C, then only one species is involved. If A became extinct, we would still only have one species. But with B gone we have two, since A and C do not interbreed.

5. The problem involves the time at which an individual becomes an effective part of the population. Is a seed a new individual at the time of zygote formation, at the time of dispersal, or not until germination? The same problem exists for animals, particularly egg-laying ones. For some purposes reproductive additions to a bird population may be counted from egg-laying, and for other purposes from hatching.

SUPPLEMENTARY MATERIALS

ADDITIONAL PROBLEMS

As indicated previously, the ideas concerning population change must be developed chiefly through the consideration of problems. As far as possible, problems should be given a local flavour. Those appearing below are far from adequate in number, but they may serve as a starting point.

1. What is the density of the pupil population in your English classroom compared with the density in your biology classroom?

2. On a farm of 450 hectares is a total of 1275 rabbits. Studies indicate the following rates for this population.

Mortality	2225/year
Natality	3400/year
Emigration	775/year
Immigration	150/year

Is the population increasing? Decreasing? At what rate? Predict the population at the end of four years. What is likely to happen to the population of producers in this area during the four years? [*Increasing at a rate of 550 per year. Producers likely to be decreased.*]

3. In a field of wheat the population of weeds (that is, all the plants other than wheat) is estimated at 35 per m^2. Half the field is treated with chemical A, half with chemical B. The density of living weeds in the A half:

End of first week	$22/m^2$
End of second week	$13/m^2$
End of third week	$8/m^2$
End of fourth week	$6/m^2$
End of fifth week	$7/m^2$

The density of living weeds in the B half:

End of first week	$32/m^2$
End of second week	$26/m^2$
End of third week	$18/m^2$
End of fourth week	$7/m^2$
End of fifth week	$5/m^2$

During the five weeks, what is the net rate of decline in the density of the weed population in each half of the field? Compare the rates of decline in the two halves during the first week after treatment. What population determiner is dominant in this situation? What population determiner is probably operating in the A half of the field between the fourth and fifth weeks? [*Decline in A: 5.6 weeds per week. In B: 6.0 weeds per week. Chemical A is initially more effective as a weed killer but does not act over as long a period as B. Natality surpassed mortality between the fourth and fifth weeks in the A half. Many more questions can be based on these data.*]

4. Many plants, such as couch grass or strawberries, spread themselves by runners which extend out and form roots and leaves at intervals. All the 'plants' may remain connected to one another. (*a*) Would you call a patch of one of these a single plant or a population? (*b*) How would you set about measuring the density of such plants? [*Where the individual cannot be picked out easily, as with couch grass, a convenient unit of vegetation can be used to reflect the density of the species, i.e. erect shoots, flowering stems or even whole clumps.*]

AUDIOVISUAL MATERIALS

It is probable that more is to be gained by concentration on problems than by much use of time for films and filmstrips.

Population Ecology. Encyclopaedia Britannica Films, available on hire from the National Audio-Visual Aids Library and the Scottish Central Film Library. 16 mm, colour, 21 min. (Probably the best film available on the subject.)

TEACHER'S REFERENCE BOOKS

LACK, D. *Population Studies of Birds*. Oxford University Press, 1966. (Birds probably provide the best existing data for studies of natural population dynamics.)

MALTHUS, T., J. HUXLEY, and F. OSBORN. *On Population: Three Essays*. Mentor, 1964. (Provides the most ready access to Malthus. Since human population is not stressed in this chapter, the Huxley and Osborn essays may be postponed until study of Chapter 5 of *MHE*—the fifth book of the series.)

SOLOMON, M. *Population Dynamics*. Arnold, 1971.

WILLIAMSON, M. *The Analysis of Biological Populations*. Arnold, 1971.

3

COMMUNITIES AND ECOSYSTEMS

MAJOR IDEAS

1. A biotic community consists of the sum of all interactions among the species populations at some particular time and place.
2. Various kinds of community interactions can be recognized. Basically these relationships are concerned with the exchange of energy between species.
3. The significance of a community interaction is best determined by measuring its effect upon the population densities of the species involved.
4. The delimiting of communities is a subjective process. In nature there is always continuity in both space and time.
5. Every biotic community is part of an ecosystem, which is a complex of all biotic and abiotic interactions within a definable unit of space and time.
6. At any particular place a transition of ecosystems occurs through time. This is ecological succession.

PLANNING AHEAD

The fieldwork of Investigation 3.1 may present opportunities to assemble specimens required for the investigations in Chapters 1 and 2 of *DLT*. Many small invertebrates can easily be kept alive for several weeks, as can mosses, lichens, algae and fungi. Leaves for Investigation 2.1 in *DLT* must be collected now, if not already done.

Order cultures of *Serratia marcescens* and *Sarcina lutea* for Investigation 3.2, and a mixed culture of bacteria for use in Investigation 3.3, all in *DLT*.

If you have not already done so, become acquainted with the scheme of classification used in the Appendix to *DLT*.

GUIDELINES

The break between Chapters 2 and 3 is one of convenience. The sequence from individual to ecosystem extends through both chapters. The pupil should come to see this continuity clearly.

It is essential that the pupil gain direct experience with communities. Within the classroom simple laboratory communities can, of course, be maintained. Aquaria and terraria are old standbys. In many classrooms, pupils, their lunches, their viruses, houseflies and mice form a community that the alert teacher will use, especially in illustrating parasitism and commensalism. Work on Investigation 3.1 in *PLW* could be started at this stage.

In the view of the writers, all this is still inadequate. In city schools especially, it is important to get beyond the walls of the classroom, to show pupils that life is not entirely an *in vitro* phenomenon—that community interactions are not confined to test-tubes, glass tanks and laboratories.

In addition to having first-hand experience with a local community, pupils should become acquainted with several other communities. Pp. 73–75 are a step towards this. Pictures—including slides, and films particularly—can be useful. In every case, attention should be directed towards the ecological relationships that bind the species in a community together. Mere enumeration of species does not depict a community.

The text of Chapter 3 fills only 29 pages. It does not lend itself well to short assignments.

Pp. 71–75 should be read before the fieldwork for Investigation 3.1 is done; the rest afterwards.

With classes that can be trusted to glean for themselves the more obvious ideas from the text, the teacher can take time to stress some of the more abstract features of the chapter. First, that the terminology of community relationships offers an opportunity to extend the discussion (begun in Chapter 2) of definitions in science. Second, that the paradox of change versus constancy—of steady state—is an idea that recurs frequently. It came early in Chapter 1 and again in Chapter 2. Third, that the continuity of communities and ecosystems in time and space without real boundaries is merely a special example of the unity of biology.

Chapter 3 completes the series of ecological units begun in Chapter 2 and also marks the end of the book. Therefore, it should be brought to a close with a careful review of the series: individual, population, species population, community, ecosystem. This may be the time for a test but the pupils must understand that it is only an interim summary and ends nothing.

TEACHING NOTES

The Biotic Community (pp. 71–76)

The concept. pp. 71–72. Do not confuse a community with a society. In the usage adopted in *WL*, a community involves interspecific relationships; a society involves intraspecific relationships only. In a beehive there is a society, not a community (unless species other than bees are also considered).

pp. 73–75. Use this river community to stimulate discussion of similar communities in your neighbourhood. You should not expect pupils to memorize the specific animals and their relationships in the Tees community.

pp. 75, Fig. 3.4. The question in the caption does not at this point require the use of the terminology developed later in the chapter, though you may want to come back to this illustration later.

INVESTIGATION 3.1

STUDY OF A BIOTIC COMMUNITY (pp. 76–82)

Every investigation should be carefully planned, but one that takes pupils out of the school building demands extra care. Unless both teacher and pupils know exactly what is to be done, this investigation is liable to lack educational value. Given time and attention, however, it can be one of the most rewarding experiences of the year.

For reasons given on page 76 of the pupil's text, it is impossible to prescribe a single procedure for this investigation. Nevertheless, enough information has been given to the pupil to enable him to take an active part in the planning. The teacher, of course, will know considerably more about the methods of community study than is provided in the pupil's text. *The Ecology of Land Plants and Animals*, a BSCS Laboratory Block by E. A. Philips (D. C. Heath & Co., 1963), is also a valuable guide. In addition, see references at end of chapter for more information.

General Procedures

Selecting a study area. You, if not your pupils, should be thoroughly acquainted with the potentialities of your region. It is desirable that the area to be studied is as close to the school as possible so that repeated visits may be made. The desirability, however, of repeated visits may in some cases be outweighed by other considerations. If the resources in the immediate vicinity of the school are poor, and if administrative conditions permit, an excursion to a desirable site at a greater distance may be preferred.

In no school is this investigation impossible. Biotic communities exist even in the most urban situation. Look for waste ground, vacant sites, the area around a hoarding, even cracks in cement and tarmac—all these contain plants, insects, nematodes. Although the producers may be miles away, the city environment also contains larger organisms—mice, rats, cats, pigeons. Urban biology teachers will find the book by R. Fitter, *London's Natural History* (Collins, 1945) useful.

Having chosen the area for study, *visit it in advance* together with a group of pupils, if possible. With the area under observation, its possibilities can be sketched out.

Organizing the work. Next, on the basis of the directly observed possibilities and the suggested procedures (pp. 77–81), the class can proceed to make detailed plans. The plans must fit the number of pupils in the class. Each pupil must have a definite place in the plans: he must have something to do, he must know what his responsibility is, and he must know how to carry it out. Each group leader must know the overall plan for his group. Fig. T-3 (p. 38) is an example of a form (used in a woodland study) that provides the teacher and group leaders with a way of checking on individual responsibilities.

Written instructions are essential. Forms for recording data in the field may be devised. These, of course, must vary with the nature of the area to be explored. Some teachers have found it desirable to go through a 'dry run' with their group leaders to check out directions and forms for data.

If at all possible, Investigation 3.1 should be a *comparative* study. For example: the border of a wood may be compared with the interior; a grazed pasture with a mowed field; a well-trodden area of the school grounds with a less disturbed area. Consider also the possibility of using studies by classes in successive years as bases for comparisons that can give first-hand meaning to the concept of ecological succession.

Population densities of several selected species may be obtained by counting individuals on measured quadrats. The mean densities in one habitat can be compared with the mean densities in another. If a large number of quadrats are used, the difference between the means can be treated statistically against a null hypothesis but at the level of sophistication appropriate for the third-year pupil. It is probably sufficient in most classes to let the pupils decide subjectively whether or not the degree of numerical difference is 'significant'. Choice of species depends upon the habitats being compared. For instance, in comparing a well-trodden lawn with an out-of-the-way one, not grass but dandelions and plantains should be counted.

Collecting the data. Do all the fieldwork while the weather permits. It is not necessary to process the data immediately. If a delay is necessary, take some 35 mm colour slides of the study area. These may be useful later, when the pupils' memories need refreshing.

Some teachers think it a good idea to use organisms collected during the investigation as a basis for considering diversity among living things—a concept to be stressed in *DLT*. Collecting, however, is not an aim of Investigation 3.1 and care must be taken that the enthusiasm pupils often display for collecting does not interfere with the community study.

Specific Procedures

Gathering plant data. The dimensions to separate trees from saplings are arbitrary and can be varied.

Searching. Small specimens are best collected in small tubes containing dilute formalin, made by mixing 1 part of commercial methanal (formaldehyde) solution (37–41 per cent) with 7 parts of water or 70 per cent ethanol (propan-2-ol is satisfactory).

Examining the litter and soil. When collecting nematodes from the soil, remember that the populations vary greatly, depending upon whether the sample is from a rich organic soil, from an area poisoned with insecticide, etc. The Berlese funnel works best if there is a considerable temperature gradient in the soil; thus the bottom of the soil should be cool and the top warmed by the light. Anything that weakens the gradient lowers the efficiency of the apparatus.

Studying the Data

If at all possible, assemble a labelled collection of the species most likely to be encountered. This may be difficult when you do the investigation for the first time, but the collection can be increasingly comprehensive in later years. Use this collection as a basis for pupils to identify their organisms. Although specific identification is not necessary to obtain acquaintance with a community, pupils love to have names for things, and this increases their sense of accomplishment. Do not let identification become the goal of the work.

Discussion of the questions is more effective if delayed until the section 'Community Structure' (pp. 82 et seq.) has been studied.

Either before or after the class discussion of the questions, each pupil should write his own report on the work as a whole. It should include:

1. The purpose of his investigation.
2. A brief account of the methods used.
3. A summary of the data.
4. His own detailed interpretation of the data in the form of a description of the community structure and relationships (most important of all).

Make reference to the results of this fieldwork wherever possible when dealing with *DLT* and *PLW*.

Community Structure (pp. 82–91)

p. 83, Fig. 3.9. The lion is a predator, the giraffe prey, the vultures commensals. Compare with Fig. 1.2, p. 2.

p. 84, Fig. 3.10. The wheat is a host, the smut a parasite.

JOB SHEET

GROUP ———— GROUP LEADER ————

PREPARATIONS

Staking out the area:

General description:

DATA COLLECTION

Trees:

Shrubs and saplings:

Herbs and seedlings:

Searching:

Litter and soil samples:

GROUP EQUIPMENT

Stakes, 8	Hammer, 1	Rubber bands, 4
String, 50 m	Old magazines, 6	Plastic bags, 2
Metre sticks, 2	Collecting bottles, 4	Wire circles, 2
Rulers (metric), 4	Forceps, 2	Trowels, 2

Fig. T-3

p. 85, Fig. 3.11. Fleas of mice are not likely to parasitize a hawk (parasites are usually host-specific); but if they weaken a mouse, they could make predation by the hawk easier. If they carried infectious organisms that killed off enough mice, they could reduce the number of mice available to the hawk.

p. 87, Fig. 3.12. The *Hydra* obtains food produced by the algae; and the algae obtain a habitat, protection and presumably some raw materials. This relationship can be called mutualism.

p. 89, ¶ 1. You may want to point out—if some pupil does not do so—that throughout this chapter micro-organisms have been neglected.

p. 89, ¶ 4. The relationship between high species diversity and high stability has been observed often enough to establish it as a fact. What is not certain is whether high species diversity *causes* the stability, or the reverse (great environmental stability *causes* high species diversity). In other words, while the correlation is observed, the cause and effect relationships are not well known.

p. 91, ¶ 4. A climax community is often referred to as a 'stable' community. But stability is a relative matter: a lichen community may persist for a very long time and, as the next paragraph indicates, climax communities are liable to change.

Ecosystems (pp. 92–95)

p. 94, ¶ 1; p. 95, ¶ 1. Use Fig. 1.17. The ideas expressed here constitute a deliberate repetition of ideas from pp. 29–31. Do any of your pupils recognize this? Those who do are really studying biology well; those who do not are perhaps merely reading pages in a book.

pp. 94–95, Fig. 3.16. All the energy in the food eventually becomes heat energy. Before that happens it appears, for example, as muscle movement and electrical forces in human bodies. Some of it passes through the human body still in chemical form (food) and enters the life processes of saprovores in sewage disposal systems.

INVESTIGATION 3.2

ABIOTIC ENVIRONMENT: A COMPARATIVE STUDY (pp. 96–97)

Procedure

This investigation not only concerns the ecosystem concept but is related to matters of tolerance and ecological distribution that arise in Chapter 2 of *PLW*. One group per class may be sent to do the fieldwork.

Differences in temperature among the three habitats are obviously more pronounced in early autumn than later. For best results vegetation should not have become dormant. The work should be done on a sunny day. The three habitats should be as near to each other as possible so that topographical differences are minimal.

In this investigation whirling hygrometers are difficult to use, and at the 0 cm height impossible. However, if available, they do hold the two thermometers (wet- and dry-bulb) conveniently in the same place. Use distilled water to soak the bulb sleeves. (See Fig. T-4, p. 40.)

Studying the Data

If you have several groups, gather data into a table drawn on the blackboard.

A–K. These depend upon the results obtained in the investigation. On a sunny day, bare ground is usually the warmest and driest habitat at all levels; vegetative habitats are cooler and moister below the top of the vegetation than above it. In general, these measurements illustrate the modifying effect of vegetation, the conversion of radiant energy of sunlight to heat on contact with the earth, and, finally, the idea of micro-climatic variations.

L, M. Micro-climates are generally of importance to an organism whose size does not transcend the extent of one distinguishable micro-climate (beetle and horse-fly) but of little importance to an organism whose size encompasses the whole range of micro-climatic variation (cow).

N. Shade is a micro-habitat factor important to producers. Air currents, related (along with relative humidity) to the evaporation rate, may also be important.

PROBLEMS (pp. 98–99)

1. In caves where there is no light, producers cannot live; but organic substances enter the caves, mainly with flowing waters, and serve as food for the saprovores that are the base of cave food webs.

2. The point of the contrast between aquatic and terrestrial Antarctic communities is a comparison of abiotic conditions. The terrestrial habitat is so extreme that few species can survive; yet in the water, where the violent fluctuations in atmospheric conditions are modified, a greater number of species can find niches.

3. All three of these communities are similar in that they depend upon a continual input of foods.

4. This problem can lead the pupil in many directions. It is a good preliminary to the

RELATIVE HUMIDITY (percentage)

Difference between dry-bulb and wet-bulb readings	Dry bulb temperature (°C)																				
	10	11	12	13	14	15	16	17	18	19	20	21	22	23	24	25	26	27	28	29	30
0.5	94	94	94	95	95	95	95	95	95	95	96	96	96	96	96	96	96	96	96	96	96
1.0	88	89	89	89	90	90	90	90	91	91	91	91	92	92	92	92	92	92	93	93	93
1.5	82	83	83	84	85	85	85	86	86	87	87	87	87	88	88	88	88	89	89	89	89
2.0	77	78	78	79	79	80	81	81	82	82	83	83	83	84	84	84	85	85	85	86	86
2.5	71	72	73	74	75	75	76	76	77	78	78	79	80	80	80	81	81	82	82	82	83
3.0	66	67	68	69	70	71	71	72	73	74	74	75	76	76	77	77	78	78	78	79	79
3.5	60	61	63	64	65	66	67	68	69	70	70	71	72	72	73	74	74	75	75	76	76
4.0	55	56	58	59	60	61	63	64	65	65	66	67	68	69	69	70	71	71	72	72	73
4.5	50	51	53	54	56	57	58	60	61	62	63	64	64	65	66	67	67	68	69	69	70
5.0	44	46	48	50	51	53	54	55	57	58	59	60	61	62	62	63	64	65	65	66	67
5.5	39	41	43	45	47	48	50	51	53	54	55	56	57	58	59	60	61	62	62	63	64
6.0	34	36	39	41	42	44	46	47	49	50	51	53	54	55	56	57	58	58	59	60	61
6.5	29	32	34	36	38	40	42	43	45	46	48	49	50	52	53	54	54	56	56	57	58
7.0	24	27	29	32	34	36	38	40	41	43	44	46	47	48	49	50	51	52	53	54	55
7.5	20	22	25	28	30	32	34	36	38	39	41	42	44	45	46	47	49	50	51	52	52
8.0	15	18	21	23	26	27	30	32	34	36	37	39	40	42	43	44	46	47	48	49	50
8.5	10	13	16	19	22	24	26	28	30	32	34	36	37	39	40	41	43	44	45	46	47
9.0	9	9	12	15	18	20	23	25	27	29	31	32	34	36	37	39	40	41	42	43	44
9.5		5	8	11	14	16	19	21	23	26	28	29	31	33	34	36	37	38	40	41	42
10.0				7	10	13	15	18	20	22	24	26	28	30	31	33	34	36	37	38	39
10.5					6	9	12	14	17	19	21	23	25	27	29	30	32	33	34	36	37
11.0						6	8	11	14	16	18	20	22	24	26	28	29	31	32	33	35
11.5							5	8	10	13	15	17	19	21	23	25	26	28	29	31	32
12.0									7	10	12	14	17	19	20	22	24	26	27	28	30
12.5										7	9	12	14	16	18	20	21	23	25	26	28
13.0											6	9	11	13	15	17	19	21	22	24	25
13.5												6	8	11	13	15	17	18	20	22	23
14.0													6	8	10	12	14	16	18	19	21
14.5														6	8	10	12	14	16	17	19
15.0															5	8	10	12	13	15	17
16.0																	5	7	9	11	13
17.0																			5	7	9
18.0																					5
19.0																					
20.0																					

Fig. T-4

concept of infectious diseases developed in Chapter 1, *PLW*. Chestnut blight is an example of a host–parasite relationship that has not evolved into a steady state. Immunity has not developed and the parasite has decimated the population of the host organism. On the other hand, in whooping cough host immunity has developed—hence there is low fatality to the host. In addition, artificial immunization of the host can be used so that the disease is no longer a population hazard.

5. No single answer can be given to these questions. The important thing is for pupils to realize how much energy is imported into the city and how much the biotic community is dependent upon man. Be sure the commensals of man—such as house mice, cockroaches and silverfish—are not forgotten.

SUPPLEMENTARY MATERIALS

ADDITIONAL PROBLEMS

1. A croft on a Hebridean island can be said to be self-sufficient. What does this mean in terms of biotic communities? [*Most of the biotic energy is captured and expended locally; food is raised and consumed on the croft.*]

2. Why may it be reasonable to include the fishing grounds of the Dogger Bank, the orchards of the Isle of Ely and the potato fields of County Kerry in the biotic community of London? [*A modern city may be thought of as a biotic community, but its energy system cannot be understood unless study is extended far beyond the city boundaries.*]

3. In Fig. 2.20 the smoothing of a graph line between points is shown. To explain this further,

the teacher may present other population data plotted on a grid and have the class draw smooth curves for each set of data.

4. 'Thrushes eat several kinds of fleshy fruits. The seeds of these fruits pass through the digestive tract and are distributed throughout the countryside, and, surprisingly, are better able to germinate than seeds that have not been eaten by thrushes.' Discuss the interrelationships illustrated by this quotation and formulate an experiment that might help to classify the type of interspecies relationship. [*The plant species gain in two ways from this mutualistic relationship: the germination rate of the seeds is increased by the action of the thrush's gizzard which abrades the tough seed coats; the dispersal of the seeds is increased by the wandering of the thrushes. The thrushes gain nutrients from the fruits and from some of the seeds.*]

5. Cattle stomachs contain enormous numbers of micro-organisms, some of which digest the cellulose which is abundant in the plants eaten by cattle, and in so doing produce substances that are useful nutrients for the cattle. Evaluate the relationship between cattle and these micro-organisms. [*The micro-organisms get a place to live and easily digested amounts of food; the cattle benefit by having cellulose digested into substances they can use.*]

ADDITIONAL INVESTIGATION

COMPETITION BETWEEN TWO SPECIES OF PLANTS

Introduction

Many problems of human ecology involve such artificial biotic communities as gardens and cultivated fields. In this investigation intraspecific competition between species of garden plants that do not naturally occur together is the basis of an experiment. Pupils should set up a hypothesis that seems reasonable for the suggested procedure.

Materials (for each group)

Soil (a good loam), enough to fill three seed boxes
Trowel
Wood block
Wooden boxes, approximately 55 cm × 33 cm × 10 cm, 3
Sharp pencil
Tomato seeds, about 450
Radish seeds, about 450
Sheet glass, enough to cover 3 boxes
Scissors
Paper tissues
Balance, sensitivity to 0.1 g

Procedure

Prepare the soil by removing stones and breaking all lumps. Place in the boxes and smooth down to a level surface. Use a block of wood to press the soil down firmly, but do not pack it tight. Use enough soil to make a firmed layer at least 7 cm deep. Water and allow to stand for twenty-four hours.

With a sharp stick or pencil, draw furrows on the surface of the soil, parallel to the long sides of the box, about 2 mm deep and 5 cm apart. In one box, place tomato seeds about 1 cm apart along each row. In a second box, place radish seeds about 1 cm apart along each row. These boxes are 'pure cultures'. In the third box, place tomato and radish seeds alternately 1 cm apart along each row. This is a 'mixed culture'. Use the wooden block to firm the soil again along the lines of planting. This will barely cover the seeds with soil.

To reduce evaporation from the soil until after the seeds germinate, cover the boxes with sheets of glass and place in a warm, shaded place. When germination begins, the boxes should be moved into the light. Later the glass should be removed. By this time the plants will be large enough to be watered gently without being disturbed. Keep all the boxes equally moist. If the boxes are kept on the classroom windowsill, turn them daily so that the plants on all sides of the boxes receive about the same amount of light over a period of days.

The experiment may be brought to a conclusion in 35–50 days, i.e. when the plants are fairly large but before they begin to topple over. At that time all the tomato plants in the mixed culture need to be trimmed (with scissors) close to the soil and weighed together. Then clip off and weigh all the tomato plants in the pure culture. Pull up all the radishes in the mixed culture, carefully wash the soil from the fleshy roots, blot dry on paper tissues, and weigh together. Follow the same procedure with the radishes in the pure culture. Since there were only half as many radish seeds and tomato seeds in the mixed culture as there were in the pure cultures, it is necessary to divide the weights of the plants from the pure cultures by 2.

Studying the Data

Represent the data in the form of histograms, with *weight* on the vertical axis and *type of culture* on the horizontal axis. Arrange the bars along the horizontal axis, from left to right, in the order in which the cultures are weighed.

A. Is there a difference between the total weight of the tomato plants grown in pure culture and those grown in mixed culture? If so, how do you account for the difference?

B. Is there any difference between the weight of the radish crop grown in pure culture and those grown in mixed culture? If so, how do you account for the difference?

C. Does the growing of tomatoes and radishes together have more effect on the tomatoes or on the radishes? Or is the effect equal? Or is there no effect?

D. Attempt to explain how the effects you noted may have occurred. The effects of competition are being judged by the weight of the 'crops' developed—a quite reasonable method from the viewpoint of an agriculturalist. On pp. 86–88 the basis on which ecologists prefer to judge the effects of community relationships is discussed.

E. Explain why the weights of the crops may be considered *indirect* evidence of competition.

For Further Investigation

Carry out the same experiment, using other kinds of plants. Spacing between plants and the depth of planting may have to be varied to suit the kinds of plants used.

For the Teacher

This is a simple experiment that can easily be turned over to a small group. The work can be done at home if sufficient window space is available. However, growth of the plants in the laboratory arouses interest and discussion that should make the group's report of the results more meaningful to the class.

The seeds can be treated with fungicide, as described in Investigation 1.2.

The procedure assumes that the roots of the tomato plants represent a negligible weight factor. If pupils object to this assumption, it is possible to harvest the whole plants by gently washing the soil away from the root systems.

A–C. These questions follow directly from whatever data are obtained.

D. If differences are found between the pure cultures and the mixed cultures for either species, they can perhaps most readily be explained on the basis of more effective absorption of water and/or nutrients by one species in the presence of another. At later stages, however, one species may outgrow the other and thus reduce its supply of radiant energy.

E. Since these species do not reproduce vegetatively, the effect of competition depends upon the quantity and viability of the seed produced. Any conclusions based on the results of this experiment depend upon the assumption that the size of plant, as indicated by weight, is related to its ability to produce viable seed.

Some pupils may want to investigate intraspecific competition, though it is more closely related to the idea of crowding (discussed in Chapter 2) than to community study. Such an investigation should parallel Investigation 3.2. One species of seeds should be used, but they should be planted in several boxes. The intervals between seeds should be different in each seed box. The effects of spacing can be measured by weight. If peas, which are self-pollinating, are used, the effects of spacing can be correlated with the quantity of seeds produced. There is value in allowing pupils to see that pertinent data may be obtained from an experiment in a variety of ways.

AUDIOVISUAL MATERIALS

The concept introduced in Chapter 3 lends itself rather well to visual instruction. Many relevant films and filmstrips are available, but most are pitched at a level that should be too superficial for 15–16-year-old pupils.

Filmstrips

Symbiosis—Strange Partners in Nature. ('Darwin's World of Nature', Part 8.) *Life* Filmstrips. Colour. (The pictures are good, but the captions are full of teleology and anthropomorphism. The term 'symbiosis' is used in the restricted sense, i.e. equivalent to 'mutualism'.)

The Ecological Succession. McGraw-Hill, 1960. (Contains accurate terminology, though rather more than necessary. The picturing of bare-rock succession parallels our description.)

The City as a Community. McGraw-Hill. (Useful to give city pupils some ideas for conducting a community study.)

*Films**

Life in a Wood. National Film Board of Canada, available through CHFL. 16 mm, colour, 17 min. (Shows the factors that affect the dynamic balance of populations in a woodland community through the cycle of the seasons.)

The Cave Community. Encyclopaedia Britannica Films, available on hire from NAVAL and SCFL. 16 mm, colour, 11 min. (A good example of the flow of energy through a community.)

Succession—From Sand Dune to Forest. Encyclopaedia Britannica Films, available on hire from NAVAL and SCFL. 16 mm. (In depicting succession on the south-east shore of Lake Michigan, this film shows many good examples of community relationships. An alert teacher can exploit these more effectively than the narrator does.)

Plant–Animal Communities: Physical Environment. Coronet Films, available through Gateway Educational Films. 16 mm, colour, 11 min. (The slant of this film leads to the subject matter of Chapter 2 in *PLW*. It could be used to pick up the thread of ecological thought in beginning that chapter.)

TEACHER'S REFERENCE BOOKS

DARLINGTON, A.—Editor. *Woodland Life.* Blandford Press, 1966.

MACAN, T. T. *A Guide to Freshwater Invertebrate Animals.* Longman, 1960.

NATURE CONSERVANCY. *Nature Trails.* Warne, 1968.

OVINGTON, J. D. *Woodlands.* The English Universities Press, 1965. (Woodland reviewed as a living entity of plants and animals in a delicate balance with climate and soil.)

*CHFL = Canada House Film Library
NAVAL = National Audio Visual Aids Library
SCFL = Scottish Central Film Library

DIVERSITY AMONG LIVING THINGS

During the twentieth century all the major developments in biology have tended to strengthen the concept of a fundamental unity in life processes. But the diversity of organisms remains an obvious and inescapable fact. Biologists—at least *some* biologists—must deal with it, and any biology course having primarily cultural aims ought to deal with it. This diversity should be considered early in the course, because pupils need a mental map on which to arrange the numerous organisms that must be used to illustrate the workings of biological principles and processes.

The chief intellectual burden of this book is *not* the characteristics of the numerous groupings into which taxonomists have sorted organisms. All such groupings involve a large element of subjectivity; they are mutable, if not ephemeral. Much depends upon still-accumulating paleontological evidence, and much depends upon the way in which the groups are defined. 'The alligator belongs to the class Reptilia' is a kind of statement frequently heard—as if the class Reptilia were an *a priori* entity to which alligators must conform. Today such a statement is merely a matter of convenience; to the modern biologists it does not have Aristotelian implications. But it certainly will have such implications for pupils unless they are cautioned to interpret the statement thus: 'Alligators have characteristics that allow us most conveniently to place them in the Reptilia—as that class is defined by most zoologists.' With such an interpretation, the commitment to memory of the proper disposition of organisms into one of the many possible schemes of classification becomes absurd.

What, then, is the essential content of this volume? It lies in three abstract ideas: first, the purposes and the nature of biological classification (Chapter 1); second, the scheme of biological nomenclature (Chapter 2); third, the difficulties inherent in attempts to fit the facts of nature into a conceptual mould (Chapter 3). Whatever time you allow for class discussion should be devoted primarily to these three ideas. Because the number of major ideas in this book is fewer than the number of pages might indicate, a schedule should be adopted that is in proportion to the density of ideas rather than to the number of pages.

DIVERSITY AMONG LIVING THINGS

1
ANIMALS

MAJOR IDEAS

1. Despite the bewildering diversity of animal forms, some major patterns of structural characteristics can be discerned in the animal kingdom. These patterns form the basis for the hierarchical ordering of taxonomic groups.
2. Because the taxonomist often finds it convenient to work with dead specimens—and, in the case of extinct species, must do so—the structural characteristics of organisms have been most frequently used in classification. But other kinds of characteristics are increasingly employed by modern taxonomists.
3. In different species, structures having obviously similar basic components vary in detail; these variations appear to fit any given species for efficient functioning in a particular environment. This is the concept of structural adaptation—an important element in all evolutionary theories, but equally applicable to a theory of special creation.
4. Since the time of Darwin, the similarities among organisms have been ascribed to common evolutionary development. By means of levels of classification, the modern taxonomist seeks to express the varying degrees of evolutionary divergence; but simultaneously he seeks to provide biologists with convenient means of grouping organisms.

PLANNING AHEAD

Considerable time may be needed to mount the leaves to be used in Investigation 2.1. Once done, however, most of the mounts should last for several years. Save some dried leaves for use in Investigation 1.4, *PLW*. Be sure you have a few twigs and dried insects for the same investigation.

Try to accumulate as many kinds of living plants as possible for display in your laboratory during work on Chapter 2. In gathering these, keep in mind the assortment of kinds desirable for Investigation 2.2; employ living specimens so far as is practicable.

If you have not ordered the cultures of bacteria needed for Investigation 3.2, do not delay any longer. While ordering these, you should also consider acquiring a culture of *Agrobacterium tumefaciens* for Investigation 1.2 of *PLW*.

You will also require potted bean, tomato, or sunflower plants for Investigation 1.2, *PLW*. If you have to grow them yourself, start now.

GUIDELINES

Diversity of form among animals is the pervading idea in this chapter. To the greatest degree possible, pupils should have opportunity to see this diversity in living animals. Investigations 1.2 and 1.3 serve to some extent. Visits to a zoo or an aquarium will help. Also, pictorial means may be used to give pupils a still broader view of the diversity of animal forms. The pictures in the pupil's book should be fully exploited. In addition, notice boards, filmstrips, and slides should be employed. For notice board materials *National Geographic Magazine*, *Life* magazine, and *The World of Wildlife* are good sources. Pupils vary greatly in their acquaintance with animals; you must gauge the background experience of your classes and then select materials accordingly.

Chapter 1 begins (pp. 1–5) with the abstract concept of classification. There follows a long part (pp. 10–34) concerned with descriptions of major animal phyla strung together on the theme of structural adaptation—a theme important later, especially in Chapter 4 of *PLW*, and Chapter 3 of *MHE*. The concluding part (pp. 39–42) returns to abstraction with an exploration of the meaning of biological classification.

The most logical first assignment is pp. 1–5 (guide questions 1–3, p. 45), during which Investigation 1.1 can be carried out as homework and then discussed in class. However, if reading difficulties are anticipated, this investigation might better be carried out in class with immediate teacher direction after the study of pp. 1–5 has been completed. If the basic concept of a hierarchical classification is difficult for pupils, you may find it helpful to have them first discuss a classification for a group of not unduly diverse manipulable objects—an assortment of bolts, screws, and nails of various sizes and types, for example.

The next block of text can be conveniently divided into two reading assignments: the first, pp. 10–19 (guide question 5), about chordates; and the second, pp. 10–34 (guide questions 6–13), beginning with the arthropods and continuing through the rest of the animal phyla that have been chosen for discussion. This is not material for exhaustive study; the facts are plentiful but are not intended to be committed to memory. Support this with as much illustrative material as possible. Investigation 1.2 then serves as a summary of some ideas concerning animal organization while it also introduces the idea of taxonomic keys. The remainder of the chapter can then be covered in one reading assignment, pp. 39–42 (guide question 15), to be immediately followed by Investigation 1.3, which brings the pupil back from abstraction to the concreteness of living animals.

Finally, you should make frequent reference to the Appendix, 'A Catalogue of Living Things', though several such references occur in the pupil's book. Call attention to the introductory paragraphs on p. 106 and point out that the sequence of Protists, Plants, Animals, is different from the sequence in the text.

TEACHING NOTES

The Principles of Classification (pp. 1–5)

p. 1, ¶ 2. Many so-called primitive peoples of the present day have large numbers of common names for the organisms surrounding them, and these are by no means restricted to organisms that are immediately harmful or beneficial.

p. 3, Fig. 1.1. The technical form of each group name has been used in the chart, but in class discussions you are urged to employ the English forms whenever possible. Thus, say 'chordates', 'arthropods', 'mammals'; leave 'Chordata', 'Arthropoda', and 'Mammalia' for formal use. Pupils should know technical forms exist, but they do not need to burden themselves with difficult spelling and pronunciation (unless they individually wish to do so).

p. 4, Fig. 1.2. To most pupils, wolf and coyote look most alike. If some pupils think that either resembles the fox more closely than the other does, call attention to the fact that the fore-limbs of the fox are more lightly constructed and the whole animal is relatively smaller. Most of the structural differences between *Canis* and *Vulpes* that are important to taxonomists are not visible in the drawings.

INVESTIGATION 1.1

THE LEVELS OF CLASSIFICATION (pp. 5–10)

This investigation is quite feasible, with a few hints from you, as an independent home assignment. However, for some classes various degrees of assistance must be given—to the extreme of working out the entire investigation in class with your direct guidance.

Procedure

In some cases *explicit* information concerning the group in which a specific animal is classified is not given in the Appendix. This will no doubt lead to some degree of healthy frustration on the part of the pupil and also to a more careful and searching scrutiny of this appendix and maybe other suitable texts—resulting, it is hoped, in a better understanding of the dimensions of animal diversity.

Specific difficulties that may arise are:

p. 6, Fig. 1.3. E. If the past four items have led pupils to expect chimpanzee and gorilla to be differentiated from man on all counts, this item may be puzzling. The number of incisors is the same for all three species. Although both chimpanzee and gorilla are illustrated in the Appendix, they are not explicitly identified as apes.

p. 7, Fig. 1.4. A. No picture is required for this. C. Some inference is required here, because in the figure the clavicle is only indicated in black, not named. D, E. Identity of canines and incisors should be carried over by the pupil from Fig. 1.3. Although no dog (order Carnivora) is shown in the Appendix, the necessary information is included in Fig. 1.1. Neither is a cat (genus *Felis*) shown, but the cheetah is sufficiently catlike for most pupils, by inference, to locate the order of the cat.

p. 8, Fig. 1.5. A. Your laboratory should contain a specimen of a frog for pupils to examine. B. See note on A for Fig. 1.4. C. You may want to briefly explain the difference between 'warm-bloodedness' and 'cold-bloodedness' or to refer to p. 14, paragraph 2, and p. 16, paragraph 1. The necessary information on dogs is available in Fig. 1.1. An illustration of a tree frog is in the Appendix.

p. 8, Fig. 1.6. C. Some pupils may call the brain of a bird small, but point out that the job here is to make comparisons and that, compared with the brain of the crayfish, the bird brain is large, even considering relative body size. D. This is another case where attention is called to a point that does not distinguish; all have paired appendages, though some pupils may call attention to a difference in numbers. E. All pupils recognize the embryonic similarity between man and bird. Some may, nevertheless, complain that they cannot definitely answer the question with respect to the crayfish. They are right, of course.

Some difficulty may be experienced in establishing the phylum to which the crayfish belongs, because only blue crabs and water fleas represent the Arthropoda in the Appendix. If no pupil does so, call attention to the close similarity between a lobster and a crayfish; then Fig. 1.1 resolves the difficulty.

Discussion

A—E. Pupils sometimes try to make these questions more complicated than they are, attempting to give reasons for the classification by repeating the structural features noted in the charts instead of merely citing the classification levels. This results from misreading: the question is not concerned with the evidence itself but the way the evidence is expressed in the hierarchy of classificatory levels.

F. Species 3 and 4 are more closely similar than species 1 and 2. On the basis of the evidence given, it is impossible to make any other general statement.

Be sure during discussion that details do not obscure the basic point of the investigation: that the hierarchy of classificatory levels is an expression of degrees of likeness—the greater the likeness between two organisms, the lower the level at which they are grouped together. At the species level, organisms to be grouped together must be so very much alike that they are capable of interbreeding.

The Animal Kingdom (pp. 10–34)

This discussion is confined to animals living today and excludes (as does the Appendix) animals known only from fossils. This simplifies (and, of course, somewhat distorts) the picture of animal diversity.

p. 10, Fig. 1.7. The caption question is rhetorical but some pupils may very well come up with some ideas such as 'two eyes', 'mouth', or even the more penetrating (and ultimately the more important) 'backbone'. Probably very few will think of 'bilateral symmetry' unless they have read ahead. A short discussion on the figure, without teacher comment, is a good motivational device at the time the reading assignment is given.

p. 11, Fig. 1.8. This is a photograph of a cleared specimen.

p. 12, ¶ 2, line 2. Note the wording 'are placed'. No organism 'belongs' in a taxonomic group except on the basis of some set of characteristics specified by someone making a classification. This is a basic point throughout *DLT*, and it needs constant emphasis by the teacher. We regret that we have not ourselves always been able to avoid the misleading 'belongs' terminology.

p. 12, Fig. 1.10. Five pairs of pharyngeal openings (gill slits or clefts) are present in the dogfish.

p. 14, ¶ 2. Many of the characteristics used by vertebrate taxonomists are in the skeleton. The importance of this becomes evident in Chapter 4 of *PLW*. You might want to mention a few such mammalian characteristics: (*a*) the much smaller number of skull bones as compared with reptiles, (*b*) the formation of the auditory arch by three bones instead of one, as in reptiles, and (*c*) the clear differentiation of tooth types (with a few exceptions).

p. 14, ¶ 4. Most pupils should be able to mention many structural adaptations in mammals—beginning with themselves. Also refer to Fig. 1.11.

p. 14, ¶ 5. What is a feather? This is a good reference question for an interested pupil. Interest in birds is often high, and there are many other questions that might be pursued. Among the multitude of bird books, one that can be recommended is: *The Readers Digest/AA Book of British Birds* (Collins, 1969).

p. 15, ¶ 2; Plate I (p. 56). Except for fishes, probably no group of vertebrates shows as much colour diversity as birds.

p. 15, Fig. 1.13. Spoonbills sweep their distinctively shaped beaks back and forth through the water to separate out small marine organisms that form their principal food. The peculiar beaks of crossbills enable them to pry out the seeds from pine and spruce cones very efficiently; many other birds must expend a great deal of effort or wait until the cones have opened before their food supply is available to them. The strong, hooked beaks of eagles are well adapted for tearing the flesh of the relatively large animals they prey on for food. The relatively narrow and sharp beaks of robins permit them to grab and hold earthworms pulled from burrows and to capture soft insect larvae. The pouch-like beaks of pelicans enable them to collect and manipulate small fish, eliminating excess water from the catch, and also enable them temporarily to store the fish before swallowing them or feeding them to young.

p. 17, ¶ 2, line 6. Movement through a narrow burrow, not much wider than the animal living in it, would be impeded by appendages extending out from the animal's body. Friction between the body surface and the burrow walls, plus the squirming motions characteristic of many burrowing animals, makes efficient locomotion possible in a very confining space.

p. 17, ¶ 5, and Plate IV (p. 88). In general, fast-swimming fish have a slender, tapered, streamlined body form. In slower-swimming species the body is usually much less elongated, and movement of pectoral and pelvic fins is of greater importance in propelling the fish through the water. In some fish, trunkfish for example, the body is quite rigid and fins play the major role in locomotion. Locomotion of fish is actually quite a complex matter. A reference that may be of help to pupils interested in this problem is: F. D. Ommanney and the Editors of *Life*, *The Fishes* (reference p. T53).

Structural adaptations of bottom-living fish may include flattened body, shift in position of eyes, alteration of fundamental patterns of body symmetry, modification in shape, and position of fins.

p. 19, Figs. 1.16 and 1.17. For background information consult N. J. Berrill, *The Origin of Vertebrates* (Oxford University Press, 1955).

p. 22, ¶ 1. A useful reference is: Immes, A. D., *Insect Natural History* (Collins, 1971).

p. 22, Fig. 1.21. In flies, the posterior wings are represented by balancing organs only (halteres). The relatively small but powerful anterior wings are well adapted for hovering and for rapid flights of short duration. Normal cruising speed in still air is approximately 8 km/h. The large wings of the type of butterfly shown in the illustration (a swallowtail) are well adapted for short periods of relatively active flight, alternating with periods of gliding on favourable air currents. Speed of flight is approximately 20 km/h. The long, slender wings of dragonflies, unlike those of butterflies, beat independently, the forward ones rising as the rear ones fall. This permits not only rapid forward flight but also well-coordinated hovering manoeuvres. Speed of flight is approximately 40 km/h. In beetles the anterior pair of wings form thick, hard structures called elytra or shards that cover and protect the membranous posterior wings. In many cases, these anterior structures reduce the effectiveness of the posterior pair of wings, the only ones used in flight. Many beetles fly for only short periods of time. The proximal halves of the anterior wings of shield bugs are thickened somewhat, as are those of beetles, though not to the same extent. They probably supply some motive power, but the posterior, wholly membranous wings are probably more effective. The delicate, ornate wings of lacewings are relatively poor flight organs.

p. 23, Fig. 1.22. *Mosquito:* Mouthparts form a piercing organ with a groove through which saliva (containing an anti-clotting substance) is injected and the blood of the victim is sucked. *Honeybee:* Paired mandibles are used for shaping and crushing wax used in comb building. A long, tube-like proboscis is employed in sucking nectar from flowers. *Grasshopper:* The strong, hard, paired mouthparts are used for biting and chewing vegetation. *Butterfly:* The long, coiled proboscis is used for sucking nectar from flowers. It can reach deeper into tubular flowers than can the proboscis of a honeybee. *Ant:* Paired mouthparts are used for seizing, biting, and chewing a wide variety of food, depending upon the species.

p. 24, ¶ 4, line 5. Other zoologists might use the following characteristics to support a classification which placed centipedes and millepedes in the same class. Millepedes and centipedes have many structural features in common. The basic body plan includes a head and an elongated trunk, made up of many similar leg-bearing segments. Both groups have a single pair of antennae, and the mouthparts show basic similarities. Both groups are often included in a single class, the Myriopoda. Yet many taxonomists, including the ones responsible for the Appendix, are more impressed by the differences.

p. 24, ¶ 6, line 4. These crabs which breathe by gills have adaptations which allow them to live on land. Terrestrial crabs are commonly nocturnal and live in burrows; both modes of behaviour reduce exposure to dry air. In coconut crabs the gills are reduced; most respiratory exchange takes place through the surfaces of highly vascular folds of epithelial tissue which hang from the top of enclosing branchial chambers. All these adaptations reduce rate of water loss to the point where the animal can absorb sufficient oxygen without becoming excessively dehydrated.

INVESTIGATION 1.2

STRUCTURAL CHARACTERISTICS IN THE IDENTIFICATION OF ANIMALS (pp. 35–39)

Much of the advice given for Investigation 1.1 in *WL* applies equally well to this investigation.

Materials

Since the keys in the second part of the investigation are very simple, they do not take into consideration exceptional representatives of the groups. Therefore, in choosing specimens, take care to select those that will 'key out'. Suggestions:

For Table 1: rat or mouse, dogfish, frog, bat, canary or budgerigar, snake, turtle, goldfish, lizard, lamprey.

For Table 2: earthworm, crayfish, jellyfish, *Hydra*, butterfly or moth, starfish, oyster, beetle, spider, grasshopper, planarian, snail (but avoid slugs), millepede, centipede, tick.

Living animals should of course be used as far as possible.

The tables are rather cumbersome to draw. Some teachers find it convenient to duplicate the tables and have pupils paste the copies in their lab-books. The number of columns in the tables depends upon the number of animals to be observed.

Procedure

At the beginning of observation, two minutes should be allowed at each station. This time can be reduced as pupils become familiar with the tables, but it is unlikely that more than twenty specimens can be examined in a single period. If diversity is to be evident despite this limitation, a judicious choice of specimens is important.

Although most of the terminology in the tables has been encountered by the pupil during his study of previous pages, some explicit definitions may have to be developed as pupils work through this investigation. Those most likely to cause difficulty are, for convenience, noted below.

Table 1. Bony skeleton: a skeleton in which most of the parts are hard and relatively rigid, because of the hard mineral matter they contain.

Cartilaginous skeleton: a skeleton in which all the parts are tough but flexible, because they are composed of cartilage—a substance that does not contain hard minerals.

Table 2. Exoskeleton: a skeleton on the outer surface of an animal, enclosing the animal.

Radial symmetry: a plan in which the parts are arranged in a circular manner around a central point or region—as in a bicycle wheel.

Bilateral symmetry: a plan in which the parts occur in pairs along the right and left sides of a line running from one end of the object to the other—as in the body of a bus.

Body segmentation: a structural pattern in which the body is divided into a series of more or less similar sections, the boundaries of which are usually indicated by grooves encircling the body.

Tentacles: slender, very flexible structures usually capable of being lengthened or shortened and usually attached near the mouth.

Antennae: slender structures that can be waved about but cannot be changed in length, usually attached to the head.

The use of a key can be assigned as homework. However, in some classes it is wise to assist in working several examples through the keys before leaving pupils to independent work. It is important to call special attention to the cautions in the Note on p. 39.

For further experience with taxonomic keys, see Investigation 2.1 and Teacher's Guide material for Investigation 5.2 of *LO*.

For Further Investigation

Any vertebrate with feathers can immediately be placed in the class Aves. Any invertebrate which has jointed appendages can immediately be placed in the phylum Arthropoda. Any arthropod without antennae can be placed in the class Arachnida. Any adult arthropod with three pairs of legs on the thorax can be placed in the class Insecta.

The Meaning of Biological Classification (pp. 39–42)

Chapter 3 of *MHE* is entitled 'Evolution', but the writers of this series regard the theory of evolution (considered distinct from the theory of natural selection) no less fundamental to the science of biology than the molecular theory is to the physical sciences. The idea that organisms change through time has been implicit from the beginning. It is still implicit in this section of Chapter 1—the word 'evolution' does not occur here—but a few questions lead to that idea as a simple way to account for a vast wealth of facts. Without that idea one is left in chaos; there is no scientific meaning to the facts.

p. 40, ¶ 3. This is an analogy. An analogy is useful for explanation if it is not carried too far. A trick of sophists is to substitute analogy for demonstration or evidence. Teachers are prone to this fallacy!

Plate II (between pp. 56–57). Refer your pupils to Plate II, A–F. Do not overlook the caption of this plate. All the organisms shown—except the Portuguese man-of-war, a hydrozoan (Coelenterata)—can be readily found in the Appendix (either by direct comparison with a figure there or by reference back to this figure).

p. 41, ¶ 4. Another analogy!

INVESTIGATION 1.3

DIVERSITY IN THE ANIMAL KINGDOM: A COMPARATIVE STUDY (pp. 42–45)

This investigation is not intended to be an abbreviated 'type study' of animal phyla. Its primary purpose is to sharpen the pupil's observation of living animals. Secondly, through it the pupil becomes acquainted with five major patterns of animal structure; these are organisms that are good examples for many purposes and are used frequently. And thirdly, it provides first-hand evidence for the important theme of structure–function relationship (see p. T1).

Materials

The principal teaching problem is one of logistics. It is sometimes difficult simultaneously to assemble all the animals in a healthy, active condition. All the species are worth maintaining as permanent inhabitants of the laboratory. Therefore, some attention to culturing these animals on a permanent basis is justified. If cultures are routinely maintained (see Teacher's Reference Books at the end of this chapter), the problem of timing orders from suppliers is eliminated. You may wish to add or substitute other animals. Suggestions: *Daphnia, Tubifex*, grasshoppers. Of course, if substitutions are made, you must revise the specific directions.

Procedure

Note that reference to the pupil's lab-book is dropped here and in most subsequent investigations. It is assumed that the use of the lab-book for all laboratory records is now an established habit.

The five stations should be as far from each other as the plan of the laboratory permits. Movable tables and peripheral facilities allow the best arrangement, but adaptations can be made in other situations. Three to six pupils per station is perhaps ideal, but more can be accommodated. Each group should be permitted about eight minutes at each station. Therefore the observation time occupies all of a double class period. Consequently, directions for observation must be thoroughly studied before the laboratory period; and everything must be in readiness at all stations when the class arrives.

It is unlikely that the procedure can be followed exactly as written. You may direct pupils to omit some of the general questions or to add others. (But be sure that any added questions can be answered *from the material available*; this is not the time to send pupils rushing to an encyclopaedia.) Some of the directions for specific animals may have to be bypassed—feeding the frog probably causes the most difficulty.

Discussion

Concentrate first on the structure–function idea in the 'Summary'. Then return to specific observations if time allows.

GUIDE QUESTIONS (p. 45)

2. The second part of this question can be answered in part on the basis of material in the first reading assignment (pp. 1–5), but it becomes more understandable at the end of the chapter (p. 41, paragraph 2). Return to it when you reach that point.

PROBLEMS (p. 46)

2. This problem incidentally illustrates the principle of adaptive convergence discussed on pp. 135–6 of *PLW*. (*a*) Lampreys lack jaws and bony skeletons. (*b*) Snakes have three-chambered hearts and amniote eggs. (*c*) Salamanders lack scales, claws, and amniote eggs. (*d*) Armadillos have mammary glands and sparse hair. (*e*) Bats have hair and mammary glands and lack feathers.

4. Taxonomy will continue to develop because: (*a*) additional fossil forms will continue to be discovered; (*b*) new kinds of evidence indicating different relationships will undoubtedly be discovered; (*c*) new species of organisms will continue to appear as a result of evolution.

5. See Simpson and Beck (reference: p. 47) for a discussion of patterns of symmetry in relation to motility in animals.

6. The pupils should consult zoology textbooks that have a phylogenetic approach, e.g. Buchsbaum (reference: p. 47).

7. This one should be easy, but it also gives a hint of an approach to Problem 2.

8. The class Monoplacophora (p. 117) had already been established, but it was known only from fossils. *The Galathea Deep Sea Expedition, 1950–1952* (reference: p. 47) tells the story of the discovery of a living member of the class *Neopilina galathea*. Another example of the same kind of event was the discovery of *Latimeria* in 1938. But probably most major changes in classification during the past century have resulted from the discovery of fossils rather than of living organisms. If the problem is allowed to include such discoveries, then it is a very wide one. Just one example: the discovery of the toothed birds in 1872.

AUDIOVISUAL MATERIALS

Slides

Of materials for projection, the best for this chapter are 2×2 slides. Large stocks of such slides, offering illustrations of almost all animal phyla, are available from the principal biological suppliers: T. Gerrard & Co. Ltd, Harris Biological Supplies Ltd.

Note: For full addresses see Appendix C.

From these stocks the teacher can select slides showing animals that are unavailable as living specimens.

Filmstrips

Except for human physiology, no topic in biology has been more abundantly treated in filmstrips than has classification. For the most part, however, filmstrips that the *Environmental Approach* writers have seen are unsuitable for our purposes, either being too greatly involved with anatomy and cellular detail, or approaching classification from an authoritarian viewpoint. As indicated above, the teaching of animal diversity is better accomplished with slides; you are then free to discuss alternative systems of classification.

Films

The comments on filmstrips apply in some degree to films. The following are relatively free of the disadvantages mentioned above.

Introducing Insects. National Film Board of Canada, available through NAVA. 16 mm, colour, 17 min. (Illustrates well the diversity among insects and discusses classification. Includes some good slow-motion and time-lapse sequences.)

Echinoderms. Oxford Biological Films, distributed by Ealing. Super-8 film loop, 4 min. (The dimension of motion is especially desirable in presenting this phylum, with which most pupils can have little first-hand acquaintance.)

What Is a Fish? What Is an Amphibian? What Is a Reptile? What Is a Bird? What Is a Mammal? Encyclopaedia Britannica Films. 16 mm, colour, various lengths. (A good series if your laboratory is deficient in living specimens and if opportunities for your pupils to visit aquaria and zoos are few.)

Between the Tides. British Transport Film Library. (Gives a very interesting and colourful introduction to marine life of the shore and also rock pools.)

TEACHER'S REFERENCE BOOKS

CARR, A., and the Editors of *Life*. *The Reptiles*. Time, Inc., 1963. (Excellent pictures. The authoritative text goes rather far beyond the scope of the chapter but it includes much on adaptations.)

CARRINGTON, R., and the Editors of *Life*. *The Mammals*. Time, Inc., 1963. (Contains excellent illustrations.)

FARB, P., and the Editors of *Life*. *The Insects*. Time, Inc., 1962. (Contains excellent illustrations.)

LERWILL, C. J., *An Introduction to the Classification of Animals*. Constable, 1971.

NEEDHAM, J. G. *Culture Methods for Invertebrate Animals*. Dover, 1959. (An old standby for information on culturing laboratory invertebrates.)

OMMANNEY, F. D., and the Editors of *Life*. *The Fishes*. Time, Inc., 1963. (See remarks under Carr above.)

PETERSON, R. T., and the Editors of *Life*. *The Birds*. Time, Inc., 1963.

ROTHSCHILD, LORD. *A Classification of Living Animals*. Longmans, 1961. (Perhaps one of the best systematic reviews of the animal kingdom.)

YOUNG, J. Z. *The Life of Vertebrates*. Oxford University Press, 1962. (Good background for teacher rather than pupil.)

2
PLANTS

MAJOR IDEAS

Classification and structural adaptation continue to receive emphasis in Chapter 2. Additional major ideas are:

1. Binomial nomenclature developed during the age of world exploration, when the need arose for an orderly system that would provide distinctive names for each of the multitude of newly discovered organisms.
2. Through the application of a few simple rules, the binomial system has resulted in relative stability and clarity in biological nomenclature for nearly two hundred years.
3. The great majority of familiar land plants can be grouped together in one phylum. This phylum, the Tracheophyta, is characterized by a vascular system, through which liquids are conveyed.
4. Plants of other phyla either are aquatic or grow best in rather moist habitats.

PLANNING AHEAD

You may want to set up Investigation 3.1 before you complete Chapter 2 (see note to p. 70, paragraph 3, p. T58).

If you are unfamiliar with techniques for handling microbes, you should try out those required in Investigations 3.2 and 3.3. before attempting to guide pupils You may find the following film loops helpful: *Sterilization Techniques*; *Bacteriological Techniques: Preparing and Dispensing*; *Bacteriological Techniques: Innoculating*; *Bacteriological Techniques: Serial Dilution and Pour Plate*; *Bacteriological Techniques: Staining* (Ealing Scientific Ltd., 23 Leman Street, London E1.)

If you have not ordered cultures of *Serratia marcescens* and *Sarcina lutea* for Investigation 3.2, do so immediately. Check the quantities of glassware available for Investigation 3.2; the result will determine the size of your groups. Also, prepare the glass tubes needed to set up Investigation 3.4.

Some form of incubator will be useful for Investigation 1.1 of *PLW*. Securing *Agrobacterium tumefaciens* sometimes involves some letter writing. See p. T74 for the names and addresses of agencies which will supply to schools and colleges.

Seeds for Investigation 2.1 in *PLW* are difficult to obtain locally in the autumn. If you do not have seeds on hand, order some from a biological supply company.

GUIDELINES

It has been customary to wrap the whole of taxonomy in a neat package of abstractions that are quickly and superficially covered, quickly and completely forgotten. The authors believe that taxonomy both explicitly and implicitly permeates all areas of biology. Moreover, they believe it to be an intellectual achievement, an important facet of human cultural development, and therefore a matter of importance in a humanistically oriented secondary school biology course (p. T5). The result of these beliefs is an attempt to avoid locking taxonomy into a compartment by itself. Experience indicates that taxonomy is more clearly understood—and (we think) better appreciated as an intellectual achievement—when it is divided into parts and taught as several appropriate points than is the case when it is presented in one massive lump.

So the concept of species was presented first (Chapter 2 of *WL*); classification followed (Chapter 1, *DLT*); and now in Chapter 2 comes nomenclature.

For assignment, the obvious divisions of the textbook chapter are pp. 48–53 (guide questions 1–4), pp. 53–68 (guide questions 5–11), and pp. 68–79 (guide questions 12–16). The first of these is rather short, but, as has been mentioned previously, number of pages is not necessarily a good guide to number of ideas. In this chapter nomenclature requires a major share of discussion time.

Investigation 2.1 can be introduced immediately following completion of the first study assignment. Pupils have already become familiar with dichotomous keys through their work in Chapter 1. They should therefore have little real difficulty in working through this investigation, which does not really depend, except for its title, on any study of pp. 53–59. However, it is best to defer Investigation 2.2 until the study assignments have all been completed. It provides a good means of reviewing many of the ideas and much of the information presented in the chapter.

Refer to the 'Guidelines' of Chapter 1 for suggestions concerning the treatment of the diversity concept. If you follow the pupil's book, you will avoid becoming entangled in alternation of generations—the traditional mental bogey of any pilgrimage through the plant kingdom that should not be allowed to deter the pupil from gaining a panoramic view of plant diversity. In due time reproductive cycles will be dealt with—Chapter 1, in *MHE*.

TEACHING NOTES

Plant Classification (pp. 48–49)

p. 154, ¶ 3. A pupil report on the life of Linnaeus might be useful. Encyclopaedia accounts are dull, though Linnaeus himself certainly was not. But even an average pupil of 15–16 ought to be able to contrive an interesting report from *The Prince of Botanists, Linnaeus* by N. Gourlie (Witherby, 1955).

A Problem: Nomenclature (pp. 50–51)

p. 51, Fig. 2.2. The Latin name is translated: 'simple wild pink, sweetly blushing'. Note that the term 'caryophyllus' was used for pinks in addition to the term 'dianthus'; Linnaeus combined the two in designating the carnation binomially (p. 51, paragraph 2).

p. 52, ¶ 2. Pupils may ask, 'What prevents duplication of names?' Basically the answer to this is the rule of priority. The first person to publish a name establishes that name as valid. An example is Linnaeus' name for the carnation. If someone now published another name for it, that name would become a synonym. This has happened quite frequently because it is difficult to know about all publications of names. For vascular plants alone it has been estimated that there are three times as many names as species. The system *is* simple, as the text says, but regulating it is complex; this is the business of several international organizations of taxonomists, who do not always agree, but who have a better record, on the whole, than international political organizations.

p. 52, Fig. 2.3. For some groups of organisms that were poorly known in Linnaeus' day, other beginning points for the rule of priority have been agreed upon. For example, most nomenclature of fungi dates from Fries's *Systema Mycologicum* of 1821.

p. 52, ¶ 3. Some expansion of the discussion may be interesting to pupils.

p. 53, ¶'s 1, 2. Further examples of short generic epithets are *Pica* (magpies), *Sus* (pigs), *Acer* (sycamore), *Bos* (cattle), *Apis* (bees), *Erica* (heaths), *Ficus* (figs), *Ilex* (hollies), *Chen* (snow geese), *Anas* and *Aix* (ducks), *Geum* (avens). On the other hand, the layman's view has some justification; consider *Strongylocentrotus* (sea urchins) and *Dolichocephalocyrtus* (beetles).

Some further examples of generic epithets far from classical Latin and Greek are *Fothergilla, Forsythia, Cunninghamia, Koelreuteria, Torreya, Kickxia, Muhlenbergia,* and of course *Welwitschia* (Fig. 2.11)—all these are plants named after botanists; *Lama* is a mammal with a name derived from an American Indian language, and *Ginkgo* is a gymnosperm tree with a Chinese name.

Tracheophytes (in part) (pp. 53–59)

p. 53. *Tracheophytes:* Establish the habit of using anglicized forms of names: 'tracheophytes' rather than 'Tracheophyta'. Where completely English words are available, use them: 'liverworts' rather than either 'Hepaticae' or 'hepatics'.

p. 54, ¶ 2. 'Why are stamens and pistils believed to be modified leaves?' a pupil who has begin to acquire the spirit of this course may well ask. A full answer, involving paleontological evidence, is difficult to give at this point, but you can show that stamens often merge into petals in 'double' cultivated flowers such as many roses and begonias.

p. 55, ¶ 4. In addition to the illustrations in Plate V you should try to have on hand an assortment of flowers for pupils to examine for first-hand appreciation of diversity. If wild flowers are not readily available, visit a florist and ask for sprays of flowers that are not fresh enough to be sold.

p. 57, Fig. 2.7. *Poppy:* The mature fruit or capsule has a ring of openings or pores at its top and is borne at the tip of a long slender stalk. When the capsules are blown by the wind, the tiny seeds are shaken out through the pores, a few at a time, and are carried away.

Sycamore: The paired, winged fruits (samaras) are well adapted to being carried away from the parent plant by the wind.

Touch-me-not: In this species the mature fruit is a fleshy capsule. If a fruit is pinched, or even touched when ripe, the capsule wall suddenly splits into a number of segments that roll up with considerable force, thereby scattering the seeds.

Cocklebur: The paired pistillate flowers are enclosed in a protective coat composed of hooked, spine-like, involucral bracts. When the fruit is mature, this outer covering serves to attach the fruits to the fur of mammals that brush up against the plant. If the animal later dislodges the fruit, a very efficient adaptation of seed dispersal has operated to transfer the seeds of the parent plant to a new location, often at a considerable distance from it.

Dandelion: Each single-seeded fruit is equipped with a 'parachute' (pappus), actually made up of delicate filaments derived from the calyx of the flower. When the fruit is mature, winds or air currents detach the fruits from the disc of the inflorescence and carry them away from the parent plant.

p. 58, Fig. 2.8. Much of the monocot seed shown here is endosperm, but this part is not labelled because presence or absence of endosperm is not a feature that distinguishes between monocots and dicots.

p. 59, Fig. 2.9. *Sweet pea:* The leaflets of the compound leaves of this legume are developed as tendrils that curl around any available supporting structures and enable the shoots of the plant to maintain its position in space. Most of the photosynthetic activity is carried on by the large, paired stipules attached at the base of the leaf.

Pitcher plant: The cup-like, highly modified leaves become partially filled with rain water. The waxy, inner surface of a cup is provided with hairs, all pointing towards the base. Insects that fall into the cups find it difficult or impossible to climb out. They eventually drown and are digested by the bacteria and enzymes in the water in the cup. The outer surfaces of the leaf carry on normal photosynthesis.

Venus flytrap: The terminal part of each leaf in this plant consists of two toothed halves that are attached to one another by a median hinge. Three sensitive hairs (invisible in this picture) project from the inner surface of each leaf half. If an insect touches any two of these hairs in succession or a single hair twice, the two halves quickly snap shut and the insect is imprisoned between them. Glands then secrete enzymes which bring about the digestion of the prey. Eventually the two halves separate, and the leaf is ready to capture its next victim.

Both the pitcher plant and the Venus flytrap are exotic species which are obtainable from horticulturalists. They are adapted to live in soils deficient in soluble nitrogen-bearing nutrients. By digesting and absorbing nitrogenous compounds in captured insects, the species are presumably better able to survive in such soils than other species that lack such insect-capturing structures.

Victoria regia: More correctly, this plant should be called *Victoria amazonica*. A species whose leaves are just a little smaller is *Victoria cruziana*. The leaves of *Victoria amazonica* may exceed 2 m in diameter. The petioles extend upward from a tuberous rhizome embedded in the mud a metre or more below the surface of the water. Stomata located only on the upper surface of the leaf blade, air spaces in the leaf mesophyll, and numerous perforations, are all adaptations permitting the photosynthetic leaves of this water plant to function as effective organs at the surface of the water. The plants which are indigenous to the Amazon can be seen at leading botanic gardens, such as Kew and Edinburgh, and at Glasgow *Victoria cruziana* is grown.

Cactus: In the species shown, an *Opuntia*, or prickly-pear cactus, the green photosynthetic tissue of the shoot is located near the surface of the flattened but thick stems. Although very small leaf-like structures are present on young stems, these do not persist. On older stems the spines, which discourage herbivores from eating the succulent photosynthetic tissues, are believed by many botanists to be highly modified leaves.

INVESTIGATION 2.1

DIVERSITY IN ANGIOSPERM LEAVES (pp. 60–61)

As the most abundant and conspicuous groups of plants known to pupils, angiosperms deserve special attention. Emphasis on diversity of leaf structure can be justified because materials for study can be so readily obtained and preserved.

Attention to this diversity through work on key construction is desirable because it continues emphasis on the mechanisms of classification and because the exercise promotes skill in using keys. As mentioned earlier, learning to use a key is a truly educational objective because it tends to free a pupil from dependence upon others for the identification of organisms.

Materials

Each pupil should individually work out a dichotomous key for the ten leaves in a set. Thus each class of 20 pupils requires 20 sets of named leaves (Set A) and also 20 sets of leaves without names (Set B) to use in checking the keys. Each set includes 10 different leaves. $20+20 = 40$; $40 \times 10 = 400$ leaves. That is a lot of leaves, particularly if they must be obtained at a time when many species are not in leaf.

Obviously you can reduce the supply problem by collecting, pressing, and mounting sets of leaves prior to the time they are needed for this investigation. Collecting leaves of 10 different species presents no problem in spring, summer, or early autumn.

Pressing, if a standard press is not available, can be quite satisfactorily accomplished by interleaving specimens in old telephone directories. Every day or every other day, depending upon the thickness of the specimens and the relative humidity, move the leaves to dry pages in the books. Completely dried specimens can then be mounted on herbarium sheets and labelled.

You can do an even better job of mounting by making use of clear transparent vinyl or acetate film, which can be obtained precoated with adhesive. Such materials can usually be obtained in paint or hardware stores and stationers. Somewhat more expensive materials are available from biological supply companies. Three methods of mounting have been used satisfactorily: (*a*) The leaf is placed between two sheets of adhesive vinyl plastic, adhesive surface against adhesive surface. (*b*) The leaf is placed on a sheet of acetate and then covered with a sheet of adhesive vinyl that bonds to the acetate. (*c*) The leaf is placed on a sheet of thin cardboard and then covered with a sheet of adhesive vinyl that bonds to the cardboard. In each of the first two methods, a slip of paper bearing a number and name (for Set A) or a number only (for Set B) is mounted with the leaf. In the third method the label can be written directly on the cardboard before the adhesive vinyl is applied. Methods (*a*) and (*b*) have the advantages that with them both sides of the leaf are visible, and the patterns of venation can easily be seen if the mount is held up to the light. But in terms of expense and time, Method (*c*) is preferred. If a hole is punched in the upper left-hand corner of each mount, sets can be assembled and then kept together by means of loose-leaf mounting rings. Such sets of leaves, when used with reasonable care, are quite durable and should remain in good condition for a number of years.

The particular species included in each set of leaves is of course not crucial, but a fairly wide range of diversity is desirable. Beech, plane, sycamore, poplar, horse chestnut, lime, hazel, oak, birch, ash, elm, willow, privet, hawthorn, cherry, and apple are all usually obtainable. Leaves do not necessarily have to be obtained from woody plants, but such leaves are usually firmer than those from herbaceous plants and more easily dried without wrinkling. Some attention should also be given to leaf size, since it is desirable that the mounts be of about the same size. A selection of ten different species from a group such as the one mentioned above will provide plenty of evidence of diversity.

Procedure

The sample key presented on p. 60 employs the same principle as that used in the keys of Chapter 1. Only the format is new, but it is so obvious that most pupils require little or no help.

You should be especially concerned with establishing a learning pattern so that pupils achieve a clear understanding of the dimensions of diversity in terms of differences between specimen and specimen and between species and species. Comparison of keys prepared by pupils in a given class almost always shows no two keys alike. Even the criteria used for constructing the keys (really the criteria used for classifying) usually differ from pupil to pupil. Finally, the need for careful observation and description is clearly demonstrated when two pupils attempt to use and then evaluate each other's keys. Thus, because the points are made in the doing, a minimum of discussion is required at the conclusion of the work.

The project suggested 'For Further Investigation' is an excellent assignment for homework immediately following completion of this investigation.

Tracheophytes (contd.) (pp. 62–68)

p. 62, ¶ 1. The distinction between gymnosperms and angiosperms is a rather more technical matter than this paragraph allows, but it is of no great importance for the purposes of this chapter, although really observant pupils may present you with a somewhat embarrassing specimen: the 'fruit' of yew, which may occur in the autumn around many schools. A little examination shows that the yew seed is rather well buried in the pulp but is not completely surrounded by it, as is the seed of a peach.

p. 64, ¶ 2. The age of a tree, particularly one living in middle and high latitudes, can be determined accurately if the annual rings, shown in a cross-section of the trunk (or in a core that has been taken by boring into the centre of the trunk), are counted. The level at which either cross-section or core is obtained should be as close to the ground as is feasible.

p. 66, ¶ 2. Pupils who want to prepare a herbarium will find that ferns are good plants to start with. Ferns are easily collected; most can be pressed without difficulty; they present a good appearance without causing the disappointment involved in the loss of colour from flowering plants; and most are not in any critical danger of extinction.

Non-vascular plants (pp. 68–75)

p. 68, ¶ 2. Because non-vascular plants are even less familiar to pupils than are tracheophytes, it is especially important to have as many specimens, preferably living, as it is possible for you to gather together.

p. 70, ¶ 2. The term 'mycelium' for the mass of hyphae has not been used because the authors could find no real need for it. You may wish to reintroduce it, but before you do so ask yourself the question, 'Are my pupils going to use 'mycelium' so often that it will have distinct advantage over the more cumbersome 'mass of hyphae''?'

p. 70, ¶ 3. It is not inappropriate to set up Investigation 3.1 at this time. Some of the 'micro-organisms' that result from it are fungi.

p. 70, ¶ 5. The terms—mould, mildew, mushroom, etc.—as commonly applied to fungi do not have any taxonomic significance. Plants bearing these common names are scattered among the fungal classes.

p. 71, Fig. 2.25. Spores of a field mushroom are located at the tips of specialized hyphae, the basidia. These project from the surface of the gills (plates) that radiate from the centre of the cap.

p. 72, Fig. 2.26. Spores of an ascomycete, such as the sac fungus shown, are formed in elongated sacs (asci) that, together with sterile, non-spore-bearing hyphae, are located in a layer on the inner surface.

p. 72, ¶ 1. The classification of species of *Penicillium* as members of the Ascomycetes is not approved by some mycologists. Since no perfect forms are known for a majority of the species included in the form genus *Penicillium*, a case could be made for relegating them to the Fungi Imperfecti (p. 74). However, their inclusion in Ascomycetes seems justified by (*a*) the close similarity between the asexual (conidial) stages of many *Penicillium* species and Ascomycete genera having identifiable sexual stages, and (*b*) the fact that most Fungi Imperfecti, upon the discovery of their sexual stages, turn out to be Ascomycetes. None of this need be brought to the attention of most pupils; but for some, this can be used as an excellent illustration of the difficulties in taxonomy.

p. 72, Fig. 2.28. The story of penicillin has been told so often that most third-year secondary pupils probably know something of it. However, a good pupil report on Fleming's discovery can do three things at this time: (*a*) relieve steady attention to taxonomy, (*b*) remind pupils that chance favours the prepared mind, (*c*) remind pupils of the complexities of ecological relationships.

p. 74, and Plates VI and VII (p. 89). The phyla of each of the genera illustrated are as follows: *Enteromorpha* and *Ulva*, phylum Chlorophyta; *Laminaria*, *Fucus* and *Ascophyllum*, phylum Phaeophyta; *Chondrus* and *Polysiphonia*, phylum Rhodophyta; and *Oscillatoria*, phylum Cyanophyta. This can be deduced from colour alone because none of the algae illustrated in this figure represents any of the numerous colour exceptions; all are also listed under their phylum names in the Appendix. *Oscillatoria* may cause some difficulty, as the Cyanophyta have now been placed in the kingdom Protista. This transfer of kingdom does not, however, alter the fact that they can be referred to as 'algae', a term that is essentially a common name without taxonomic significance. Note also that in Plate

VI identification is by generic epithet only; each of the genera contains a number of species, but the species depicted has not been designated.

p. 74, ¶ 2. A good variety of freshwater algae can be maintained in your classroom aquarium. You should have some examples of marine algae mounted on herbarium sheets. Pupils should also be allowed to scrape the sides of an aquarium to collect and examine microscopically the diatoms usually found there. Some microscope time might also be devoted to prepared slides of diatoms and desmids, which are fascinating in their diversity.

p. 75, ¶s 1–2. Refer pupils to Plate VIII and in paragraph 2 note that here is another place where your pupils are not told what to think. How are they bearing up under this treatment? How are *you* handling matters of this sort when you construct a quiz?

INVESTIGATION 2.2

THE CONCEPT OF 'PRIMITIVE CHARACTERISTICS' (pp. 75–78)

Pupils generally have far less acquaintance with plants than with animals; therefore, more effort is required to broaden the concept 'plant'. Such is the primary purpose of this investigation. The secondary purpose derives from the last portion of Chapter 1 (pp. 39–42). From the idea that classification reflects kinship (genetic) relationships, it logically follows that characteristics of organisms have diverged through successive generations in the past. That some present-day organisms may have retained more, and others fewer, of their ancestors' characteristics is but a corollary of this proposition.

Materials

The pupil is being asked to observe. Inference should be kept to a minimum. Therefore the plant materials should show as many as possible of the characteristics needed for correct scoring. For example, gymnosperm specimens should include seeds; angiosperm specimens should have fruits or flowers; mosses, ferns, and lycopods should bear spore cases. Some distinctions, as between shrub and tree, are difficult to exhibit in specimens small enough for the laboratory.

The easiest plants to obtain, of course, are just those that tend to confirm the popular stereotype. Considerable effort (and, if necessary, money) should be expended to provide a true diversity of plants; perhaps not more than three of the ten specimens should be angiosperms, and not more than one of these should be herbaceous.

Suggestions for plant materials follow. For microscope: green algae (*Spirogyra*, *Oedogonium*, *Ulothrix*); yeasts. For hand lens or stereomicroscope: moulds (*Rhizopus*, *Aspergillus*); liverworts (*Marchantia*, *Conocephalum*); mosses (*Polytrichum*, *Mnium*, *Hypnum*, *Dicranum*); lichens. For naked eye: *Lycopodium*, *Equisetum*; ferns (*Polystichum*, *Pteridium*, *Polypodium*); pine; spruce; buttercup; bluebell; wallflower; *Zebrina*; household geranium; *Antirrhinum*. If you wish to consider blue-green algae as plants, the chart allows you to do so.

As always, fresh material is preferable to that which is preserved. Most of the plants listed above are fairly easy to obtain in the autumn. To obtain bread moulds, start about ten days before the work is scheduled. If possible, use home-baked bread or rolls, since commercial bread contains mould inhibitors. Break the bread into ten pieces to avoid later handling. Spores of the mould are usually abundant in the air; after the bread has been exposed to air in the laboratory for a day, sprinkle it with water, cover, and keep in a warm, dark place.

Procedure

The chart on pp. 76–77 bears a superficial resemblance to a dichotomous key. The resemblance enables the pupil, who has had experience with keys in Investigations 1.2 and 2.1, to grasp the plan of work quickly. But of course the chart is not a key, because it does not lead to identification.

Begin the work by dividing the class into groups of two to four pupils and assigning each group to a station. Each station should be supplied with a specimen of each kind of plant. If sufficient specimens of each kind of plant are not available, some plan of rotation among stations is required. By this time pupils should be able to move from station to station readily. Work is expedited if you run through the scoring of a specimen (one not included in the investigation) before setting pupils to work.

Discussion

The numbers in the chart have been worked out so that, when summed along possible courses through the chart, they will provide low scores for plants generally considered primitive by botanists and high scores for plants generally considered advanced. For the purposes of this investigation, 'primitive' and 'advanced' are adequately explained in the 'Background Information' on p. 75. You must bear in mind and communicate to the pupils the controversial nature of the scores. For example, it is probable that many mycologists would justifiably object to the rather low score that is assigned to fungi.

A. If species were formerly fewer and simpler, then the more numerous and more highly developed species of later times would logically be more diverse—that is, there would be more kinds.

B–D. The greater the difference in score between characteristics at any one dichotomy, the more important the difference was considered to be by the maker of the chart.

E. This entails listing characteristics common to plants that scored low and high respectively.

F. It is dichotomously branching on the basis of contrasting characteristics.

G. It does not lead to an identification.

PROBLEMS (p. 80)

1. The dark spots, usually arranged in a definite pattern, are groups of sporangia called sori, which in some species of ferns are partially covered with a protective structure, the indusium. Typically, sori are more conspicuous on the older leaves of a fern, but they can be seen in various stages of development as progressively younger leaves are examined. The presence of sori—and sporangia—is of course entirely normal in a healthy fern plant.

2. Compounds containing elements in which the soil is deficient can be obtained by insectivorous plants through absorption of digested material derived from captured insects. See comments on pitcher plant and Venus flytrap in Fig. 2.9 (p. 59).

3. In helping pupils work out answers to these questions, you can avoid some confusion if you contrast a plant such as a species of *Penicillium* which produces only asexual spores, with a seed-bearing species such as a bean. 'Advantage' and 'disadvantage' should be interpreted in terms of reproductive effectiveness. Since both spore bearers and seed bearers are abundant today and both have been abundant for long geological ages, it follows that 'advantages' and 'disadvantages' have balanced in the long run. With the above as the ground rules, some pertinent characteristics are as follows.

Advantages (seed-bearing plant): (*a*) Embryos in seeds are provided with a 'built-in' food supply. (*b*) Adaptations of fruit or of seed coats may provide protection and a variety of mechanisms for dispersal. (*c*) Great variety of offspring is possible because seeds are sexually produced (not a likely idea from pupils at this point).

Advantages (spores): (*a*) Because they are very small in size, spores can be produced in tremendous numbers. (*b*) They have a low density and high surface-to-volume ratio, which are good characteristics for dispersal by winds and air currents. (*c*) Because they are asexually produced, new plants developed from them have characteristics of the parent; if the parent is well adapted, so also will be the offspring.

Disadvantages (seed-bearing plant): (*a*) Since seeds contain foods, they may be used by many animals as a source of nutrients when, unless protected by resistant coats, the seeds are destroyed. (*b*) In many cases seeds are produced in relatively small numbers.

Disadvantages (spores): (*a*) The food stored in the spore is necessarily very limited. (*b*) Since all the offspring resemble the parent closely, if they are distributed to environments quite different from that of the parent plant, the probability of any of them being able to survive is low.

In discussing this problem with pupils, keep in mind that in due course you will be discussing the microspores and megaspores produced by seed-producing tracheophytes and also that many spores are produced as the direct result of sexual reproduction. Thus the discussion, if not held within the bounds of the 'ground rules' noted above, may lead more to confusion than to enlightenment.

4. The parts of plants that we use as concentrated sources of foods are precisely those in which the plant has stored foods for its own use or for its offspring's. Food storage in seeds represents an adaptive characteristic increasing reproductive effectiveness per disseminule. Storage of foods in underground parts of herbaceous plants serves to protect the plants' food supply from animal depredation and, in certain climates, from damage by extremes of environmental temperatures or by desiccation. This protection is obviously of survival value to the species concerned.

5. Not all producers appear green, although all contain at least one of the several kinds of chlorophyll that may be involved in photosynthesis, but accessory pigments may mask the green colour of the chlorophylls. Common terrestrial examples: red cabbage, *Coleus* species, copper beeches, etc. Pupils should also remember the variously coloured algae.

AUDIOVISUAL MATERIALS

The remarks concerning audiovisual materials for Chapter 1 apply equally well to Chapter 2.

Filmstrips

Classification of Plants. Edita, distributed by Rank. Colour. (Although this filmstrip uses the ancient cryptogam–phanerogam classification, it is clearly organized in a way that can easily be related to more modern classification. The drawings are excellent.)

Great Names in Biology: Carolus Linnaeus. Encyclopaedia Britannica Films. Colour. (Good for biographical background.)

Films

Fungi. Simple Plants: The Algae. Gymnosperms. Encyclopaedia Britannica Films. All 16 mm, colour. (These three films are part of a series covering the plant kingdom. As a whole, the series places too much emphasis on phylogeny for our purposes in *DLT*; but taken individually, the films may be useful for emphasizing plant diversity with pupils who have poor experiential backgrounds.)

Carnivorous Plants. Ealing. Super-8 film loop, colour, 2 min. (Good, if your pupils do not have an opportunity to become directly acquainted with this example of plant diversity.)

TEACHER'S REFERENCE BOOKS

DAVIS, P. H., V. H. HEYWOOD. *Principles of Angiosperm Taxonomy.* Oliver & Boyd, 1965. (An advanced comprehensive survey of angiosperm taxonomic principles.)

HESLOP-HARRISON, J. *New Concepts in Flowering–Plant Taxonomy.* Heinemann, 1963. (A clear exposition of the problems facing a modern plant taxonomist.)

HEYWOOD, V. H. *Plant Taxonomy.* Arnold, 1967. (An account of the recent developments in the field of plant taxonomy.)

SWAIN, T.—Editor. *Chemical Plant Taxonomy.* Academic Press, 1963. (The first comprehensive attempt to survey the scope and usefulness of chemical plant taxonomy.)

3
PROTISTS

MAJOR IDEAS

Structural adaptations, classification, and nomenclature should continue to receive emphasis during work on Chapter 3. Additional ideas to emphasize are:

1. All that we know of micro-organisms has come to light within the era of modern science. This knowledge—representing a whole new dimension in biology—has been integrated into the older biological knowledge slowly, with difficulty, and (still) incompletely.
2. Science is an international enterprise, sustained by the efforts of many men in many countries. (This, of course, is an idea that should carry through the entire year, but it is especially appropriate for emphasis in Chapter 3.)
3. The more man learns about nature, the less easy he finds it to fit his knowledge into compartments. The boundaries between taxonomic groups seem sharp when a scheme of classification is studied in a book, but indistinct when organisms themselves are observed.

PLANNING AHEAD

With everything in readiness for the laboratory work of Chapter 3, you also have much of your preparation for Chapter 1 completed, because both chapters are concerned primarily with microbiology. However, check specifically: facilities for incubating cultures (Investigation 1.1, *PLW*, p. T73); make sure *Agrobacterium tumefaciens* has been ordered, see p. T74 (Investigation 1.2, *PLW*); availability of plants of sufficient number and size for inoculation (Investigation 1.2, *PLW*). It may also be advisable to collect soil samples for Investigation 1.3, *PLW*. Also make sure that you have the materials for Investigation 1.4, *PLW*, particularly dead insects, dead leaves, and twigs.

It is not too early to survey your pictorial and projection materials for Chapter 2, *PLW*. That chapter requires pupil experience with various biomes, and, except for the biome in which you live, that experience can be gained only through slides and films and audio recordings.

GUIDELINES

Animals are very familiar to pupils, plants (except in a general way) less familiar, and protists quite unfamiliar. Consequently, though shorter than Chapters 1 and 2, Chapter 3 presents more difficulties. Providing first-hand experience with protists is a task incommensurate with the size of the organisms. Yet, if protists are not to be mythical beasts, you must show them to pupils. Therefore the core of this chapter, even more than others, is the laboratory work.

A good beginning is made with Investigation 3.1. It has previously been suggested (p. T54) that this may be set up well before the end of Chapter 2. Many of the organisms that appear in the cultures fall into our plant kingdom and others into our protist kingdom (microbiologists have never been very careful about distinctions between kingdoms). Thus Investigation 3.1 not only forms a good bridge between Chapters 2 and 3; it also again stresses the artificiality of classificatory schemes.

If you have not begun Investigation 3.1 previously, start it as soon as possible. While the cultures are developing, see that pupils read pp. 81–83, the first of several illuminating excursions into the history of biology. Stressing the international nature of the scientific enterprise, which is a necessary corollary to a history of science, is an important aspect of teaching science humanistically. Any other time available for discussion is best devoted to the problems involved in human attempts to impose order upon the facts of nature—as illustrated by difficulties of classification at the kingdom level, for example, or by doubts concerning the status of viruses. From the first chapter pupils have been exposed to the idea that learning science is not the memorizing of a prescribed system, into which facts are to be fitted with Procrustean determination, but rather a seeking after new and better ways to order an ever widening array of facts. Said the Abbé Galiani, 'La science est plutôt destiné à étudier qu'à connaître, à chercher qu'à trouver la vérité.' (Science is destined rather to study than to know, rather to seek truth than to find it.) Chapter 3 offers an opportunity to bring this attitude to the forefront.

If pupils have studied cells in a previous course, the question of the cellular nature of protists may be raised. Are ciliates, for example, single cells, or are they organisms that have lost the cellular state—or did they ever have cells? How do we interpret the cellular state in slime moulds?

TEACHING NOTES

The Discovery of Micro-organisms (pp. 81–83)

Note: As indicated under 'Guidelines' above, this is primarily a chapter of laboratory investigations. You can start Investigation 3.4 at any convenient time. It lasts for six weeks and so does not reach a conclusion until long after you have left the rest of Chapter 3 behind.

p. 82, ¶ 2. The first scientific society was probably the Accademmia dei Lincei of Florence. It exists today, but, unlike the Royal Society, it has not been continuously active.

p. 83, ¶ 3. Much of the technology that pupils employ in Investigations 3.2 and 3.3 came from Koch's laboratory. Petri dishes were invented by one of his students. The use of agar-agar was suggested by the wife of one of his students.

INVESTIGATION 3.1

A GARDEN OF MICRO-ORGANISMS (pp. 83–85)

Investigation 3.1 allows the biology teacher to lay claim (temporarily) to some of the olfactory ill fame usually monopolized by the chemistry department. This is not necessarily bad public relations; in some schools it has called attention to the fact that there is life in the biology laboratory!

The terminological distinctions made in 'Background Information' are important. The chapter is titled 'Protists' and begins with a discussion of micro-organisms; the pupil can therefore easily fall into the delusion that protists are micro-organisms and vice versa. However, the terms 'microbes' and 'micro-organisms' *are* synonymous, though there is a slight differentiation in usage, the latter being a little more dignified and formal, perhaps.

Materials

Any glass or plastic container more than 10 cm in diameter and having more or less vertical sides at least 4 cm high may be substituted for finger bowls. Small plastic refrigerator dishes are good; but they cannot be stacked, so they require more space after being filled. It is desirable that all the containers used in any one class be alike.

The recommended media have been chosen with an eye to culturing a wide variety of micro-organisms; there are, of course, many other possible choices. If necessary the number per group may be reduced, but the media recommended for Bowls 1, 3, 4, 5, and 8 should be retained. Peppercorns (Bowl 10) are recommended because of their historical association with Leeuwenhoek's work. The kind of fruit for Bowl 1 can be varied from group to group (rose hips and small tomatoes give good results—use whole undamaged fruits and puncture skin once with a sterile needle); so can the kind of water for Bowl 3. Pupils can supply most of these materials.

Hand lenses may be used in place of stereomicroscopes, but they are not really an adequate substitute in this investigation.

Tap water may be used in setting up the cultures (except for Bowl 3); but if the water contains much chlorine, it should be allowed to stand for twenty-four to forty-eight hours in a shallow container. Chlorine can diffuse from the large exposed surface of the water.

Procedure

It is convenient to set up two or three groups of ten pupils each. Each pupil is responsible for one of his group's ten bowls. (Few classes have exactly twenty or thirty pupils; see p. T30 for suggested alternatives.)

It is important to note that all bowls must be clean at the outset (and of course should be cleaned thoroughly afterwards). Here 'clean' means not sterile, but chemically clean—that is, free of all soap or detergent. After washing the bowls, rinse them thoroughly at least four times.

If the bowls are set up on Friday, they may be observed macroscopically (and olfactorily) during the following week. Each day at the beginning of the class period, the bowls should be placed in an accessible place so that each pupil can observe all bowls, at least in a cursory way. But it is too time-consuming to have all pupils make detailed notes on all bowls; each should concentrate on the bowl he set up.

One week after they have been set up, the materials in the bowls should be ripe for microscopic observation. Pupils having like bowl numbers should work together as a group. Provide each of these (ten) groups with whatever optical equipment it seems to require. For example, a group observing organisms in Bowl 3 requires monocular microscopes, but it has little use for hand lenses or stereomicroscopes.

In a classroom situation it is not feasible to attempt identification of microbes all the way to the genus and species levels; in most instances one must be content with identification at the phylum level. Some pupils may, however, wish to pursue the matter further with the aid of: A. T. Henrici, *Moulds, Yeasts and Actinomycetes*, 2nd edition (John Wiley, 1947).

Discussion

The first step is to have the reports of observations exchanged among the pupils either orally or by means of duplicated sheets.

B. Usually moulds are found in the largest number of dishes. Partly this is because bacteria, which are probably actually more widespread, are difficult to find without special techniques (Investigation 3.3), while moulds are readily visible. And partly it is because moulds grow better at the temperatures prevailing in most laboratories. If cultures were incubated at 35–40 °C, more bacterial colonies might be observed.

C. This depends upon how long observations are continued. For example, sometimes 'rotting' begins (probably mostly from bacteria), and later the material becomes covered with visible mould. A succession of 'protozoa' might be observed in Bowl 3, but this kind of succession is more likely to be detected in Investigation 3.1, *PLW* (pp. 82–83). In Chapters 1 and 2 ecological relationships were allowed to sink into the background. In Chapter 3 ideas and terms from Chapters 1 and 2 of *WL* are being picked up again. You should be aware of this swing back towards ecology; actively exploit it, because it prepares the way for a fully-fledged return to ecological concepts in *PLW*.

D. A definite diminution of material should be noticeable, especially in Bowls 1, 6 and 8. To some extent this may be attributed to desiccation, but for the most part the food is being consumed by the organisms.

These questions present opportunities for discussion to proceed in many directions. It is particularly desirable to stress the fact that in this 'garden', *foods*—not merely inorganic nutrients—are provided for the micro-organisms, since the great majority are consumers. As most of the organisms occur in situations where decomposition is proceeding, it is obvious that most are saprovores, though there is no clear-cut way to distinguish them from other consumers. Also, the possible presence of predators and parasites cannot be excluded.

At this point the discussion might be turned to such questions as: Why the odours, and why different kinds? Where did the micro-organisms come from? (This is a good prelude to Investigation 3.4.) How can you account for the changes in a given bowl over a period of time?

The Phyla of Protists (in part) (pp. 86–88)

p. 86, ¶ 2. This is the point at which to consider the questions, What is an animal? What is a plant? These questions were deliberately bypassed in Chapters 1 and 2, but they were raised in Investigation 1.1 of *WL*. Have pupils refer to their lab-book records of that investigation when you consider these questions.

p. 86, ¶ 4. As a grouping of convenience, the protist kingdom is somewhat like the Lichens, which some botanists use as a group in the plant kingdom, or like the Deuteromycetes. No one claims that all the phyla in the protist kingdom are related to each other—not even as distantly as the phyla grouped in the plant kingdom or in the animal kingdom. We may regard the protist phyla as representing different ways in which organisms have carried a varying set of characteristics from very ancient times. The animal and plant kingdoms, then, represent two other ways in which organisms have developed from ancient ancestors. Every taxonomist knows that we will develop *better* classifications as we learn more about living things. However, the taxonomic problem is not unique in science. The door is never finally closed on any major scientific problem. At any time, a new discovery may reopen any question.

p. 87, ¶ 2. Perhaps the most obvious way in which bacterial photosynthesis differs from that of other photosynthetic organisms is its lack of oxygen production.

p. 87, Fig. 3.6. Having discussed nomenclature in Chapter 2, you should take every opportunity to use biological names—without ostentation. The names of micro-organisms are suitable for this purpose, because most lack 'common' names. Be sure pupils note the great disparity in magnifications.

INVESTIGATION 3.2

MICROBIAL TECHNIQUES: POPULATIONS (pp. 89–93)

This is a complex investigation, but it is worth the effort involved. It pays dividends more in appreciation than in specific knowledge. Moreover, it provides a good test of the extent to which pupils have learned to understand and follow directions, to work together in groups and to relate procedures to outcomes. It is also a good test of how well you have mastered laboratory logistics.

Before any other action is taken, a careful check of materials must be made. Keeping in mind the quantities of glassware available, decide upon group sizes. Groups of two require, for example, 200 test-tubes and 200 petri dishes if for instance you have four classes of twenty pupils each. Such quantities may be entirely impracticable. Using groups of three or four reduces them and still provides opportunity for each pupil to participate actively. By staggering the scheduling of the investigation for different classes, you can cut down the quantities still further.

Materials

Standard nutrient agar can be purchased, but the preparation of the medium is a good project for a special group. To prepare sixty culture tubes, the group requires:

Pipette, 10 cm^3
Test-tubes, 19 mm×150 mm, 75
Glass-marking crayons
Heat source
Distilled water, 1000 cm^3
Graduated cylinder, 500 cm^3
Beaker, 1000 cm^3
Balance
Agar, granulated, 15 g
Peptone, 5 g
Beef extract, 3 g
Stirring rod
Funnel, 10–12 cm
Ring stand and ring
Rubber tubing, to fit funnel tube
Pinch or Mohr clip
Non-absorbent cotton wool
Wire test-tube basket or other suitable container
Autoclave or pressure cooker

Using the pipette (or a small graduated cylinder), place 15 cm^3 of water in a test-tube and mark the water level with a glass-marking crayon. Pour out the water. Using this tube as a ruler, mark the 15 cm^3 level on enough tubes to provide five for each team.

Heat the water to just below boiling. (If the tap water bears no unusual amounts of dissolved minerals, it may be used in place of distilled water.) Dissolve the agar first; then add peptone and beef extract. Heat until the mixture comes to a boil, stirring continuously. For dispensing the prepared medium into tubes, attach a short piece of rubber tubing to the stem of a large funnel and place a pinch clip on the tubing. Support the funnel on a ring stand at a convenient height. While the liquid is still warm (above 55°C), pour it (a portion at a time) into the funnel. A little practice in using the pinch clip facilitates filling the marked tubes. Plug the tubes with cotton wool. Place all the tubes in wire baskets, and sterilize in an autoclave at 15 lb/in^2 (10^5 N/m^2) pressure for fifteen minutes.

With a little trial and error you can make inoculating loops that hold close to 5 mm^3 (1/200 cm^3).

Serratia marcescens is a bacterium having a red pigmentation; *Sarcina lutea* is yellow. These organisms are relatively harmless. They are safe for use by secondary school pupils and are large enough to be easily studied. Cultures can be obtained from most biological supply companies.

Procedure

Of help to those whose bacteriological background is a little 'rusty' is: A. E. Sussman, *Microbes—Their Growth, Nutrition and Interaction*, A BSCS Laboratory Block (D. C. Heath & Co., 1963). Although the bacteria being cultured are ordinarily harmless, sterile procedure should be observed throughout the work, not only because the demonstration of safe handling of micro-organisms is one of the purposes of the investigation but also because it is possible for pathogenic organisms accidentally to get into the medium and multiply. Therefore, at the end of the work *all tubes and petri dishes must be sterilized* in an autoclave or pressure cooker before the medium is cleaned out of the glassware.

Studying the Data (p. 92)

3. 1 800 000 000. The populations can also be explained by calculating in reverse order, from 1 800 000 000 live bacteria per cm^3 in the original mixed culture: Into Tube 1 we transfer 1 loopful ($= 5\ mm^3 = 1/200\ cm^3$) of this suspension. In this volume there will be $1\,800\,000\,000/200 = 9\,000\,000$ organisms. If each of these organisms develops into a visible colony, there should be approximately 9 000 000 colonies present in the pour plate prepared from this tube. Thus every colony on this plate represents 200 organisms per cm^3 present in the original culture. If we now take a loopful of agar from Tube 1, included in that loopful there should be 3000 bacteria. (We introduced 9 000 000 organisms into 15 cm^3 of agar medium: each cm^3 will as a result include $9\,000\,000/15 = 600\,000$ bacteria per cm^3. A loopful will therefore include $600\,000/200 = 3000$ organisms.) This loopful is now introduced into Tube 2. In the pour plate prepared from this tube there should develop approximately 3000 colonies. Thus each colony represents 600 000 organisms present in the original mixed culture. In a loopful of the agar present in Tube 2 there will be $1/200 \times 1/15$ or 1/3000 of the organisms present in Tube 2. $1/3000$ of $3000 = 1$. Therefore we would expect only 1 organism to be introduced into Tube 3, and only 1 colony would develop in the pour plate prepared from it. Therefore each colony in this plate represents 1 800 000 000 organisms present in the original mixed culture.

In these dilution procedures there are, of course, a number of sources of error which may be identified by alert pupils and that will serve as good starting points for discussion. One source, of minor importance, is the fact that the organisms removed from Tubes 1 and 2 by means of the loop are not present in the pour plates prepared from them. Another, of far greater significance, is the fact that organisms may not survive the rigours of the handling and dilution procedures and will therefore not give rise to visible colonies. Also, incomplete mixing of organisms introduced into the tubes leads to large sampling errors. Finally, variations in the dimensions of the loops used by different groups can result in differences in data.

6. The assumption involved is that all plates are contaminated equally.

C. *If* you lift only from a single colony, then all organisms on the loop should be of a single kind, since all in the colony descended from a single individual.

D. This would be a pure culture.

G. This is the same as the number of kinds of bacteria—assuming that all kinds produce colonies distinctively different in appearance. The assumption is probably warranted in this case of a deliberately constructed 'mixed culture'. But generally speaking, 'kinds' cannot be equated with 'species' because in many bacterial species various kinds of colonies may be produced depending upon environmental circumstances and 'strains' within species.

I. See last paragraph of notes on 'Procedure' (above).

Summary (p. 93)

C and **D.** In general, the pour-plate procedure and colony counting are concerned with the determination of population density, and the streak-plate procedure with obtaining a pure culture.

Evaluation of procedures should be a habit by this time. Have the class consider the following: (*a*) What are some possible sources of error in estimating the number of bacteria in a culture by the methods used in this investigation? (*b*) How might the kind of medium and the temperature of incubation influence the number of colonies? (*c*) How could you use the pour-plate method to compare rates of growth for two different kinds of bacteria in a mixed culture?

INVESTIGATION 3.3

MICROBIAL TECHNIQUES: MICROSCOPIC STUDY OF BACTERIA (pp. 93–95)

Materials

If your biology laboratory is not equipped with sufficient gas outlets and if alcohol burners are unavailable, you may be able to arrange an exchange of laboratories with the chemistry teacher for one period. The actual laboratory time for this investigation need not exceed thirty minutes, exclusive of time for studying directions and for cleaning up.

The peppercorn infusion from Investigation 3.1 should contain a good variety of bacteria. This may be mixed with materials obtained from the cultures of Investigation 3.2.

To prepare crystal-violet stain, dissolve 2 g of crystal violet (gentian violet) in 20 cm^3 of ethanol (95 per cent) and add 180 cm^3 of distilled water. Filter just before using.

Procedure

Though sterile technique is not critical in this investigation, flaming the loop before and after use should be practised for the sake of bacteriological principle.

Mounting the stained film by adding a few drops of glycerine and a cover slip makes the colour of the stain appear more brilliant and permits the high power objective to be used at maximum resolution—an important factor in the study of such small organisms.

Some teachers provide blank slide labels. The pupil may mark such a label with the name of the organism, the date, and his initials, affix the label, and keep the finished slide as a souvenir. (If glycerine has been used, it should be rinsed off in several changes of water.) Since pupils are often proud of these tokens of their achievement, knowledge of the activities of the biology classes may be disseminated through the school. You can make a permanent mount by placing a few drops of canada balsam or of DPX mounting medium on the dry film and adding a cover slip.

The oil-immersion lens is not essential; after all, Leeuwenhoek observed bacteria in a peppercorn infusion with a lens magnifying $\times 270$. But an oil-immersion microscope always adds some interest to the observation of stained bacteria. To illustrate the extent of magnification possible with the oil-immersion microscope, a simple comparison can be made: To the centre of the blackboard fasten a 1 mm square of white paper; around this mark off a 1 m square. The bacteria, when viewed under the oil-immersion microscope, are magnified to a similar degree. From this the pupils can get an indication of the minute size of bacteria.

Completion of Investigations 3.2 and 3.3 provides pupils with rudiments of technique that can be further developed on an individual basis. Possibilities of further investigation are endless; the two ideas at the end of Investigation 3.2 are samples. The Laboratory Block by Sussman (reference p. T70) should be put in the hands of any pupils showing interest.

The Phyla of Protists (contd.) (pp. 95–102)

p. 95, ¶ 3. Unlike the photosynthetic bacteria, the blue-green algae do give off oxygen. But the blue-green algae resemble bacteria in their reproduction by simple fission and in their lack of a distinct, organized nucleus.

p. 96, Fig. 3.14. Again (previously in connection with Plate VI) remind pupils that the use of generic epithets only does not violate the binomial system of nomenclature. It is legitimate identification at a level less precise than that of species in cases where such a level of identification is considered unnecessary (as here) or where it is not possible on the basis of available information.

p. 96, ¶ 2. Another reminder that *WL*, with its attention to ecological relationships, is not a closed matter. Often pupils' past experience has been such that they find it difficult to believe that they are expected to carry over any understanding from one 'unit' to another. This can only be overcome by constant attempts to tie past experience to the present.

p. 100, ¶ 2. If pupils consulted the Appendix to check on sporozoans, they would have found there a malarial parasite named *Plasmodtum vivax*. Now, in this paragraph, they find the word 'plasmodium' unitalicized, uncapitalized, and referring in a very general way (not taxonomically) to slime moulds. This is sure to be confusing. Help them.

p. 100, ¶ 4. If you wish to demonstrate slime-mould growth, see pp. T117–18.

p. 101, Fig. 3.21. If these photographs are to mean much, the magnifications must be noted.

And they should be compared with those in other figures, especially Fig. 3.6.

p. 102, ¶ 1. See comments concerning p. 75, paragraphs 1–2 (p. T59).

INVESTIGATION 3.4

EXPERIMENTS ON SPONTANEOUS GENERATION
(pp. 102–4)

There is no need for more than one set-up per class; the thinking is much more important than the manipulations. But if you have more than one class, replication by classes may reveal that there can be some variation in the results.

Materials

Filtering can be avoided if peptone is available; use about 8 g of peptone in place of the bouillon cube.

If flasks do not have a ground-glass area that can be written on with lead pencil, label them before sterilizing by using cardboard tags tied on with string. Do *not* use a glass-marking pencil on flasks that are to be heated or autoclaved.

Procedure

A few pupils working outside of class time may prepare the materials. Those who prepared the medium for Investigation 2.2 in *WL* can prepare the medium with a minimum of teacher supervision. The work with glass tubing is new, however, and should be closely supervised. Be particularly careful when inserting the glass tubing through the rubber stoppers; cork borers may be used to make this operation safer. Choose a cork borer with an inside diameter just large enough for the tubing to fit into it. First insert the cork borer through the hole in the stopper. Then slide the tubing through the borer to the desired distance. Finally, holding the tubing in place, withdraw the borer from the stopper.

Pre-class preparations can include mixing the medium, bending the tubing, constructing the stopper combinations, and autoclaving Flasks 4, 5, 6 and 7. (Stoppers should be loose in necks of the flasks during autoclaving.) Then the preparation group can conclude the work in front of the class by boiling Flasks 2 and 3 and sealing the tops with paraffin wax. (Be sure the stopper of Flask 3 is tilted in the mouth of the flask during the boiling.) If flasks have been labelled with tags during the preparations, they can now be relabelled with glass-marking crayon.

During the weeks while observations are being made, flasks may be picked up for examination; but they should not be violently shaken.

If all goes well, put aside Flask 7 at the conclusion of the investigation and save it for use at the beginning of Chapter 1, *MHE.*

Studying the Data

Typical results are as follows:

Flask 1 becomes turbid within a day or two. In this and the other flasks that become turbid, patches of mould often develop on the surface of the medium.

Flask 2 usually becomes turbid a day or two later than Flask 1. Before the conclusion of the experiment its contents have usually evaporated.

Flask 3 may take a very long time to become turbid; indeed, it may never become turbid. Nevertheless, though many microbes are killed by boiling, the resistant spores of some bacteria may survive and eventually produce turbidity.

Flask 4 usually becomes turbid at about the same time as Flask 2. Its contents usually disappear by evaporation.

Flask 5 with its small opening may not become turbid for many days or (if there are few air currents) even weeks. But if it is kept long enough, turbidity will appear.

Flask 6 often remains clear long after the experiment has been concluded.

Flask 7 should remain perfectly clear for as long as it is undisturbed. Some flasks of this kind have been dated and kept for years as exhibits. The design of the S-shaped tube is comparable to that of Pasteur's 'swan-neck' flask.

Pupils may doubt that the turbidity of the broth is caused by bacteria. (If the course is doing what it should, they will demand evidence.) Make a check on Flasks 2 and 4, using the methods of Investigation 3.3. Usually some good slides of bacilli and cocci will be obtained, and these can be compared with prepared slides of known bacteria.

A and **B.** If Flask 3 becomes cloudy, it usually does so much later than Flask 2, for boiling kills most if not all of the organisms originally present in the broth. Although boiling also kills organisms in Flask 2, nothing hinders the contents from acquiring new ones from the air.

C. This question should engender considerable discussion. Believers in abiogenesis argued that Spallanzani's sealed flask did not develop microbes because heating had destroyed the 'power' of the air in the flask and there was no way for new air to enter. Anaerobic organisms were not known in their day. It is instructive to note that while Spallanzani failed to make his point theoretically, the practical import of his work followed rather quickly. Much of Napoleon's success has been attributed to his use of foods preserved by canning to feed his troops.

E. Pasteur needed to meet the argument made by Spallanzani's critics.

F. The curve in the tube forms a dust trap. Dust particles bearing spores of micro-organisms do not get beyond this point even when air currents around the flask are rather strong. If the autoclaving has been properly done and the sealing has been done promptly and well, no living organisms are present in the broth; the tubing prevents new, living ones from entering, and therefore no turbidity develops. The single curve in Flask 6 has no such trap, but it is a rather effective barrier, and frequently no growth of micro-organisms occurs in Flask 6 during the time of the experiment. Pasteur went one step further; he tilted a 'swan-neck' flask until its contents flowed into the crook of the tube; soon thereafter a growth of micro-organisms appeared in the flask.

G. Growth of micro-organisms may appear almost simultaneously in Flasks 1, 2, and 4; but frequently it occurs slightly sooner in Flask 1. Presumably this results from an abundant supply of micro-organisms in the broth at the beginning, while growth in Flasks 2 and 4 depends upon micro-organisms arriving in the broth after heating.

Conclusion

Discussion can proceed in many directions. If results are not somewhat similar to those described above, the whole experiment may be inconclusive. Then the discussion should centre on reasons for discrepancies between the class's data and results reported by Spallanzani and Pasteur (and many later investigators). If results are essentially similar to those described above, pupils should be led to see that the experiment discredits the idea that micro-organisms can arise from matter that has no living organisms already within it. The ideas in the introduction may be put into a hypothesis expressed in negative terms: 'Micro-organisms cannot arise unless they have ancestors', or similar wording. You then have the opportunity to discuss the impossibility of proving a negative proposition. The best that can be done is to amass evidence making the proposition more and more unlikely. This is what has happened to abiogenesis.

GUIDE QUESTIONS (p. 105)

4. The best example and, perhaps, the oldest is the use of yeasts to make alcoholic beverages and later to leaven bread. Others: in making cheese, in making yogurt, in making silage, in retting flax, in curing tea and cacao, etc.

9. For more on this matter refer to Fig. 1.2 in *LO*.

PROBLEMS (p. 105)

1. This statement is a complex way to describe fission in combination with the principle of biogenesis.

2. We believe that the exact number is 2 361 183 241 434 822 606 848.

At 12.00 noon the population is
$2^0 = 1$ bacterium;
at 12.20 it is $2^1 = 2$ bacteria;
at 12.40 it is $2^2 = 4$ bacteria;
at 1.00 p.m. it is $2^3 = 8$ bacteria;
and so on to 12.00 noon the following day, when it is 2^{71}.

What is more important, have the pupil look at the magnifications in Fig. 3.6 and then try to imagine what a population of the size he has calculated would look like. The idea of reproductive potential is a key idea in this course, culminating in Chapter 5 of *MHE* (p. 161, paragraph 2).

3. This can be discussed at several levels. Most pupils realize that cold storage does not kill micro-organisms. The pupil is most likely to say that the micro-organisms that bring about decay of food (by consuming it themselves) do not multiply at low temperatures. This is, however, a relative matter. Between 0 °C and room temperature, moulds generally thrive better than bacteria. Few pupils at this stage are likely to point out the effect of lowering temperature on metabolic processes or to link this with rates of chemical reaction.

AUDIOVISUAL MATERIALS

The remarks concerning audiovisual materials for Chapter 1 (p. T52) apply equally well to Chapter 3.

Filmstrips

Great Names in Biology: Antony van Leeuwenhoek. Encyclopaedia Britannica Filmstrips. Colour. (Good for biographical background.)

Bacteria. Encyclopaedia Britannica Filmstrips. Colour. (A good visual survey of major bacterial groups.)

Films

Bacteria. Encyclopaedia Britannica Films. 16 mm, colour. (Emphasizes the kinds of bacteria and their life processes.)

Bacteria. McGraw-Hill Book Co., Inc. 16 mm, colour, 30 min. (A good film for background, without too many excursions beyond the concerns of Chapter 3.)

Bacteriological Techniques. Ealing. A series of four film loops: *Preparing and Dispensing*, 3 min. 25 s; *Inoculating*, 3 min. 34 s; *Serial Dilution and Pour Plates*, 3 min. 53 s; *Staining*, 3 min. 20 s. Super-8, colour.

TEACHER'S REFERENCE BOOKS

DADE, H. A., J. GUNNEL. *Class Work with Fungi*. Commonwealth Mycological Institute, Kew, Surrey, 1966.

DOBELL, C. *Antony van Leeuwenhoek and his 'Little Animals'*. Constable, 1960. (The letters Leeuwenhoek sent to the Royal Society of London, in which he described his discoveries.)

HARRIGAN, W. F., M. E. MCCANCE. *Laboratory Methods in Microbiology*. Academic Press, 1966. (A handbook containing much useful information.)

STEVENSON, G. *The Biology of the Fungi, Bacteria and Viruses*. Arnold, 1970.

SUSSMAN, A. E. *Microbes—Their Growth, Nutrition and Interaction*. A BSCS Laboratory Block. Heath, 1963. (An excellent source of ideas for microbial investigations.)

The Oxoid Manual of Culture Media. 3rd edition. 1965. Issued by the Oxoid Division of Oxo Ltd., Southwark Bridge Road, London SE1.

VICKERMAN, K., F. COX. *The Protozoa*. Murray, 1972.

PATTERNS IN THE LIVING WORLD

The break between *Diversity Among Living Things* and *Patterns in the Living World* is not a major one. The basic ideas of *DLT* require no elaborate summation; moreover, work on Investigation 3.4 must extend beyond the time of transition. Emphasis on the diversity of organisms comes to an end with Chapter 3 in *DLT*, but protists—viewed in their ecological settings—remain the centre of attention in Chapter 1, *PLW*. Therefore, in the classroom the transition from book to book need not be conspicuous.

Reading the introduction to *PLW* should be sufficient to establish in the pupil's mind the change in viewpoint that characterizes the return from taxonomic to ecological thinking. But some misunderstanding may arise from the use of the term 'patterns' in the section title. The word can be ambiguous. If we think of it as denoting something *a priori*, then of course it can have no use in modern science. But this is not its meaning here. Sir Cyril Hinshelwood pointed the way to the proper meaning when he said: 'Science is not a mere collection of facts, which are infinitely numerous and mostly uninteresting, but the attempt of the human mind to order these facts into satisfying patterns'—the 'patterns' being mutable mental constructs based on available data. A further quotation (from Henri Poincaré) may help you to guide the pupils' thinking: 'Science is built up of facts as a house is built of bricks; but a collection of facts is no more a science than a heap of bricks is a house.'

1

PATTERNS OF LIFE IN THE MICROSCOPIC WORLD

MAJOR IDEAS

1. Micro-organisms are found in all natural ecosystems; many have distributional patterns relatively unaffected by geographical features.
2. Disease is not a special curse of the human species, but a universal attribute of living things. An infectious disease is an ecological relationship between two organisms and therefore cannot be adequately understood except from an ecological viewpoint.
3. Virulence is a characteristic of pathogens; resistance is a characteristic of organisms that pathogens affect. The milieu in which pathogen and host meet is a third factor in the causation of disease.
4. Different diseases prevail in different parts of the world as a result of different ecological conditions. Among the ecological conditions necessary to the existence of some diseases is the presence of suitable vectors.
5. Soil is an ecosystem consisting of living organisms, organic remains and inorganic substances. Although the larger organisms are more conspicuous, microscopic ones play important roles in soil life.
6. The community relationships discussed in Chapter 3 of *WL* can be found in soil micro-communities.
7. Micro-organisms are sensitive to the chemical nature of the ecosystems in which they occur. Ionization, especially that which results in pH of water, is a significant chemical process in soil environments.
8. The biogeochemical cycle of the element nitrogen is dependent on the activities of soil micro-organisms.
9. Much of the return of other elements to the abiotic environment is also accomplished by soil microbes. The rate of decomposition varies with type of soil and nature of the substance being decomposed.

PLANNING AHEAD

Investigation 2.1 requires radish, vetch, tomato, and lettuce; if these were not obtained previously, they should be ordered now.

Investigation 2.2 is facilitated if you can provide pupils with grids exactly like those used in the climatograms on p. 41—that is, having twelve divisions horizontally and eighteen vertically. Such a grid can be drawn on a stencil and produced in quantity very cheaply.

To make the understanding of biomes vivid, assemble as much pictorial material as possible. Pin boards should receive heavy use during work on Chapter 2, and the displays should be changed frequently. Obtaining good ecological pictures is not easy; even the *National Geographic Magazine* tends to avoid natural landscapes. Some years may elapse before you have a really adequate file of pictures, but begin such a file now.

Collect the pond water and pond organisms needed for Investigations 3.1 and 3.2. If necessary, augment collections with species ordered from a biological supply company.

GUIDELINES

Investigation 1.1 provides motivation for an excursion into a long chapter containing few pictures and a large number of ideas. Moreover, these ideas are in some cases entirely unfamiliar to most pupils, or (and this is much harder to cope with) they are ideas involving unfamiliar ways of looking at superficially familiar things. All of this calls for a slowing of the rapid pace that should have characterized the work in *DLT*.

The reading material is most logically broken into three assignments: pp. 2–16 (guide questions 1–12, p. 32), pp. 19–22 (guide questions 13–17, p. 33), and pp. 24–30 (guide questions 19–22).

Investigation 1.2 requires a long time to bring to completion, as does Investigation 1.4. Each one will run beyond the time needed for study of the chapter. Meanwhile, observations on Investigation 3.4 in *DLT* should be continuing. Because of this overlapping of several investigations, maintaining interest and preventing confusion requires much attention from the teacher.

Curriculum-makers universally regard disease prevention and soil conservation as important topics. If, however, the pupil is ever to progress beyond the rote repetition of 'right' prescriptions for conservation of soil and prevention of disease, he must be induced to wrestle with the concepts embodied in the terms 'soil' and 'disease'. Chapter 1 aims to bring the pupil to grips with the biological realities of these concepts.

There is, of course, need to apply the concepts in a local situation. Do you teach suburban pupils whose parents are struggling to make a lawn on subsoil hastily bulldozed into place by the departing builder? Do your pupils know soil only as a commodity that is sold in plastic bags at the garden shop? Are the farmers and gardeners of your area coping with inadequately drained soils, acid soils, alkaline soils, sandy coastal soils lacking humus, upland clay soils?

The Appendix of the pupil's book was frequently referred to during *DLT*. Continue to call attention to it. As various groups of organisms are mentioned in Chapter 1, they should be placed in the framework of the classification scheme.

A final suggestion: Make the investigations central to the presentation of this chapter. Discussions of laboratory work will provide the teaching opportunities for most of the major concepts.

TEACHING NOTES

INVESTIGATION 1.1

MICRO-ORGANISMS IN SCHOOL ENVIRONMENTS (pp. 1–2)

If the class did a thorough job on Investigation 3.2 in *DLT*, then Investigation 1.1 can be done easily—especially if you can retain the groups of pupil assistants who learned the techniques of medium preparation and sterilization. This investigation often generates a considerable amount of interest outside biology classes.

Materials

Each plate should contain 15 cm^3 of nutrient agar. The following quantities provide materials for 66 plates:

Water (perferably distilled), 1 litre
Peptone, 5 g
Beef extract, 3 g
Agar, 15 g
An incubation temperature of 37 °C is desirable.

Procedure

Ideally each pupil should expose a plate, but groups of any size can be used. If enough groups or individuals are involved, several plates may be exposed at each location. Then the results from each location can be averaged; this reduces chance variation.

Besides a total count of all colonies, the data may be classified into several categories: number of bacterial colonies, number of mould colonies, number of kinds of bacterial colonies, number of kinds of mould colonies, and number of colonies of each particular kind. Let the pupils decide how the different kinds of mould and bacterial colonies can be distinguished.

The type of medium and the temperature of incubation influence the development of colonies on the exposed plates. The medium used here is a general one on which many populations will grow. Certain specialized media (e.g. blood agar) are required to culture most pathogenic species.

Caution: Nevertheless, pathogenic organisms can be picked up and cultivated on plain nutrient agar. Therefore, after exposure the plates *must be taped closed.* Counting the colonies must be done without opening the dishes. If glass dishes are used, only *after dishes have been sterilized* should they be opened for washing. Disposable petri dishes are ideal for this experiment. Plastic and glass dishes may be opened when completely immersed in a strong solution of lysol (approximately 5 per cent) and should be left overnight to sterilize.

Discussion (p. 1)

A. Results will vary from one school to another and even within one school from one year to the next.

C. This harks back to the idea of species diversity. In general, plates exposed in environments with small populations of many different kinds are more likely to pick up micro-organisms of disease, because it takes only one or a few bacteria to start a colony—and the greater the number of kinds of bacteria in the environment, the greater is the probability of 'trapping' a pathogen. In nature, human pathogens are rare among the total species of micro-organisms. Of course, if you deliberately biased your sampling—by exposing plates in a sickroom, for example—the generalization would not hold.

A, D, E. These items all refer to methodology. It is important to standardize the duration of plate exposure to obtain a more uniform census. Secondly, it is important to have replicates from the several locations, so that the results will more reliably represent the populations available. The caution above answers **E**.

Microbes and Disease (pp. 2–16)

This section is an attempt to provide materials with which pupils can think about disease scientifically. Pupils are always eager to recount their own symptoms or those of all their friends and relations and to testify to miraculous cures. To cut off *all* of this stifles interest; you must instead use it to guide pupils into discussions of principles. The book assumes that pupils already know something about many human diseases. To emphasize the biological universality of disease, try to mention animal and plant disease as often as possible.

p. 5, Fig. 1.2. For other examples of infectious plant disease, see Fig. 3.10 in *WL*, as well as Fig. 1.7 in *PLW*.

p. 6, ¶ 1. In some ways, allergies are similar to infections: in both, organisms react to the invasion of foreign substances. But in an allergy the foreign substance is not reproductive.

pp. 6–10. If you wish, this is an opportunity to discuss venereal diseases in general—which, almost by definition, are contagious. But emphasize that syphilis and malaria are extremes of a continuum; many diseases are neither as strictly contagious as syphilis nor as strictly vector-borne as malaria.

p. 8, Fig. 1.5. Eliminate the vector.

p. 9, Fig. 1.6. Each dot is a colony of bacteria or mould started from an individual organism that dropped from one of the fly's feet as it walked.

p. 10, ¶ 7. Note that virulence and resistance are both relative terms; to gauge either, the value of the other must be kept constant. This is a difficult idea for most pupils, who live in a simple world of 'good' or 'bad'.

p. 12, Fig. 1.8. Use this cartoon as a catalyst for discussion or for pupil reports on other clashes between attitudes and innovation.

pp. 11–14. The terminology for kinds of immunity is unsettled. That used here follows for the most part Frobisher, *Fundamentals of Microbiology*, 8th edition (Philadelphia: W. B. Saunders Co., 1970).

p. 12, ¶ 3. Note that cowpox is rather unspecific, occurring in both cattle and men. The story of vaccination is a good example of empiricism in science; the practice long preceded the modern theory of immunity and the germ theory.

p. 16, Fig. 1.11. Tsetse flies are tropical insects; they are most likely to spread to other tropical regions outside Africa. Tropical South America is probably most endangered, because there is considerable air traffic across the South Atlantic.

INVESTIGATION 1.2

INVESTIGATING AN INFECTIOUS DISEASE (pp. 17–19)

The writers regard this as an extremely important investigation. It clearly demonstrates disease transmission, a topic that traditionally has been 'talked to death' instead of demonstrated. Moreover, it applies the skills practised by pupils in Investigations 3.2 and 3.3 in *DLT*.

Materials

Agrobacterium tumefaciens is not an animal pathogen. It is not dangerous to pupils. However, it should be handled carefully and with full attention to sterile technique, for it is dangerous to plants, and plants are also important. The bacilli are large enough to be studied under the 4 mm high power objective of the microscope. Cultures of *Agrobacterium* are available from: National Collection of Industrial Bacteria, Department of Trade and Industry, Torry Research Station, PO Box 31, 135 Abbey Road, Aberdeen, Scotland; and National Collection of Plant Pathological Bacteria, Plant Pathology Laboratories, Hatching Green, Harpenden, Hertfordshire.

The bacteria obtained from these sources is freeze dried and in ampoules. Instructions for recovery of bacteria are issued with all samples purchased. The following is a modified extract: To open the ampoule make a file cut across the tube at the mid-point of the cotton wool plug and crack the tube by touching the cut with a red-hot glass rod of small diameter. Allow air to enter slowly through the crack before removing the top. If this is not done the plug may be blown to the bottom of the tube. The plug, which may contain some dried culture, should be removed and replaced by a small sterile plug. With a Pasteur pipette add aseptically to the dried culture—including the numbered paper strip—a small quantity of sterile distilled water and mix well.

You are obligated to *sterilize* all soil, pots, and plants at the conclusion of the work. Emphasis on careful handling of pathogens is an essential part of biological education.

Plants must be started at least a month before this exercise is set up. When ready for use, they should have three or four true leaves; 7 cm pots are large enough, or half-pint milk cartons may be used. Tomato is a good species to use. Pot labels are not essential; the pots themselves may be marked with a soft pencil.

For the crystal-violet solution, see Investigation 3.3, *DLT*.

The scalpel must be quite sharp; a razor blade is preferable.

To prepare culture tubes of dextrose agar, you need:

Peptone,	5 g
Beef extract,	3 g
Agar,	15 g
Glucose (dextrose),	10 g
Distilled water,	1000 cm^3

Heat water (just below boiling) and dissolve the agar in it; then add peptone, beef extract and dextrose, stirring the mixture well. Pour 30 to 40 cm^3 into cotton-plugged tubes. Sterilize tubes in autoclave or pressure cooker, and cool them at a slant of about 30 degrees.

Procedure

Investigation 1.2 *can* be handled by one group for each class. Group members can then present their research report in seminar style.

Galls may be removed when about 5 mm in diameter. Incubate dextrose agar slopes containing juice from crushed gall tissue at 25 °C.

Note that reference to lab-books as a place for recording laboratory observations is being dropped. Use of the lab-book should now be a habit.

Again, *Caution:* At the end of the work, be sure to sterilize in an autoclave or pressure cooker all plants, pots, and soil. An alternative method is to bake plants, pots and soil in an oven for 1 hour at 150 °C.

Discussion and Conclusions (p. 18)

C. In an experimental procedure all set-ups must be treated exactly alike except for the variable under investigation. If the control plant (A) is not punctured with a needle, the way is open to ascribing the gall development to the needle wound rather than to the bacillus.

E. This leads to the idea that morphology alone is not sufficient for species identification in micro-organisms (p. 88, paragraph 2, *DLT*). Biochemical characteristics and physiological reactions are used to distinguish among species of similar morphology.

F–H. This investigation merely associates in one instance a suspected pathogen with a particular symptom in the host. This is far from proof of causation. To carry out Koch's postulates, one should begin with extensive investigation of organisms in naturally occurring galls. The pupils have carried out only No. 3 and part of No. 4 of Koch's postulates.

The Soil Ecosystem (in part) (pp. 19–22)

If any work was done with soil organisms during Investigation 3.1, *WL*, be sure that pupils recall it now.

p. 20, ¶s 3, 4. It is not necessary to go deeply into the chemistry of ionization. But by all means use any knowledge that pupils may have gained in chemistry classes concerning atomic structure.

INVESTIGATION 1.3

A CHEMICAL CHARACTERISTIC OF SOILS (pp. 22–24)

This investigation provides encounters necessary for the meaningful assimilation of some basic ideas of acidity and alkalinity. Although some use of indicators is possible on a very simple

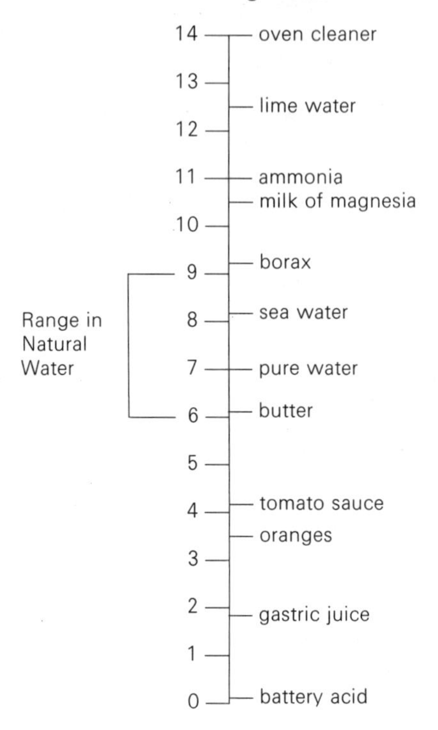

Fig. T-5. pH of some common substances.

basis (as, for example, Investigation 1.5, *WL*), biological experimentation on any level must remain quite unsophisticated unless the pupil learns some method of measuring pH. Figure T-5 provides some examples of the range of pH in some substances with which pupils may be acquainted.

Materials

Hydrochloric acid and sodium hydroxide should be 0.1 M solutions.

A pH paper with a range from 3 to 9 will cover most needs.

The greater the number of soil samples, the better. If many are available, different groups can measure different samples. However, several groups should measure each sample to obtain more than one reading. Samples should come from a variety of habitats, such as bogs, burned-over areas, woodland, heath, hedgerows, and gardens.

The '10 g' of soil should be estimated, not weighed.

Studying the Data

The ranges of the recommended indicators are as follows:

Methyl orange	pH 3–4.6
Phenol red	pH 6.6–8.0
Phenolphthalein	pH 8.2–10.0

B. Examples: bog soils are usually acid; many garden soils are close to neutral; soil from limestone and chalk regions may be alkaline.

C. Decomposing organic matter usually releases acids; therefore, soils with high humus content are often acid.

D. Because the soil water is diluted in the procedure, the pupil is not measuring the true pH of the soil.

For Further Investigation

If the pH of a soil is modified, composition of the microbial population of that soil may be modified. This can be demonstrated by plating a soil on agar adjusted for pH 5, 6, 7, and 8 and noting the differences in bacterial counts.

The Soil Ecosystem (contd.) (pp. 24–30)

p. 25, ¶s 2, 3. Pupils might be asked to fit these community relationships into the framework of Chapter 3, *WL*. Antibiosis might be considered a negative–neutral relationship.

p. 26, ¶ 6. The words 'good' and 'bad' are being used in the time-honoured absolute sense that we are trying to avoid. If the insects attacked were honeybees, we would be inclined to label the nematodes 'bad', especially if we were beekeepers. But the scientific questions remain: How much damage is done to bees compared with damage done to other insects? And how important are the bees compared with the other insects? Even from a strictly anthropocentric viewpoint, 'good' and 'bad' are terms that are relative ecologically.

p. 28, Fig. 1.17. Note that substances are printed in capitals, agents of change in lower case. The diagram concerns the nitrogen cycle in the biosphere as a whole. In a particular place large amounts of nitrogen compounds may be lost from a soil by leaching, but they are not lost from the biosphere. Likewise, nitrogen compounds may accumulate and remain stationary in the cycle for long periods of time, as in the guano islands of Chile, but from a geological viewpoint these compounds are temporary. Fig. 1.17 should be compared with Figs. 1.12,

1.13, and 1.14, *WL*, which, however, are not as detailed as 1.17.

p. 29, Fig. 1.18. In the situation depicted here, there is definite evidence that the vascular plant benefits; but in the situation of Fig. 1.15, the evidence is clear that the vascular plant is harmed.

p. 30, ¶s 2, 3. In principle the *Rhizobium*–legume relationship is somewhat similar to the mycorrhizal situation.

INVESTIGATION 1.4

DECOMPOSING ACTION OF SOIL MICROBES (pp. 30–31)

This experiment shows the decomposing power of moulds and bacteria (including the actinomycetes) and also ties in with what the pupil has learned about the chemical composition of plant and animal structures. It is not expected to generate any great enthusiasm. It is neither colourful nor spectacular—and it may seem interminable to some. Yet its purposes are important. And its length can help make the point that biological processes are often slow and the biological investigator must be patient.

Ask the pupil to set up hypotheses concerning the relative rates of decomposition of the substances and the rates of decomposition of the two soils.

Materials

Lids of standard-size petri dishes fit snugly inside 10 cm flowerpots. Cans may be substituted for flowerpots, but be sure to punch several holes in the bottoms.

Obtain builder's sharp sand and wash it well; the fewer qualities of a true soil it has, the better. On the other hand, the garden soil should be a humus-rich loam.

Old tights are good sources of nylon.

The aluminium dishes in which frozen pies are packed make excellent substitutes for deep dishes if 10 cm flowerpots are used.

Procedure

Not all members of the class need be involved in setting up the experiment. Two or three set-ups per class are quite sufficient for purposes of comparison; two to four pupils can work at each set-up. However, the whole class should understand the procedure and participate in observing and recording the progress of decomposition.

The surfaces of the pots should not be watered; they must be kept moist by capillarity from the dishes in which the pots sit. Therefore the dishes must be kept filled with water. Since observations occur at infrequent intervals, one pupil from each group should be delegated to check the water each day.

The observation times listed on the chart (p. 31) may be varied.

Studying the Data

A. Most of the substances have been mentioned in the text. Cotton string is mostly cellulose. Cellulose also occurs in a twig or a leaf; there is a small amount in rolled oats. The twig and the leaf contain considerable amounts of lignins, and the leaf also contains some pectins. The rolled oats are mostly starch, with small amounts of protein and fat. An insect, too, contains proteins and fat, but its exoskeleton is mostly chitin. Nylon is included as an example of a man-made material; the elements in it are similar to those in natural organic substances, but no micro-organisms ever had a chance to use nylon as a food until man added it to the environment.

B to D. The general rate of decomposition depends on temperature. Four weeks is scarcely enough time, even in well heated buildings; eight weeks is closer to the average. Probably the first evidence of decomposition will be mould hyphae on the oats. Later, moulds may appear on other items. Still later, it may be possible to detect the typical earthy odour of actinomycetes and the putrid odour of bacterial decay.

Conclusions

B. Rates of decomposition on sand may or may not differ from rates on soil. Usually there are differences, decay of most items proceeding more slowly on the sand because of its smaller initial population of decomposers—but the facts should not be forced to fit the theory. In discussing results, stress the production of humus and the return of minerals to the soil, where they are used again as plant nutrients. In the economy of nature, the decay of food, fibre, and wood is no different from the decomposition of other plant and animal remains. 'Desirable' and 'undesirable' decomposition are only expressions of man's point of view.

PROBLEMS (pp. 33–34)

1. (*a*) Disease causation is a complex interaction between host, parasite, and environment. A pathogen may or may not lead to disease, depending on its virulence, the degree of immunity of the host, and the nature of the environment. (*b*) No. A fly may carry typhoid bacilli mechanically on its body surface without being infected by the bacillus. (*c*) The lower the host-specificity of a disease, the wider the range of experimental organisms available to the researcher. (*d*) This is a straightforward library research question. Raise the further question whether the same immunizations are required for travel to all parts of the world. A good resource book for (*d*) is: A. C. Turner, *The Traveller's Health Guide* (Tom Stacey, 1971).

3. 'Asian flu' involves strains of virus that may differ in virulence. An attack of the disease confers some immunity, although it is of short duration. The virus spreads from man to man by moisture droplets—probably both directly and indirectly.

4. One explanation of antibody formation is based on the idea of complementary molecular shapes. For example, gamma-globulin is changed by contact with the active part of a toxin. These 'template' gamma-globulin molecules combine with and inactivate other toxins of the type that produced the template.

5. The following is a partial list of pathogenic micro-organisms that are most likely to be spread by sewage and of the diseases they produce: *Salmonella typhosa* (typhoid fever), *Salmonella paratyphi* and *Salmonella schottmuelleri* (two forms of paratyphoid fever), *Salmonella enteritidis* (enteritis), *Shigella dysenteriae* (bacillary dysentery), *Vibrio comma* (Asiatic cholera), *Endamoeba histolytica* (amoebic dysentery). Some infectious diseases with alternate hosts, such as typhus (fleas on rats), might also be favoured. Obviously the incidence of noninfectious diseases would be little affected.

6. (*a*) Many aspects of this subquestion can be investigated. Soil texture influences air, water, and nutrient content, the nitrogen content, rate of decomposition, and rate of leaching. (*b*) Acid soil water may retard the work of nitrifying bacteria and of all forms of nitrogen-fixing bacteria. Acidity influences leaching of cations, many of which are important plant nutrients. Secondary effects, such as lowered aeration in clay soils with low pH, are numerous. (*c*) Very little leaching occurs in desert soils, so inorganic nutrients are likely to be readily available. (*d*) The purpose of the subquestion is to evoke reasonable hypotheses from the pupils. There are many: for example, (1) destruction of root systems, (2) introduction into a soil of unfavourable pH, (3) lack of mycorrhizal fungi in the new soil. (*e*) For each decrease of 10 °C in mean annual temperature, an increase in humus content of two to three times has been shown to occur; in warm, moist, well-aerated soil decomposition takes place at a very rapid rate.

7. During the first year the nutrients in plot A were tied up in the rye, and therefore the potatoes in plot B did better. However, the second year the nutrients were released from the rye, and the rye improved the drainage of the soils; and so potatoes in A did better.

8. Some biological examples are practices relating to (*a*) immunization, (*b*) crop rotation, (*c*) pesticide control of insects, (*d*) animal and plant breeding. This problem emphasizes empiricism; a 'scientific method' is not always necessary for technological advance.

AUDIOVISUAL MATERIALS

Films

Micro-organisms—Harmful Activities. University of Indiana Audiovisual Centre, available from Boulton-Hawker Films. 16 mm, colour, 18 min. (A good film to back up this chapter. Discusses specificity of infection, types of immunity, antibodies, and Koch's postulates. Shows part of the technique used in Investigation 1.2.)

World at Your Feet. National Film Board of Canada, distributed by Guild Sound and Vision Ltd. 16 mm, colour, 23 min. (Contains good material on soil structure and soil physics, but is inclined to place poetics above ecological principles.)

The Nitrogen Cycle. Rank. Black and white.

Life in a Cubic Foot of Soil. Coronet Films, available from Gateway Educational Films Ltd. 16 mm, colour, 11 min. (A good introduction to the components of a soil.)

TEACHER'S REFERENCE BOOKS

BOYCOTT, J. A. *Natural History of Infectious Disease*. Arnold, 1971. (Describes some of the factors responsible for the phenomena of infectious disease.)

BURNET, F. M. *Natural History of Infectious*

Disease. Cambridge University Press, 1962. (Excellent background reading.)

NOTTINGHAM UNIVERSITY SCHOOL OF AGRICULTURE. D. K. M. KEVAN—Editor. *Soil Zoology*. Butterworth, 1955. (Excellent for the economically important nematodes. These small animals that inhabit soils and often play an important role in the microcommunity have been much neglected. This book contributes to overcoming such neglect.)

2

PATTERNS OF LIFE ON LAND

MAJOR IDEAS

1. In any given area the naturally occurring organisms are those that can survive and successfully reproduce under the environmental conditions (abiotic and biotic) prevailing there.
2. Over large areas of the Earth, organisms and environment produce characteristic landscape patterns—biomes.
3. Rainfall and temperature (both ultimately determined by the global pattern of solar radiation) are the major abiotic environmental factors that determine the distribution of biomes.
4. In general, as one approaches the equator from the poles, one passes through biomes of increasing diversity of species, increasing complexity of community interrelationships, and increasing depth of vegetation.
5. In the low and middle latitudes an additional sequence of biomes may be distinguished along a gradient characterized by a diminishing availability of water.
6. The geographical distribution of many species cannot be explained on ecological grounds alone.
7. The present distribution of a species and the distribution of its fossils and of the fossils of its putative ancestors are keys to the distributional history of the species. A similar statement can be made concerning higher levels of classification—genera, families, orders, etc.
8. Organisms vary in ability to disperse; therefore, to explain the distribution of any species one must consider its structural and physiological characteristics in relation to the nature of the physical barriers that it may have had to cross in spreading from its area of origin.
9. Man has greatly influenced the distribution of many organisms by transporting them—sometimes deliberately, sometimes accidentally—across barriers that they had not themselves been able to surmount.
10. Man is an increasingly important element in determining the characteristics of landscapes. He has now succeeded in completely transforming large portions of some biomes. Most completely man-dominated are urban and suburban areas, where even the effects of climate are so highly modified that the eye of a trained geographer is needed to detect them, and where the major portion of biotic energy is derived from distant regions.

PLANNING AHEAD

Pond-water for Investigation 3.1 can be collected at any time. Water from a quietly flowing stream can also be used. If collected more than a week before use, the unsterilized water should be kept in an aquarium. If it is too late to collect the larger organisms and if you do not maintain them in an aquarium, order them immediately from a biological supply company. Order or collect the organisms to be used for Investigation 3.2 also.

The grid for Investigation 4.1 (p. 127) should be drawn on to a stencil and duplicated.

Start onion-root growth for Investigation 1.3, *LO*.

GUIDELINES

The most logical division of this chapter for study is indicated by the positions of the investigations; the first assignment would be pp. 35–38 (guide questions 1–3); the second, pp. 40–58 (guide questions 4–18); the third, pp. 61–68 (guide questions 19–23). The second assignment is long, but most of its substance is descriptive.

Emphasis in this chapter should be placed upon the biome in which the pupil lives. However, one of the objectives of all true education is to broaden the pupil's view of his world, so other biomes should not be neglected. Direct experience with the local biome can be either assumed or secured, but experience with other biomes must usually be developed by means of visual aids. The clearer and more immediate the pupil's view of his local biome, the clearer will be his understanding of other biomes.

Even becoming familiar with the local biome may be somewhat difficult, especially in 'inner city' schools. A field excursion may be the best means of achieving such familiarity. If the teacher himself is well acquainted with his region, he will undoubtedly be able to arrange an itinerary for a half- or full-day excursion that will display the salient characteristics of the biome.

Of course the original 'climatic climax' of the local biome may long since have disappeared from any place within feasible travelling distance, but excellent examples of the influence of man on the biotic landscape and of successional stages should be available. With a properly planned sequence of observations and a series of sharp, attention-fixing questions, the teacher can organize the immediate experiences that pupils will gain from a field trip into an adequate acquaintance with the local biome. For those who are using these books as a series, some acquaintance will have been acquired during Investigation 3.1, *WL.*

Unlike that fieldwork, however, a field excursion for Chapter 2 in this book does not require growing season conditions. The aspect of the landscape in winter is quite suitable for the study of biome characteristics unless deep snow prevails.

A biome is a biotic expression of a climate. To understand biomes, then, the pupil must have some knowledge of (*a*) the atmospheric factors that, when statistically summarized, constitute climate, and (*b*) the astronomical and geophysical phenomena that determine the distribution of climates. As in the case of distant biota, an understanding of distant climates is best developed against a background of familiarity with the local climate.

Many biology pupils are already likely to possess some background in climatology. However, for those teachers who wish to use it, some basic information concerning world distribution of climates is provided on pp. T88–93.

An excellent source of information about distant biomes and climates is often overlooked, that is pupils who have lived in other biomes. Such pupils often have photographs as well as an eagerness to talk about their former surroundings. Teachers in schools located near installations of the armed forces or near large industrial plants that frequently shift their personnel are usually aware of this resource, but other teachers might discover unsuspected possibilities by looking into their pupil records or by direct questioning.

TEACHING NOTES

Meeting the Environment (pp. 35–38)

p. 35, ¶ 1. Inevitably this chapter must mention many kinds of organisms. But mere names are poor for arousing mental images. Therefore, almost all organisms mentioned have also been pictured. The indexes should be used to help guide pupils to illustrations elsewhere in the five volumes.

p. 36, ¶ 2. Unlike the questions in paragraph 1, the questions in this paragraph are not entirely rhetorical. Some speculation on these points is in order.

p. 36, ¶ 4. The term 'tolerance' was introduced incidentally in Chapter 1 (p. 22). It is further illustrated by Investigation 3.2. You may wish to use that investigation now, if you expect to omit Chapter 3.

INVESTIGATION 2.1

A STUDY OF ENVIRONMENTAL TOLERANCE (pp. 38–39)

Seeds are a convenient form of organism to use for experiment.

Moisture is, of course, a basic distributional factor, but it is difficult to maintain and measure in experimental situations. Effects of light and temperature are more easily investigated and are also important factors in distribution of organisms.

This investigation makes use of some technical skills developed in Investigation 1.2, *WL.* If pupils are made aware of this, most will strive to show that they did indeed learn something.

Groups of five or six pupils each are suitable. These are not too large to provide each pupil with a task. One pupil may be assigned as a group leader. Extra pupils can distribute materials and coordinate the allocation of dishes to their proper environments.

Materials

The plant species selected for this investigation provide a variety of responses to the conditions of the experiment. Other seeds may be substituted as follows:

For *tomato:* beet, carrot, sunflower.

For *radish:* flax, bindweed.

For *lettuce:* onion, evening primrose, willow herb.

For *vetch:* celery, larkspur, columbine, plantain, shepherd's purse.

If sufficient petri dishes are not available, satisfactory substitute containers can be made from cardboard milk cartons cut down to a depth of about 1.5 cm. Or dishes and covers may be made by moulding circles of heavy aluminium foil around glass petri dishes.

The strips of cardboard used as dividers should be as wide as the petri dishes (or other containers) are deep. A good grade is that used in shirt boxes. The tabs at the ends of the dividers may be fastened to the dishes with rubber cement.

The plastic of which the bags are made should be thin (1 or 2 mm), to allow free diffusion of gases. The plastic sheeting used by cleaners to protect clothing can be used in place of bags. The dishes should be wrapped in the plastic and the loose ends fastened with string or a rubber band.

For covering the seeds to be grown in darkness, coffee tins can be used in place of cardboard boxes, or the dishes may be wrapped in aluminium foil.

A conventional refrigerator may be used to establish the cold environment. However, because the refrigerator light normally goes off when the door is closed, modification is necessary to ensure continuous light. Such light can be provided quite simply by running an extension cord, with lamp attached, into the refrigerator. With this cord in place, the rubber gasket around the door will still permit normal opening and closing and at the same time prevent loss of cold air. An alternate method is to run a bypass wire across the light switch in the door frame. This will allow the regular refrigerator light to burn continuously. It may be necessary to remove part of the frame in order to gain access to the switch. Because it produces heat, an incandescent lamp should not be used as the light source. Small fluorescent lamps (8 to 15 watts) and fixtures for them are available commercially. But even fluorescent lamps generate small amounts of heat, so it is wise to check the refrigerator a few days prior to the experiment to make sure that a temperature of 10–12 °C is maintained.

Teachers who have access to a commercial incubator need only make provision for a continuous source of light to provide the proper environment for Dishes 3 and 4. Use an extension cord, as in the case of the refrigerator. If the incubator door will not close when the cord is in place, it may be necessary to remove the thermometer and run the extension cord through the thermometer hole. As a safety measure, wrap tape around the cord at the point where it enters the incubator.

If an incubator is not available, a simple heat box can be constructed from the following materials:

Metal box of suitable size
Fluorescent light tube (same wattage as that used in refrigerator)
100 W incandescent bulb
Small egg incubator thermostat rated at 23–50 °C
Thermometer

The equipment is to be assembled as shown in Fig. T-6. A metal partition separates the heat source from the petri dishes. Since the incandescent lamp serves as the heat source, it should be wired in series with the thermostat; the fluorescent light should be wired parallel. It might be necessary to experiment with a slightly higher or lower wattage to obtain the desired temperature.

Home-made or modified equipment is often better than none, but schools should be encouraged to provide standard laboratory equipment obtained from commercial supply companies. Care should be exercised in building or modifying equipment. *Before* constructing new electrical equipment or modifying existing equipment, check with the school authorities concerning fire codes and regulations.

Tomato seeds germinate most rapidly in complete darkness at about 26 °C. Lettuce germinates best in light. Vetch requires a cool environment. Radish germinates well under a wide range of conditions.

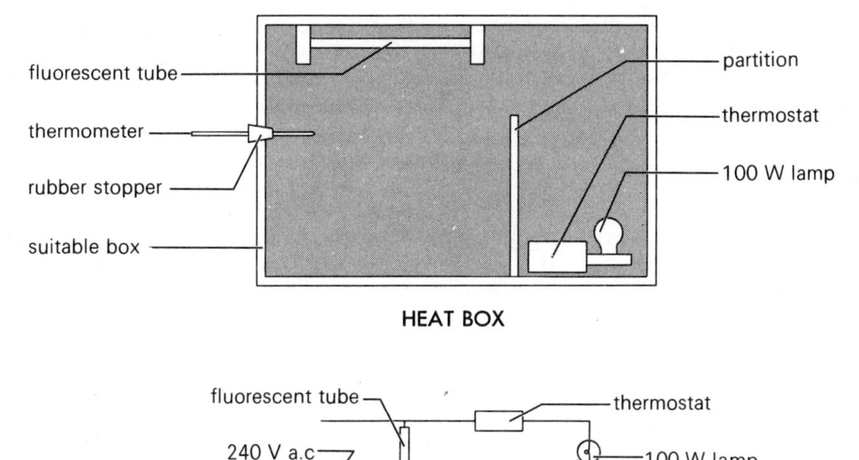

Fig. T-6

Have the pupils discuss the hypotheses before performing the investigation. It is important that they are consciously trying to test their hypotheses throughout.

Studying the Data

A. Refer pupils to Investigation 2.2 in *WL*, especially 'Studying the Data', paragraph 2.

D. Students may have difficulty separating the influence of individual factors from combinations of factors. However, this is a very common experimental design; you should lend assistance when necessary.

Conclusions

A, B. Both of these questions are speculative. The important thing to emphasize is the necessity in scientific reasoning that no conclusion be in conflict with evidence.

Ecological Distribution of Life on Land (pp. 40–58)

p. 40, ¶ 4. This paragraph echoes the quotation in the introduction to *PLW* in this Guide (p. T71). It seems important to refer frequently to the relation between science, which is seeking for order, and nature, which may be–but is not *necessarily*—orderly.

p. 41, Fig. 2.5. Pupils should notice the temperature relationships in these climatograms, when they are arranged in order of decreasing latitude.

Plate II (p. 24). This map is based on one in Odum (reference: p. 75 of *PLW*). It is quite generalized. However, much of the world simply has not been studied sufficiently from the biome point of view to make a better map possible at present. Maps of *vegetation only* are many, and some are quite detailed.

p. 42, ¶ 3. Watch for a tendency to misread 'man has modified the effects of climate'. Modification of *climate* itself does not concern us here.

p. 42, ¶ 5. 'No lack of water': Referring to Fig. 2.5, pupils may be inclined to dispute this for tundra and taiga. If so, this is the place to introduce the idea of precipitation–evaporation ratio.

p. 42, ¶ 6. Something like tundra is found on the Palmer Peninsula of Antarctica. Life on polar ice is based on producers in the seas, so it is omitted from consideration here.

p. 43, ¶ 1. It may be desirable to demonstrate the relationship between the angle of the sun's rays and the intensity of solar radiation. See pp. T88–89.

p. 43, ¶ 2. See Plate I, Figs. A, B and C.

p. 43, Fig. 2.6. Cold, dry winds constitute perhaps the chief factor in preventing woody plants from growing above the level of snow cover (thin in tundra) or protections such as the rock in this photograph.

p. 44, Fig. 2.7. This may only elicit the idea of protective colouration. You may wish to raise the question of how such conditions might have arisen, but do not try to pursue the idea too far at this point.

p. 45, ¶ 4. Of course any daily comparison must be of the same day of the year for both tundra and taiga.

p. 45, ¶ 7. Note that this is a brief description of an ecological succession. See Plate III.

p. 47, ¶ 3. The idea of community depth and the term 'canopy' may have been encountered during work on Investigation 3.1 in *WL.*

p. 25, Plate IV. Deer are browsers; buds and twigs are used throughout the year, but nourishment is more abundant when leaves are on woody plants. Snow can hinder travel.

p. 56, Plate V. In part, the general greenish tinge to the light at the floor of the tropical rain forest derives from the filtering out of other colours by foliage above.

p. 49, ¶ 3. 'Always green' may be equated by some pupils with 'evergreen' and this in turn with 'pines and spruces'. Forestall this.

p. 49, ¶ 4. 'Untold numbers' is quite literal; unknown thousands of species remain to be described.

p. 50, ¶ 1. Vegetation on the forest floor is so sparse that little food exists for herbivores.

p. 50, ¶ 2. The idea of limiting factors may need to be reviewed from Investigation 2.2, *WL.*

p. 50, Fig. 2.14. In all biomes, local conditions may modify the biota characteristic of the overall climate. Here water available along a stream in a valley allows growth of trees in a grassland biome.

p. 51, Fig. 2.16. Several other first-order consumers (herbivores) are mentioned on p. 52. Seed-eating birds (granivores), mostly present in winter, are also first-order consumers.

p. 52, Fig. 2.17. Early explorers described the voice of this ground squirrel as a 'bark'.

p. 54, ¶ 3. The thorns and spines on desert plants probably are protection against first-order consumers in a land where their food is sparse. More clearly demonstrable is the relation between reduction in leaf surface and retardation of water loss (see Plate VI).

p. 54, Fig. 2.21. Competition among root systems for water is probably the best explanation. But many desert plants are known to produce substances that are deleterious to other plants—a kind of antibiosis.

p. 56, Plate VII. The dead leaves on the ground and the bare branches indicate the dry season. Dr Kuchler says the picture was taken in the spring of 1966 and shows forest partly in relatively good condition, partly damaged through cutting and burning, a characteristic of most tropical deciduous forests.

p. 56, Plate VIII. The principal large predators are lions, leopards and cheetahs.

p. 56, ¶ 6. Note that Plate II does not distinguish this biome from taiga.

p. 56, ¶ 9. This is termed 'maquis' in parts of the Mediterranean, 'broad-sclerophyll woodland' by some botanists and 'chaparral' in America.

p. 56, Plate IX. The mediterranean biome is most verdant in the season between the winter rains and the beginning of the summer drought.

p. 56, ¶ 10. Such leaves retain moisture well during the long, dry summers.

p. 57, ¶ 2. The first sentence may require some explanation if pupils have not yet grasped the relation between radiant energy and heat.

p. 57, Fig. 2.24. The constant moisture favours dense growth of plants, especially epiphytes, mosses, and lycopods.

p. 58, Fig. 2.25. Note that no uniform scale is used in these diagrams.

INVESTIGATION 2.2

TEMPERATURE, RAINFALL, AND BIOME DISTRIBUTION (pp. 58–61)

Climatograms are a common means of summarizing the two most important climatic variables; they are used especially by geographers. The representation (on p. 53) of precipitation by vertical bars is a violation of the general principle that a continuous variable is best shown by a line, but it is justified by the need to distinguish easily between temperature and precipitation. This point is worth making in class discussion, because even those pupils who are familiar with the mechanics of graph making often do not know how to select a form suitable for a given set of data.

Materials

If the graphs are drawn on a variety of grids, comparisons are difficult. Therefore it is well to furnish all pupils with graph paper of the same kind. The best practice is to duplicate grids that are exactly like those in the pupil's book—12 blocks wide and 18 high. On this grid the April data for Moshi, Tanzania, go two blocks above the top, but these can easily be added.

Local climatic data can be obtained from the nearest meteorological office and from the geography departments of some schools and colleges.

Procedure

After some of the climatograms in *PLW* have been studied, the graph-making phase of Investigation 2.2 may be assigned as homework. The number of graphs to be prepared by any one pupil is a matter of choice. The minimum would seem to be three: one based on local data, one on a set of data from Group 1, and one on a set from Group 2.

When added to the six graphs in the textbook, the four graphs drawn from the data in Group 1 provide the pupil with representative climatic data from ten major biomes.

Studying the Data

Each graph should be discussed from the viewpoint of possible relationships between the climatic data and the characteristic features of the biome. Enlarged copies of the ten climatograms may be pinned up where they will be visible to all pupils. These and the discussion based on them may then serve as a background for making predictions from data in Group 2.

During discussion of the biome descriptions, attention should be focused on the relationship of the climatic factors (as graphed in the climatograms) to the biota. For example, in the climatogram for the tundra (p. 41), the graph indicates that the average monthly temperatures are above freezing—and not far above—for only three months of the year; during most of the year the average monthly temperatures are far below freezing. Furthermore, the precipitation is quite low in all months of the year. From these facts we can conclude that the producers of the tundra biome are actively producing food during only a fraction of the year and that the density of consumers must be low. But the implications of the data presented by the climatogram need to be explored. The amplitude of the yearly cycle of monthly average temperatures implies a high latitude; this, in turn, assures a long daily period of sunlight during the season when temperatures are high. In addition, the low temperatures imply a low rate of evaporation. Thus conditions for plant growth during the brief summer season are not as unfavourable as they seem at first glance. The food-production conditions suggest that a large migratory summer population of consumers may be possible. Such reasoning should be applied to all the climatograms.

Pupils should be kept constantly aware of the limitations of the data with which they are working. Some significant variables not indicated in the climatograms are mentioned on p. 61, item 2. It may also be as well to discuss the way in which the data have been derived. The temperature data are means derived from 'daily means', which are not really means in the usual statistical sense but, instead, are the mid-points in the range of hourly readings over a twenty-four hour period. The precipitation data, on the other hand, are true means for monthly precipitation over a period of years. The number of years of observation varies from station to station, of course. These data do not indicate the monthly *range* in precipitation and temperature. And the range of these variables is a significant factor for organisms, as are the extremes.

Having noted limitations in the climatic data, pupils must realize that a climatogram does not summarize precipitation and temperature for a biome as a whole. Any one climatogram merely shows the data for these variables at one station within a biome. The stations have, of course, been chosen as carefully as possible to provide data that are representative—'typical'—for each biome.

Fig. T-7 shows another kind of climatogram. Some pupils may like to replot some of their data on such a grid and attempt to associate shapes with biome characteristics.

Earth History and Distribution (pp. 61–64)

p. 61, ¶ 1. Penguins do cross the equator, but only in the cold waters of the Humboldt Current on the west coast of South America.

p. 63, Fig. 2.28. Any ideas that the pupils have about this situation must be judged entirely on the basis of logic, since there is no accepted explanation.

p. 64, ¶ 2. Note that this conclusion does not *necessarily* imply evolution.

Man's Influence on Terrestrial Ecosystems (pp. 64–68)

p. 64, ¶ 6. Note the wording. The seeds of plants are not likely to arrive in an area in serial order, but some germinate more rapidly than others, some kinds of plants become conspicuous more quickly than others, and some

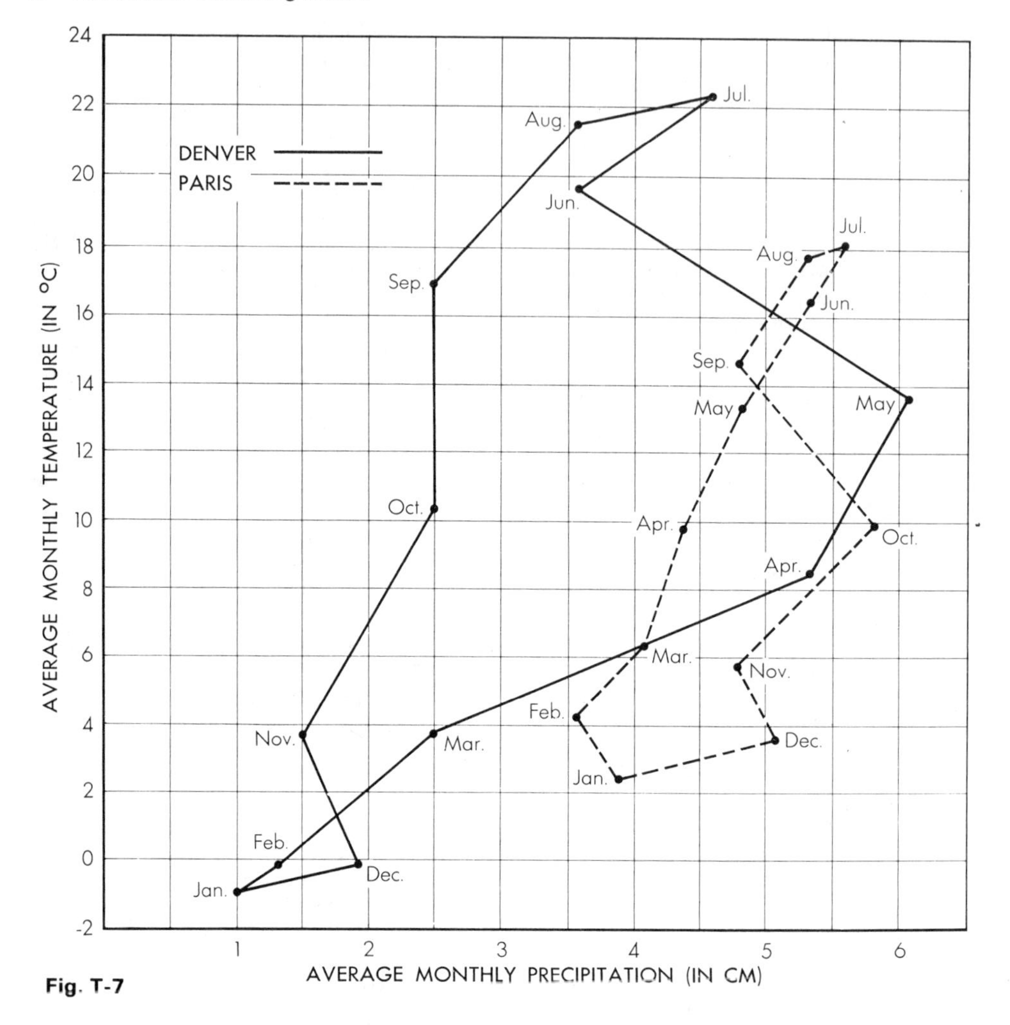

Fig. T-7

die out after flourishing temporarily. All this adds up to a succession of *adult forms*.

p. 67, Fig. 2.32. Urban pigeons are commensals of man. Without the voluntary or accidental provision of food by man, the pigeons could not exist. If anyone really wanted to get rid of the pigeons of Trafalgar Square, the ecological procedure would be clear.

INVESTIGATION 2.3

EFFECTS OF FIRE ON BIOMES (pp. 68–70)

This investigation can be worked through in a class lesson, without previous assignment. Or it can be assigned as homework and then discussed briefly in class. Or it can be assigned to certain of the better pupils for independent work, followed by a report to the class.

It is advantageous, though not essential, for the teacher to read the article on which the exercise is based: 'The Ecology of Fire' by C. F. Cooper, *Scientific American*, April 1961. Most of the illustrations used in the investigations are derived from illustrations in that article, through the courtesy of Gerard Piel, publisher of *Scientific American*.

Procedure

The answers suggested below carry the reasoning through in a logical manner. But you should be alert to reasonable alternative ideas that the more able pupils may provide. Such ideas should be welcomed, subjected to critical examination, and, if tenable, nurtured.

Fire in middle-latitude grassland

A. Mesquite.
B. Access to deep supplies of moisture during long droughts.
C. Mesquite brush with little grass.
D. Both survive.
E. Mesquite.
F. Mesquite.
G. Grass.
H. Somewhat like A or B.
I. See C.
J. Dominantly grassland.
K. Fire.
L. Lumbering.

Fire in a forest of the south eastern United States

A. (*a*) Fire damages only the needles of the longleaf pine. (*b*) Fire usually kills young deciduous plants.
B. Longleaf pine.
C. Ground fires kill the lower branches, but not the tops or the trunks.
D. Deciduous trees in the absence of fires grow faster than the pines and eventually shade them out.
E. Fire.

Discussion

A. Advantage because it maintains grass.
B. Advantage because it maintains long-leaf pine in competition with deciduous trees.
C. Burning would have to be done at a season when they were not nesting.
D. First, of course, the landowner must decide what he wants. Even so, it is impossible to predict the effects of fire without prior empirical evidence, because in all cases the variables are very numerous.

To some degree, fire is used as a management tool in all regions. Indeed, it is perhaps the most ancient tool that man has employed to change the landscape. What is not stated either here or in Cooper's article is that today there are many other such means—fertilizing, employing herbicides and the saw and axe, discing and bulldozing—and that these usually can be used more selectively than fire. Thus, though fire may have played a major part in shaping grasslands and forests in the past, ecologists are by no means agreed that fire should continue to have a large role in land management today—points worth drawing out during class discussion.

GUIDE QUESTIONS (p. 72)

4. Pupils should realize that a biome is a relative unit in a system of classifying ecosystems according to magnitude.
10. This question seeks only a superficial matching of habitat with organism, not an evolutionary or zoogeographical treatise.

PROBLEMS (pp. 73–74)

1. Some organisms that might be found on these lists are: apple, grapes, orange, plum, cherry, melon, potato, tomato, beans, corn, barley, wheat, rice, house mouse, rat, cockroach, bedbug, house sparrow, starling, pheasant. Whether or not each of these could survive in your locality without man depends in part on the climate of your region. However, some, such as bedbugs, could probably not flourish anywhere without man.
2. (*a*) Havana, New Orleans, Caracas, Boston, Winnipeg, Anchorage, Rio de Janeiro. (*b*) Brisbane, Hobart, Singapore, Manila, Little America, Tokyo, Vladivostok. (*c*) Nairobi, Tananarive, Cairo, Cape Town, Madrid, Copenhagen, Murmansk. (*d*) Day length and cloud cover particularly.
3. (*b*) Consider the influence of permafrost. (*c*) Aestivation is usually in response to dryness. (*d*) An ecotone is a transition zone between two ecosystems. See Kendeigh, *Animal Ecology* (reference: p. 74). (*e*) The most important relationship is between precipitation and evaporation: the hotter the climate, the greater the evaporation. Therefore, in two regions with equal precipitation, the hotter one is the more arid. Tundra introduces a special factor: there the permafrost prevents loss of moisture from the surface by the downward percolation that occurs elsewhere.
4. This is often explained on the basis of a climate warming trend. But increased sources of food from the presence of man may also be involved. Pupils may evolve explanations that have equal intellectual validity.
5. This problem will test the pupil's reference-using skills as well as his ecological thinking. No one reference will serve for all parts; perhaps the best single one would be that much-abused tool, the encyclopaedia.

Clues that may aid in assessing the pupil's success: (*a*) cattle; (*b*) (steppes of Asia): old world bison, horses, wild ass; (pampas of Argentina): guanaco, pampas deer: (veldt of South Africa): gnu and other large antelopes; (*c*) reindeer, caribou, musk ox; (*d*) the euphorbias, which have succulent stems, are thorny, and live in arid climates of South Africa; (*e*) the sunbirds, which are small, have long bills, and feed on nectar.

6. Contraction of geographic ranges might be due to: (*a*) climatic changes, (*b*) evolution of tolerances towards a more specific or limited range, or (*c*) results of species interactions (e.g. predation, parasitism, competition).

7. Some good evidence now exists for this theory. The filmstrip *Geological Oceanography*, by Encyclopaedia Britannica Films, gives an excellent presentation of it.

8. Work on this problem is similar to that for problem 5. Both should be voluntary undertakings unless you have personally checked the availability of pertinent references.

10. The answer to this question can become extremely sophisticated if the pupils consider effects of gravity, various types of solar radiation, temperature, lack of air, etc.

SUPPLEMENTARY MATERIALS

MATERIALS ON CLIMATE

The background material presented below may be used in any way you wish. Two suggestions: Use it as background for a lecture illustrated with a globe and maps and with charts presented on an overhead projector. Or duplicate the material and distribute it to pupils for reading.

Climates

Because climate has such an overwhelming effect upon the organisms in any landscape, we should understand something of the way in which it varies from place to place on the surface of the Earth. If you already understand the distribution of climates, you may treat this topic as a review. If the topic is new to you, it forms a necessary background in physical science for your biological study.

Distribution of radiant energy

Energy enters the biosphere in the form of radiation from the sun. This energy is changed into chemical form through the process of photosynthesis. From the organic substances thus produced, the energy for all the activities of living things is derived. The amount of solar energy received on any particular part of the Earth's surface is therefore a most important environmental factor for organisms.

The shape of the Earth and its position in relation to the solar system affect the distribution of solar radiation on the Earth's surface. Different places receive different amounts of radiation. Through the year the regions near the equator receive most; the polar regions—north and south—receive least.

This situation results from two factors. (As you read on, refer to Fig. T-8). First, to reach the Earth's surface, solar energy must, of course, pass through the Earth's atmosphere. The atmosphere absorbs some of the solar energy. The greater the thickness of the atmospheric blanket that the energy passes through, the greater the amount of energy absorbed. Rays striking the Earth at a right angle (vertically) travel through less atmosphere than do those striking the Earth at an acute angle. Hence parts of the Earth on which light falls vertically will receive more energy than other parts.

Second, radiation striking the Earth's surface at an acute angle is spread out. A torch may be used to show this. When its light falls vertically on a surface, we see a bright circle. When its light falls at a slant, we see a dimmer oval; the light is dimmer because it is spread over a greater area.

If we look at the matter from the viewpoint of an observer on Earth (where the organisms are) instead of from outer space, we can see that the higher the sun is above the horizon, the greater the amount of energy received. In any one day at any one place, the sun is highest above the horizon at noon (noon local time, not 'zone time' or 'day-light saving time'). And any place where the sun is *directly* overhead at noon will receive the maximum possible amount of solar radiation.

As the Earth revolves around the sun, the tilt of its axis causes a shift in the latitude at which a ray falls vertically at noon (Fig. T-9). At noon on 21 June the vertical ray will strike the Earth 23° 30′ (23 degrees and 30 minutes, or 23.5 degrees) north of the equator (along a line called the *Tropic of Cancer*), and on 21 December, 23° 30′ south of the equator (along the *Tropic of Capricorn*). The sun is never directly overhead at any point north or south of these latitudes; at the poles it is never more

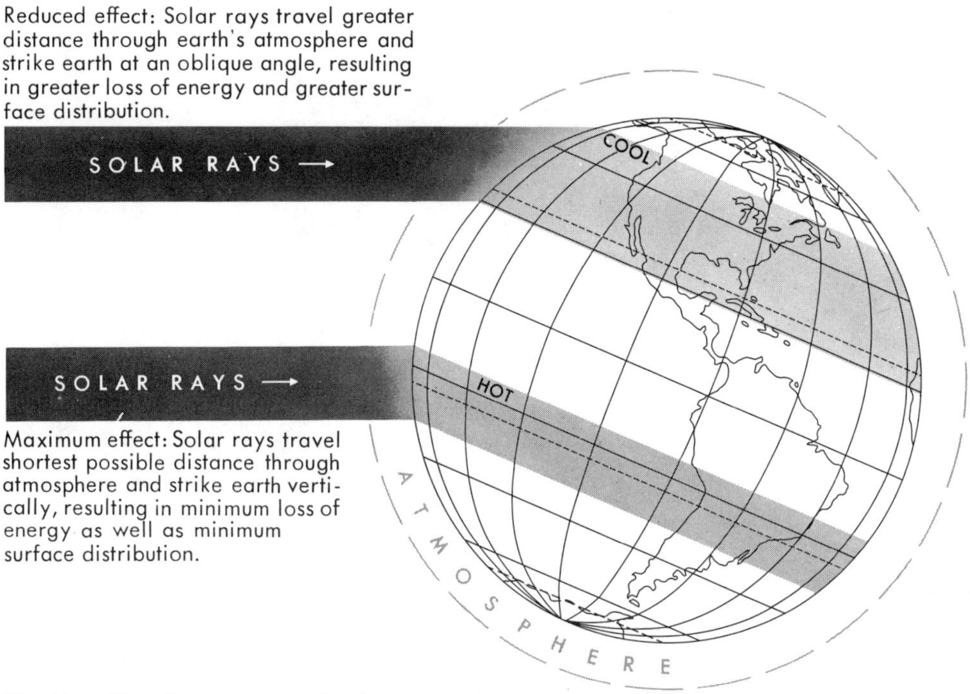

Reduced effect: Solar rays travel greater distance through earth's atmosphere and strike earth at an oblique angle, resulting in greater loss of energy and greater surface distribution.

Maximum effect: Solar rays travel shortest possible distance through atmosphere and strike earth vertically, resulting in minimum loss of energy as well as minimum surface distribution.

Fig. T-8. Showing how the Earth's atmosphere and the angle of contact affect the amount of solar radiation received on the Earth's surface—here, during a period of seasonal extremes.

than 23° 30′ above the horizon. In general, then, the solar radiation received by the Earth is greatest between the tropics and decreases rapidly toward the poles.

So far we have been considering solar radiation received in an 'instant' of time—the *intensity* of radiation. But the time unit most important from the point of view of a photosynthesizing plant is probably twenty-four hours—a day. Thus we must consider not only the intensity of the solar radiation but also its *duration* on the basis of a twenty-four hour period.

Duration is another result of the tilt of the Earth in relation to the sun. By examining Fig. T-10 carefully, we can see that the parts of the Earth beyond 66° 30′ north or south will have at least one twenty-four hour period during the year without any sunlight. We can also see that the period between sunrise and sunset is always twelve hours long at the equator, and it does not vary greatly within the tropics. On the other hand, at the poles the sun shines for six months at a stretch; twilight is measured in weeks, and night lasts until dawning begins, about five months later.

Thus there is great variation over the Earth's surface both in the duration of daily solar radiation and in its intensity. At the poles, the sun shines for six months, but the intensity is always low; within the tropics, days are never greatly prolonged, but the intensity of radiation is high. We can conclude that the rate of photosynthesis over the surface of the Earth will vary according to geographical position and time of year, with corresponding effects on the whole biotic community.

Distribution of heat

Up to this point we have discussed solar radiation as if it were only light. Actually it contains many wavelengths besides those we see (*WL*, p. 19, Fig. 1.9). But it does not contain the energy

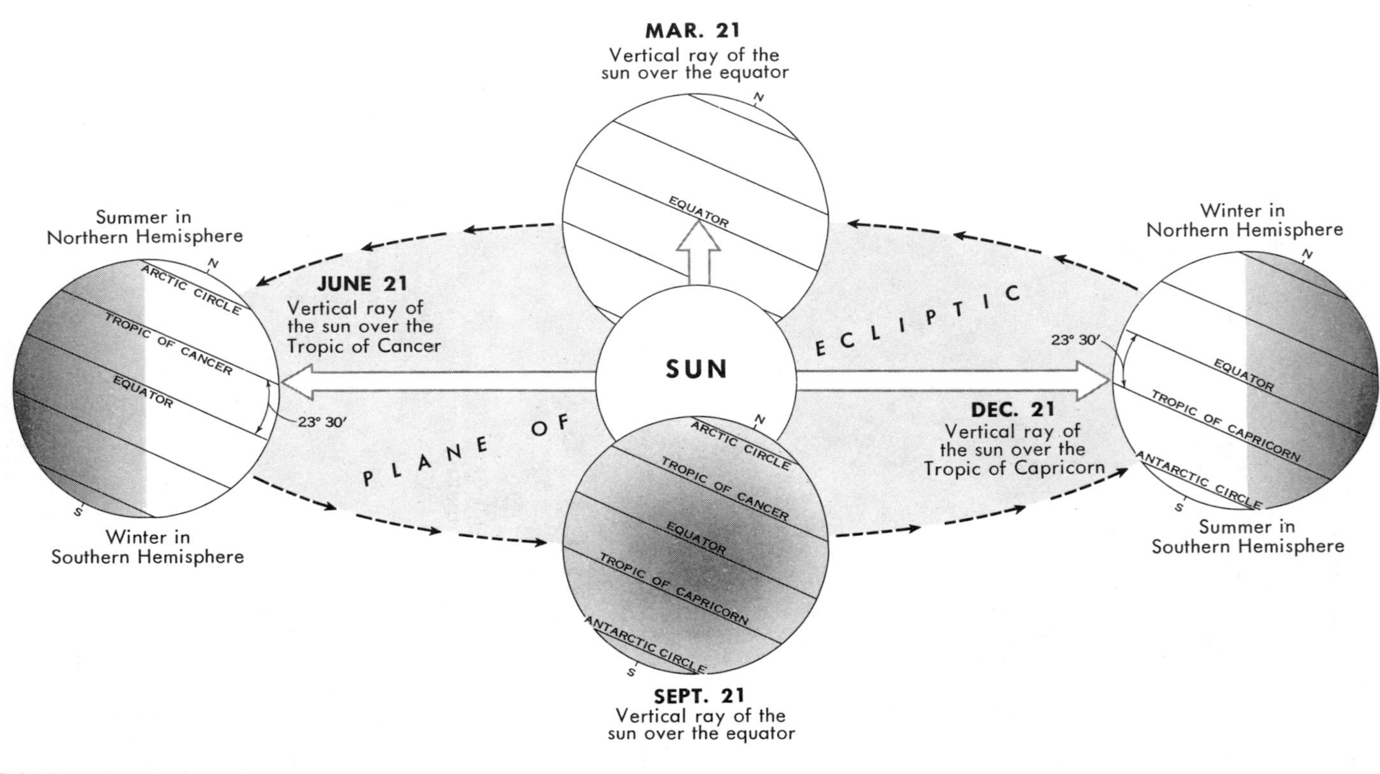

Fig. T-9. Showing the relationship between the Earth and the sun during the annual cycle.

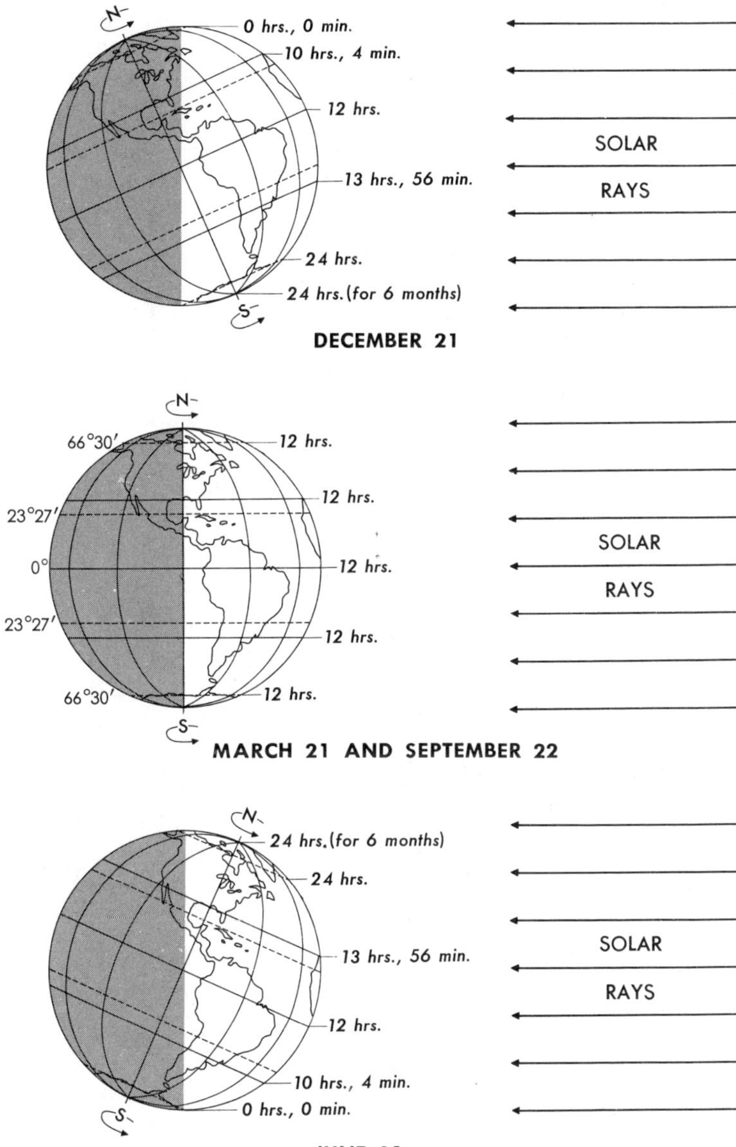

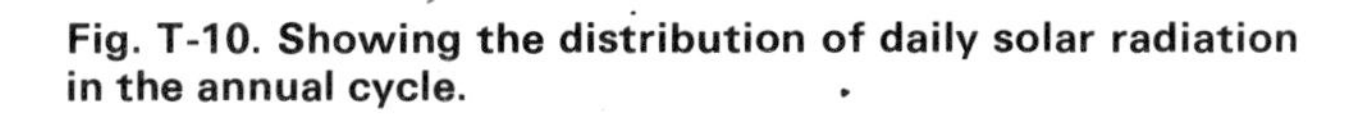
Fig. T-10. Showing the distribution of daily solar radiation in the annual cycle.

we call heat. The warmth of the sun's rays, which we feel on a summer's day, is solar radiation transformed into heat as the light strikes our bodies, the air around us, or other material things. (Thus, in interplanetary space, where radiation is plentiful but matter is not, there is very little heat). Light is the source of energy for the process of photosynthesis, but heat—some degree of it—is important for *all* biochemical processes.

The fact that the sun's 'warmth' comes only from the interaction of solar energy with matter leads to a remarkable conclusion: The Earth's atmosphere is heated mainly from *below*. When solar radiation strikes the upper atmosphere, very little heat results, because the molecules that compose the upper atmosphere are few and far apart. At lower levels, where the air is denser, more solar energy is changed to heat. But the greatest change occurs when the light hits land and water. Therefore, despite the fact that the sun shines above us, its heating effect comes from below.

When a fluid (liquid or gas) is heated from below, the warmer molecules rise. The warmer fluid spreads up and over the cooler portion. In this way heat is distributed throughout a container of water over a fire. Air is a fluid, and over the year it receives more heat in the region of the equator than anywhere else. A basic pattern of atmospheric circulation results.

This massive circulation is brought about by the unequal distribution of solar radiation; it transports heated air over the Earth's surface, as shown in Fig. T-11. Solar energy supplies the force, but gravity also plays a part. Cool air is denser—that is, heavier per unit of volume—than warm air. As a result, air descends over cooler regions and rises over warmer regions. Since the air near the equator is, in general, warmer and lighter than elsewhere on the globe, the air movement there is upward. As the warm

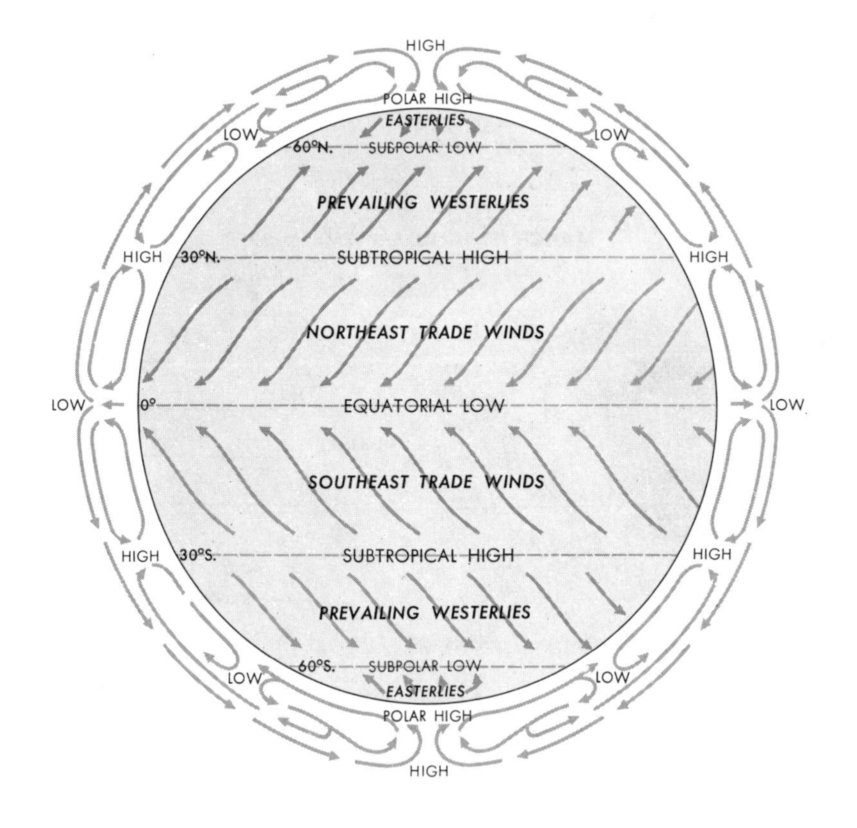

Fig. T-11. Showing the basic pattern of atmospheric circulation on the Earth's surface.

air rises, it is replaced by cooler air from both the Northern and the Southern Hemispheres. This horizontal movement of air results in constant winds, the *trade winds*. The entire wind pattern shifts with the seasonal movement of the vertical ray of the sun—northwards in the period from January to June, southwards from July to December.

Another factor affects winds—the rotation of the Earth. Because of this rotation, cooler air moving into the region of the *equatorial low* (the 'doldrums' of mariners) does not move directly from the north in the Northern Hemisphere or from the south in the Southern Hemisphere (see Fig. T-11). Instead, in the Northern Hemisphere the currents flowing towards the equator are turned to the right, resulting in winds that blow from the northeast; in the Southern Hemisphere there is an opposite deflection, to the left, resulting in winds from the southeast.

The circulation of warm air is away from the tropics; cooler air from the *subtropical highs* (the 'horse latitudes' of mariners), two regions of descending air, is constantly being brought back for reheating. This means that the equatorial regions of the Earth are cooler and the middle latitudes warmer than they would be otherwise. In other words, the region between the tropics exports heat. The air that sinks in the regions of the subtropical highs does not lose all its heat there, however. Some of it swirls out into the higher latitudes—northwards in the Northern Hemisphere, southwards in the Southern—and mixes with cold air moving toward the equator from the poles. These movements spread heat still farther towards the poles, so that even those extreme regions are warmer than would be expected from the small amount of radiation they receive.

Distribution of moisture

The oceans are the principal sources of moisture for the land. Moisture is carried to the land by circulating air. Therefore our understanding of atmospheric circulation is basic to our understanding of both temperature and moisture conditions on the Earth—and these are, in turn, the two most important factors in understanding the distribution of organisms.

Warm water evaporates faster than cold water. Warm air holds more water vapour than cold air. When air passes over warm ocean water, it picks up a lot of water vapour. When the air is cooled, the water vapour precipitates as rain or snow. It may precipitate by passing from a warmer ocean area to a cooler land area. This accounts for the relatively humid (moist) climate of western Europe. Water vapour may precipitate if it meets a cooler air mass.

Along the northern coast of Chile air passes from a cool ocean, where it picks up rather little water, to a warm land, where there are no cool air masses. The result is one of the driest deserts in the world (Plate II). But there is another way in which air may lose its moisture. We have already seen that the higher atmosphere is cooler. An air mass that is blown against a mountain will rise and, becoming cooler, lose its moisture. In this way the high Andes in Chile receive the moisture that fails to precipitate on the coast.

AUDIOVISUAL MATERIALS

There is an abundance of visual materials generally related to the content of this chapter, but by and large they tend to be quite elementary. There is a great need for new materials pitched on the senior secondary school level, materials that feature biological principles and avoid anthropomorphism, teleology and moralizing. The items listed below are among the better ones available.

Slides

Perhaps the best method of illustrating this chapter—at least for the teacher with a good background—is with slides. These can be placed in any desired sequence, and the 'captions' can be supplied by the teacher, largely in the form of questions directed to the pupils. The quantity of good ecological slides is small compared with the number of taxonomic ones, but all the following suppliers have ecological lists: T. Gerrard & Co. Ltd., Harris Biological Supplies Ltd.

Filmstrips

The World We Live In: The Tundra. Life Filmstrips. Colour. (Excellent pictures, good organization, some attention to ecological principles—and an unfortunate abundance of anthropomorphism and teleology.)

The World We Live In: The Woods of Home. Life Filmstrips. (The title refers to the middle-latitude deciduous forest. See comments for the filmstrip above.)

The World We Live In: The Rain Forest. Life Filmstrips. (Concentrates on the tropical rain forest of South America. See comments for first filmstrip.)

Films

Water and Desert Plants. BSCS Inquiry Film Loop, available from John Murray. (Through discussion the pupils can use the data in this film to discover some of the anatomical and physiological adaptations to the desert biome.)

Water and Desert Animals. BSCS Inquiry Film Loop, available from John Murray. (The physiological adaptations and responses of desert animals to the limiting factor of water can easily be observed in this film.)

Mountain Trees—An Ecological Study. BSCS Inquiry Film Loop, available from John Murray. (An excellent opportunity for pupils to apply their knowledge of ecological principles to the problems of plant distribution.)

High Arctic: Life on the Land. Encyclopaedia Britannica Films. 16 mm, colour, 22 min. (A film on a specific location in the tundra—Queen Elizabeth Islands, Canada.)

Distribution of Plants and Animals. Encyclopaedia Britannica Films. 16 mm, colour, 16 min. (The various factors that influence the distribution and survival of selected organisms in a given geographic area are discussed.)

The Temperate Deciduous Forest. Encyclopaedia Britannica Films. 16 mm, colour, 17 min. (The annual cycle of changes in the middle-latitude deciduous forest of North America is well depicted.)

The Desert. Encyclopaedia Britannica Films. 16 mm, colour, 22 min. (Adaptive mechanisms rather than community relationships are featured in this film about deserts in general.)

The Tropical Rain Forest. Encyclopaedia Britannica Films. 16 mm, colour, 17 min. (A good integration of abiotic and biotic factors in this biome.)

TEACHER'S REFERENCE BOOKS

BUXTON, P. A. *Animal Life in Deserts, a study of the fauna in relation to the environment.* Arnold, 1923.

CLOUDSLEY-THOMPSON, J. L.—Editor. *Biology of deserts.* Institute of Biology, 1954.

EYRE, S. R. *Vegetation and Soils—A World Picture.* Arnold, 1968.

GEORGE, W. B. *Animal Geography.* Heinemann, 1952.

GOOD, R. D. *The Geography of the Flowering Plants.* Longmans, 1953.

LEOPOLD, A. S. and EDITORS OF *Life. The Desert.* Time, Inc., 1961.

MILNE, L. J., MARGERY MILNE and EDITORS OF *Life. The Mountains.* Time, Inc., 1962.

RICHARDS, P. W. *The Tropical Rain Forest.* Cambridge University Press, 1955. (An excellent description of the tropical rain forest biome by a perceptive naturalist.)

WENT, F. W. 'The ecology of desert plants', *Scientific American*, April 1955. Pp. 68–75.

3

PATTERNS OF LIFE IN THE WATER

MAJOR IDEAS

1. Aquatic environments encompass almost three-quarters of the total biosphere, and in some past geological periods the proportion was even greater.
2. Current, translucency, turnover, and chemical characteristics are the most important abiotic factors in the classification of inland-water ecosystems.
3. In aquatic communities the greatest proportion of food production is accomplished by microscopic plants and protists. The diversity and abundance of aquatic consumers are indications of the vast amounts of energy made available by the aquatic-producer system.
4. The chemical characteristics of the sea are to a considerable degree the result of the activities of marine organisms, just as the chemical characteristics of the atmosphere are to some extent the result of the activities of terrestrial organisms.
5. Recent explorations have revealed that life occurs even on the abyssal floor of the oceans.
6. Man has influenced inland-water ecosystems almost as much as he has influenced the surface of the land. However, except for slight disturbance of the margin, man has as yet had little effect upon the seas. Indeed, the ocean depths are a frontier of exploration at least as rich in possibilities as the depths of solar space.

PLANNING AHEAD

If you have not yet duplicated grids for Investigation 4.1, this should be done now.

For Investigation 1.1, *LO*, you will need a few living frogs. If you do not keep frogs in your laboratory, either collect or order some now. The same investigation (and also Investigation 2.5, *LO*) requires *Elodea*. This should be always available in your classroom aquarium; if it is not, order it also. *Elodea*, however, is frequently obtainable locally from pet shops. If you intend to demonstrate slime moulds, order or collect sclerotia.

The non-living materials and the equipment required by the investigations in Chapters 1 and 2, *LO*, are probably to be found in most school laboratories. However, it would be as well to check through the lists now. Note particularly: cellulose tubing and Clinistix (for Clinitest tablets), (Investigation 1.2, *LO*); Carnoy's fluid, aceto-orcein solution, slides of onion-root tip (Investigation 1.3, *LO*); pea seeds, glass beads, potassium hydroxide, volumeters (Investigation 2.1, *LO*); vacuum bottles (Investigation 2.3, *LO*); propanone (acetone), petroleum ether (Investigation 2.4, *LO*).

GUIDELINES

In Chapter 2 (Patterns of Life on Land) principles of tolerance, limiting factors, biogeography, biomes (and man's influence on them), are treated in the context of terrestrial life. These principles are further exemplified in Chapter 3 (Patterns of Life in the Water). Additional concepts in Chapter 3 are those pertaining specifically to aquatic life. These involve only a few new terms however, so the vocabulary load is quite light.

It is undeniable that aquatic biology is important, but in view of the overlap of concepts in Chapters 2 and 3, the latter may be considered for omission. But even if Chapter 3 is omitted, some pupils who have interests in aquatic biology or who complete their work rapidly might undertake it. Such individual or small-group efforts might include the setting up of Investigation 3.1 as a demonstration for a class. Further, it may be desirable to ask all pupils to read 'The Ocean Environment' (pp. 90–95) when the subject of the origin of life arises in Chapter 4.

Some teachers do not want to omit Chapter 3 but find that in the regular sequence of chapters it comes at a season inconvenient for fieldwork. If this problem arises, the chapter may be shifted to the spring term. Quite logically it may be inserted as an interlude between Chapters 3 and 4 of *MHE*, where it can serve as a summary experience for the whole course up to this point. A field excursion can focus the pupils' attention once again upon the individual organism and its adaptations, upon species populations, and upon the interaction of both with environment. A study of man's involvement with aquatic ecosystems serves as a bridge to Chapter 4 and 5 of *MHE*. If Chapter 3, *PLW*, is studied any time after *LO* has been completed, it is possible to introduce many ideas that are unsuitable earlier on—for example, differences between fresh-water and salt-water environments in conditions for water diffusion.

The chapter easily divides into three assignments: pp. 76–81 (guide questions 1–4), pp. 83–90 (guide questions 5–10), and pp. 90–100 (guide questions 12–16).

TEACHING NOTES

Inland Waters (in part) (pp. 76–81)

p. 73. If the ideal introduction to this chapter—a field excursion—is impossible, a good beginning is to show slides illustrating the description of a pond.

pp. 77–81. Remind pupils to use *DLT* and its Appendix. Practically all organisms mentioned in the pond description are illustrated somewhere in the pupil's book.

p. 79, Fig. 3.1. Pupils should be able to recognize these as arthropods.

p. 81, Fig. 3.6. This is an integral part of the consideration of pond ecology. Do not allow it to be overlooked.

INVESTIGATION 3.1

SUCCESSION IN A FRESH-WATER ECOSYSTEM (pp. 82–83)

This investigation, which may be carried out during the winter months, allows some insight into fresh-water biology without a trip outside the classroom. It is particularly concerned with succession and provides means for deepening an understanding of that concept.

Spend at least half a period considering the 'Background Information' and 'Purpose' through discussion and audiovisual aids. Ask the pupils to suggest hypotheses for the two stated purposes. Then remember during the post-laboratory discussions to ask the pupils to relate their experimental data to their hypotheses.

Materials

For wide-mouth jars use large jars, obtainable from the school canteen or from a restaurant. Clean the jars thoroughly. Rinse them a number of times to remove all traces of soap or detergent; mere traces of these kill many aquatic organisms.

Pond-water may be sterilized in an autoclave, but boiling is sufficient. Rainwater may be collected and used in place of pond-water. If nothing but tap-water is available, condition it before using. This is done by allowing the water to stand for at least forty-eight hours in shallow glass vessels. (Large baking dishes are good.) Distilled water should, if possible, be obtained from a *glass* still; some organisms are quite sensitive to traces of metallic ions, especially copper.

Procedure

It is desirable that the set-ups be replicated. But having even one set-up for each class may present space problems—especially since the investigation should be continued for six weeks.

Jars should not be placed in direct sunlight for more than an hour at a time, especially around noon.

Studying the Data

A. Jar C.
B. Jar A.
C. Jar B.
D. Mostly from the twigs, stones, etc.

E. From dust in the air. Cf. Investigation 3.4, *DLT*.

F. Organisms usually do disappear, but it is difficult to generalize on causation. Logical speculation by pupils should be encouraged.

Conclusions

A. Succession should be most evident in Jar A.

B. Most rapid succession would probably occur in a 'new' body of non-flowing fresh water that formed where debris from a previous body of water existed.

Additional questions for discussion: Why is pond-water preferable to distilled water for this investigation? [*Sterilization kills organisms but does not remove minerals and possibly organic substances important to organisms.*] Why is distilled water used to replace water lost by evaporation? [*Since quantities replaced may differ among set-ups, distilled water provides the least change among set-ups.*] What is the source of energy in the three jars? [*Some may be in organic materials dissolved in the water, some in the debris (Jar A), some in the organisms (Jar C), but most in light.*]

Inland Waters (contd.) (pp. 83–90)

p. 84. ¶ 2. Oceanic mixing is also brought about by the great current systems which are powered, as is atmospheric circulation, by differential interception of solar radiation. Cf. material on climates, pp. T88–93.

p. 84, ¶s 1–3. The problem of Lake Baikal and Lake Tanganyika can be treated as an invitation to inquiry before pupils have read the section 'Lakes'. They can then evaluate their hypotheses as they read.

p. 84, ¶ 5. The larger and deeper the lake, the less, proportionally, can rooted, shallow-water plants contribute to the big energy budget of the ecosystem.

p. 85, ¶ 1. In flowing waters unattached or non-motile organisms (plankton are both) are swept away.

p. 85, Fig. 3.7 Note the melting snow in the background.

p. 87, ¶ 2. You might discuss water tables and the effect of drainage on them.

p. 87, ¶ 3. See Plate X.

p. 88, Fig. 3.10. Many miles of river course are compressed into this diagram.

p. 90, ¶ 1. No mention of health hazards in pollution occurs because this receives adequate attention in news media. But you should be sure the pupils are aware of it.

p. 56, Plate XI. This photograph merely calls attention to the existence of estuarine biology. If your school is near an estuary, you may wish to pursue the topic further.

The Oceans (pp. 90–100)

p. 91, Fig. 3.11. Pupils may require some assistance in interpreting this graph. The percentages on the horizontal axis are cumulative.

p. 92, Fig. 3.12. Unlike the major constituents (Na^+, Ca^{2+}, K^+, Mg^{2+}, SO_4^{2-}, Cl^-, and CO_3^{2-}), the trace elements vary greatly in concentration from place to place and from time to time.

p. 92, ¶ 2. Plate XII (p. 57) illustrates life on a coral reef.

p. 96. Fig. 3.16. The US Navy's *Sealab* project has shown that man can live and work on the continental shelf at depths of from 60 to 120 metres.

p. 96, ¶ 4. Some further comparison of the ecology of caves and ocean depths should be undertaken in class discussions.

pp. 99–100. Plate XIII (p. 57) illustrates a tidal pool on a rocky shore.

INVESTIGATION 3.2

EFFECTS OF SALINITY ON AQUATIC ORGANISMS (p. 101)

This investigation is concerned with the idea of tolerance, which was discussed in Chapter 2. Tolerance to variations in concentration of dissolved substances is of basic importance to all aquatic organisms. In a more general view, the investigation is concerned with osmoregulation. When the concept of diffusion is encountered in Chapter 1, *LO*, reference to the observations in this investigation will help to make the abstract molecular process more vivid.

Devote some time to a pre-laboratory discussion. Be sure that each pupil has developed a hypothesis and understands how to use it in directing his experimental procedure.

Procedure

Each pair of pupils should test the effects of all concentrations of sodium chloride on a particular organism. A variety of different aquatic organisms may be used. The following are suggested largely on the basis of availability: *Amoeba*, *Artemia* (brine shrimp), *Vorticella*, *Spirogyra* or other filamentous algae, *Euglena*, *Daphnia* or *Cyclops*, *Hydra*, rotifers.

If pupils have forgotten the method for drawing a liquid beneath a cover slip, refer them to Fig. 1.8, in *WL*.

You can provide a basis for comparison by having all pupils observe the effect of a 5 per cent salt solution on *Elodea* cells. The 'shrinking' of the cytoplast away from the wall is very dramatic and easily observed.

A. Several assumptions could be suggested, but the principal one is that the technique actually replaces the salt solution with relatively salt-free water. This is not likely on a single try; the salt must be flushed out with several applications of culture water.

Studying the Data

All the questions asked in this investigation depend upon the observations made during the Procedure. A great deal of variation is to be expected. Pupils should always be kept aware of the sources of observational variation, and this investigation is a good point at which to renew this concern. In addition, there are variations in the ways that pupils carry out the Procedure—especially with respect to timing—and variations in tolerance among different individual organisms. For microscopic fresh-water organisms, individual variation may be greater than variation between species. However, if *Artemia* is used, it should provide marked contrast to most other species in its ability to tolerate even a 5 per cent salt solution.

C. Failure to recover may be merely the result of failure to extend the period of observation far enough to note recovery.

D. Of the organisms listed above, *Artemia* is most tolerant to high salt concentration.

For Further Investigation

If marine species are used, it is appropriate to vary the concentrations *downwards* from about 3.5 per cent.

GUIDE QUESTIONS (p. 102)

11. A pupil should state the type of small organism he is referring to when answering this question. A general statement covering all small organisms is hardly possible.

PROBLEMS (pp. 103–4)

1. Pupils should compare such things as origin, size, depth, organisms, food webs, stage of succession.

2. Cloudiness, by reducing intensity of radiant energy, decreases photosynthesis. But windy conditions circulate phytoplankton, gases and nutrients; this increases photosynthesis because more phytoplankton is exposed to sunlight, the nutrient supply near the surface is augmented, and a rough water surface reduces reflection of light.

3. When a pond with a favourable food supply is stocked with a single species of fish, such as trout, rapid reproduction results in a dense population (cf. Investigation 2.1, *WL*); but because of intraspecific competition, individuals are stunted. If at least one prey–predator combination is included, the predator tends to consume excessive prey offspring, and the average size of remaining fish increases.

4. Acid water is associated with bogs, swamps, and many streams emptying such places. Streams draining areas having abandoned coal mines are often intensely acid. In waters of low pH, plankton, bottom organisms and fish are usually scarce.

Alkaline water is characteristic of areas having limestone or marl substrates. Alkaline waters generally have a greater diversity of species and greater productivity. Most large lakes vary in pH between 6.0 and 9.0, but extreme values of 1.7 and 12.0 have been observed.

5. Answers to these questions are all related to the idea that the density of water is greater than that of air. (*a*) Resistance in water is greater than in air and increases at a greater rate with increased rate of movement. (*b*), (*c*) Birds that catch their food by pursuit in air correspond to fish that catch their food by pursuit in water. In both, streamlining is associated with predation niches. (*d*) Plankton organisms, being drifters, encounter no resistance in moving. (*e*) More energy must be expended to walk in water owing to the increased resistance, but less energy is required to support the body because of greater buoyancy of water than air.

6. The essence of the matter is: Estuaries contain a gradient from salt water to brackish water to fresh water. This fluctuates back and forth with the tides. Hence organisms must have wide tolerances for salinity.

7. (*a*) Among nations Japan has perhaps made the greatest use of marine resources. (*b*) Food from the sea includes whales, fish, crustaceans (shrimps, lobsters), molluscs (scallops, squid, octopuses, snails, clams, oysters), and the larger algae ('seaweeds'). (*c*) It is difficult to harvest marine producers, for they are mostly microscopic, planktonic forms. (*d*) Some marine biologists believe that man is even now harvesting nearly the maximum that can be obtained on sustained yield from the sea. The supply of minerals such as nitrates and phosphates appears to limit production in much of the sea. (*e*) By the time we get organisms in edible sizes, the materials have gone through several food levels, with great loss of energy (Fig. 1.10, *WL*).

8. Here the focus is upon large areas of the seas, such as the Sargasso Sea, which are deficient in nitrates and phosphates and therefore produce very little. This is in contrast to areas with upwelling currents which circulate minerals from the depths to the surface, and to areas of rather shallow waters near the mouths of streams that contribute run-off minerals from the land. In those areas, marine fisheries are extensive, and the fishing fleets of many nations may congregate there.

9. Pollution of the seas can occur through dumping of garbage, oil (cf. Fig. 5.18, *MHE*) and radioactive wastes. The major argument for the use of seas as dumping grounds is that of dilution, but care must be taken with the effects of currents, stagnation in basins, and ecological concentration of materials in food webs.

ADDITIONAL PROBLEMS

1. Measurements of oxygen in a Cumberland lake produced the following data:

DEPTH	OXYGEN (cm^3 per litre)		
(in metres)	Aug. 24	Oct. 11	Nov. 10
0	5.5	4.8	6.0
5	5.3	4.3	6.0
10	0	4.3	5.2
15	0	4.8	4.0
20	0	5.0	2.0

Write a short paragraph that describes the major principle illustrated by these data. How does this phenomenon influence mineral cycling? Characterize the environmental demands placed on organisms living at 20 metres. Where should the proportion of saprovores be greatest? Producers? Would the measurements of oxygen in April be most like those of August, October, or November? [*The data illustrate seasonal water turnover in middle-latitude lakes. Turnover tends to circulate minerals as well as oxygen. Any organism living permanently at a 20 metre depth must be a facultative anaerobe, and many of these are saprovores. Producers must be at depths to which light sufficient for photosynthesis can penetrate. In April the spring turnover produces oxygen distribution similar to that of early autumn.*]

2. How does temperature affect the species composition of a stream?

3. Current is one of the most important limiting factors in a stream ecosystem. Stream organisms must be adapted for maintaining a constant position. Cite at least five specific organisms showing five different adaptations to stream currents.

4. Why does the introduction of small quantities of organic wastes into bodies of water sometimes increase the size and productivity of the animal populations? Why is this not a good argument for putting wastes into streams? [*Organic wastes provide food for saprovores and nutrients for photosynthetic organisms, but in large quantities they upset the stream ecosystem by encouraging the growth of large populations of such organisms. Often this results in conditions undesirable from a human viewpoint, such as the production of undesirable odours and the decrease of fish populations.*]

5. Why do lake organisms sink to the bottom when they die? How would a lake ecosystem be changed if this sinking did not occur? [*In general, organic matter is slightly more dense than water. The second part of the problem is a matter for pupil imagination.*]

6. How does the amount of nitrogen fixation in aquatic ecosystems compare with the amount of nitrogen fixed on land? [*This problem can test the pupil's research and reasoning abilities. Measurements of nitrogen fixation are many, but no overall estimates exist.*]

AUDIOVISUAL MATERIALS

Filmstrips

The Miracle of the Sea. 'The World We Live In', Part II. New York: *Life* Filmstrips, 1954. Colour. (Contains little about living things, but is useful for establishing the dimensions of the seas and the physical processes that are important factors in the abiotic environment of marine organisms.)

Creatures of the Sea. 'The World We Live In', Part VII. New York: *Life* Filmstrips, 1955. Colour. (Excellent photographs and better-than-average paintings.)

The Coral Reef. 'The World We Live In', Part VIII. New York: *Life* Filmstrips, 1955. Colour. (Contains a good collection of photographs.)

Along the Seashore. Visual Information Service Ltd. (Provides a series of plants and animals found along the sea shore—rocky, sandy and shingle.)

Films

The Intertidal Region. BSCS Inquiry Film Loop, available from John Murray. (The concept of zonation is presented and the problem of zonation on various substrates is explored.)

Life in the Intertidal Region. BSCS Inquiry Film Loop, available from John Murray. (Primarily concerned with the structural, physiological, and behavioural adaptations to the intertidal region. Pupils have an opportunity to evaluate a controlled experiment dealing with a behavioural adaptation.)

Predation and Protection in the Ocean. BSCS Inquiry Film Loop, available from John Murray. (An investigation into the evolutionary significance of various adaptations to specific niches.)

The Sea. Encyclopaedia Britannica Films. 16 mm, colour, 26 min. (Interrelationships among marine organisms are shown against the background of conditions in the environment.)

Plankton and the Open Sea. Encyclopaedia Britannica Films. 16 mm, colour, 19 min. (Uses photomicrography to show many plankton organisms. Also develops the importance of plankton in marine food chains.)

World in a Marsh. National Film Board of Canada, available from Guild Sound & Vision Ltd., and Rank Film Library. 16 mm, colour, 22 min. (Ecological relationships in a marsh. The organisms are chiefly those of eastern North America.)

Science of the Sea. Produced by Woods Hole Oceanographic Institute, distributed by Boulton-Hawker films. (Illustrates the life cycle of the sea.)

The Estuary. Rank. (Provides a general ecological study, adaptation and the effect of man.)

TEACHER'S REFERENCE BOOKS

BARRETT, J. and C. M. YONGE. *Collins Pocket Guide to the Seashore.* Collins, 1958.

BROWN, E. J. *Life in Fresh Water.* Oxford University Press, 1955.

CARSON, R. *The Sea.* McGibbon & Kee, 1964.

CLEGG, J. *The freshwater life of the British Isles.* Warne, 1959.

DUDDINGSTON, C. L. *Beginner's Guide to Seaweeds.* Pelham Books, 1971.

ENGEL, L. and EDITORS OF *Life. The Sea.* Time, Inc., 1961.

ENGELHARD. W., AND H. MERXMULLER. *The young specialist looks at pond life.* Burke, 1964.

NEWTON, L. *A Handbook of the British Seaweeds.* British Museum, 1931.

4

PATTERNS OF LIFE IN THE PAST

MAJOR IDEAS

1. Fossils are the tangible evidence for the existence of organisms in the past.
2. From this evidence paleontologists have been able to piece together a sketchy history of life on Earth. But this history is more than a mere recital of names and descriptions of extinct organisms; it includes reconstructions of biotic communities, of climates, and even of whole ecosystems.
3. To reconstruct the past from the scattered and fragmentary fossil evidence, paleontologists begin with a thorough knowledge of the organisms and ecosystems of the present. Then, by inference and logical analysis, they extrapolate this knowledge into the past—guided by the fossils. 'The present is the key to the past.'
4. Though man's understanding of the history of the Earth is, and probably always will be, meagre, it increases as more fossils are discovered. And almost every year discoveries push the antiquity of life on Earth farther back in time.
5. In successive layers of rock, the presence of certain fossils and the absence of others enable paleontologists to construct a geological time sequence. In recent years, measurements of the products of radioactive decay have helped to tie geological time more firmly to an absolute time scale.
6. While it seems likely that life on this planet originated in the primordial oceans, paleontologists have no direct evidence of how it originated. Some biologists have been willing to speculate about the origin of complex carbon compounds that could have been biochemical forerunners of simple living systems, and recent laboratory experimentation has lent support to some of these speculations.
7. The fossil record indicates that throughout the biological history of the Earth, as environments changed, once-abundant kinds of organisms became extinct and new kinds of organisms appeared. On the other hand, in ecosystems which have shown little change through time (certain marine situations, for example), many organisms have shown great stability.
8. The fossil record indicates that with the passage of time many groups of related organisms have exhibited the phenomenon of adaptive radiation (the occupation of diverse niches by members of a taxonomic group) as well as adaptive convergence (the pursuit of a particular way of life by unrelated organisms).
9. The fossil record not only indicates the past distribution of organisms but also helps to explain present distribution. 'The past sheds light upon the present.'

PLANNING AHEAD

Because time devoted to Chapter 4 should be short, all preparations for Chapter 1, *LO*, should be quickly completed and those required for Chapter 2, *LO* should be pushed forward. See advice on p. T95. In addition, if you plan to use the demonstrations in Chapter 1, *LO*, prepare the mitosis models and start the slime-mould cultures (pp. T117–18).

Have you decided what plant material to use in Investigation 3.1, *LO*? It is probably now too late to start tomato plants from seed. Although not ideal, geranium plants can be used, and they are easy to obtain. You need rather long, unbranched stems. Plants grown on window ledges in winter naturally acquire this 'leggy' form. Evergreen woody plants are unsuitable, but woody deciduous plants which have recently burst into leaf could be used.

Pelargonium or *Coleus* may be used for Investigation 3.2, *LO*. You will need a number of potted specimens. If you are not growing these plants routinely in your laboratory or greenhouse, obtain them several weeks before they are needed so that they may adjust to the laboratory environment.

GUIDELINES

Telling the story of our present knowledge of the past requires considerable space, and the mere recounting of facts is not enough in a course that attempts to stress the investigative nature of biology. But the very nature of paleontological research makes difficult pupil laboratory investigations of a paleontological type. An attempt must be made to accomplish, by means of the textbook, some aims that might preferably be accomplished in investigations.

Early in the study of Chapter 4, pupils should have some first-hand experience with fossils. If the school is near good fossil-bearing strata, this is pre-eminently an occasion for a field excursion; at least you can lead a Saturday expedition of volunteers. But regardless of your school's location, fossils should be in the classroom. The fossil collector is a breed of hobbyist that may occur anywhere, and donations from one can usually be arranged. In your own journeyings, be alert to opportunities for acquiring specimens. Finally, there are the geological supply agencies.

Chapter 4 breaks into three sections: 'Fossils, Sediments and Time' (the nature of fossils and the geological record); 'The History of Life' (the origin of life, antiquity of life, and descriptions of paleoecosystems during Cambrian, Carboniferous, Triassic and Eocene times); and 'The Work of the Paleontologist' (studying the fossil evidence, interpreting the evidence, with some concepts which have developed out of this work). The first two sections, however, can be handled as one reading assignment (pp. 105–24, guide questions 1–13) and the third as a second assignment (pp. 125–37, guide questions 14–20). The one investigation is in the latter section. In this chapter much stress needs to be placed upon illustrations.

Just as we cannot bodily transport pupils to distant biomes, so we cannot take them to the distant past. But vivid illustration can partially overcome the difficulty. There is, however, a complicating factor in illustrating Chapter 4, a factor often overlooked but crucial to the teaching of biology as a science. When we present a photograph of a scene in a distant biome, we are presenting evidence subject only to a *narrow* margin of error imposed by the limitations of photography. But when we present a painting of a scene in the distant past, we are presenting an artifact, subject to myriad errors in translation from fossil evidence. Pupils must realize this. You yourself must keep in mind that in general the more vivid the portrayal of a scene from the geological past, the further the artist has probably departed from the strict fossil evidence.

Mindful of the caution in the last paragraph, the authors have tried to balance illustrations showing restorations with illustrations of the fossils themselves. Likewise an attempt has been made to distinguish between statements that are facts concerning the occurrence of fossils and statements that are interpretations of such occurrences. You are urged to carry through and strengthen this distinction during class discussions.

Remind pupils to frequently refer to *DLT*, and to its Appendix. Neither of these treats extinct groups of organisms, but the task of fitting extinct groups into the assemblage of modern organisms is instructive. And, of course, most of the groups on the higher levels of classification have had long histories, so that mention of them recurs frequently in Chapter 4. But perhaps the chief reason for referring pupils to the Appendix of *DLT* is to give you an opportunity to show how the arrangement of taxonomic groups reflects the attempts of taxonomists to portray phylogeny—a matter briefly discussed at the end of Chapter 1, *DLT*. Chapter 4, *PLW*, illustrates one major sector of the evidence for evolution.

TEACHING NOTES

Fossils, Sediments and Time (pp. 105–10)

p. 106, ¶ 4. Not all objects that resemble organisms and that may be dug from the earth are fossils. Some are mere chance resemblances. Often it takes an expert to distinguish true fossils.

p. 106, ¶ 7. For examples of such 'indications', refer to Fig. 4.7 (fossil casts or tubes made by worm-like organisms) and Fig. 4.14 (fossil footprints left by a reptile).

p. 109, Fig. 4.3. The absolute time scale in the right-hand column is recorded in figures that have been widely published. It is perhaps too much to say that they are generally accepted. The pupil can no doubt find others in various reference works. And as more dating is done by means of radioactive isotopes, still other figures will come into use.

p. 110, Fig. 4.4. The dark gorge at the canyon bottom (*right foreground*) in which the Colorado River lies (out of sight) is cut into pre-Cambrian schist and granite, while the rimrock is Permian in age and sedimentary in nature. Like many 'geological books', this one has many missing pages, including the middle of the Paleozoic (Ordovician, Silurian, and most of the Devonian) and everything more recent than Permian. Thus, from the standpoint of life in the past, the Grand Canyon rocks in this photograph do not represent the 'Age of Reptiles' or the 'Age of Mammals'. Pupils might guess that the 'missing pages' either were not deposited or have been eroded away.

The History of Life (pp. 111–24)

p. 111, ¶ 3. Oparin's *The Origin of Life* is available in paperback (complete reference: see p. T107).

p. 112, ¶ 1. For amino acids refer pupils to p. 45, *LO*.

p. 112, ¶ 2. Research on the synthesis of complex organic molecules is continuing. Late in 1967 Arthur Kornberg and Mehran Goulian, of the Stanford University Medical School, were successful in synthesizing biologically active molecules of DNA. This DNA, made up of some 5500 nucleotides, represented the core of a bacterial virus (X174) which normally occurs within *Escherichia coli.* It exhibited biological activities, including the abilities to infect and replicate. At this point the pupil is probably not in a position to appreciate all this, but you may want to refer to it later during work on Chapter 2, *LO*, or Chapter 2, *MHE.*

p. 113, ¶ 3. The fossil evidence supports Oparin's idea that consumers preceded producers. An abundance of free organic molecules (synthesized by the same processes which resulted in the earliest organisms themselves) may well have been a source of energy for such consumers. Presumably, by the time this energy source became inadequate, producers were evolving. Of course, some early organisms may have been chemotrophic, as the iron and sulphur bacteria are today.

p. 115, ¶ 4. The diversification of the trilobites is a fine early example of adaptive radiation, discussed on p. 134.

p. 116, Fig. 4.9. The large shovel-nosed 'fish' in the foreground is an ostracoderm, a distant relative of the modern chordate class Agnatha, characterized by having neither paired appendages nor jaws. In the upper-right background the organism that has many evident appendages is a eurypterid, an ancestor of the modern horseshoe crab (an arthropod). The finned 'fish' in the middle background (a placoderm) is an extinct distant relative of modern sharks. The palm-like stalked organisms on the left are crinoids, relatives of starfish, while at their base are spiny sea urchins, also related to starfish. Trilobites, now extinct, appear in the middle foreground, while in the right foreground is a brittle star, a member of a group related to starfish and still in existence today. Above the brittle star are brachiopods, which look superficially like clams but are in a quite different phylum. Several hundred species of brachiopods still exist in the oceans today, but this is a small number in comparison with the thirty thousand or more that occurred in the past.

p. 117, Fig. 4.11. This map may be overlooked by many pupils. However, you may find it useful for relating past geography to present. Consider, too, the question: On what kind of evidence is such a map based? The geological evidence of evolution in land masses led Darwin's thinking towards the idea of organic evolution.

p. 121, Fig. 4.15. The crocodile-like animal on the left is actually a phytosaur (the name means 'plant lizard' but is incorrect because the reptile was a carnivore). This 'false crocodile' was a primitive thecodont reptile in a group that became extinct, though very distant relatives eventually gave rise to modern crocodiles (see Fig. 4.29). To a degree, the similarity of crocodiles and false crocodiles is an example of adaptive convergence (pp. 135–6), in which two different taxonomic groups exhibit similar characteristics and occupy similar niches. The large organism in the water is one of the most primitive amphibians (*Eupelor*); in the far background on the right are two dinosaurs, descendants of thecodont stock. A different type of thecodont is in the background behind *Phytosaurus*, and still another is standing on its hind legs in the right foreground beside the pond. The thecodont group exhibited adaptive radiation (p. 134), giving rise to phytosaurs, crocodiles, reptiles with wings, and several types of dinosaurs (again see Fig. 4.29). Vegetation along the shore is made up of ancestors of our modern horsetails and on the shore at the right are tree ferns. The large trees in the background are araucarian trees, true conifers related to our modern monkey-puzzle pine. By the way, Fig. 1.17, *WL*, shows a food web for a late Mesozoic ecosystem somewhat like this Triassic one.

p. 124, ¶ 2. There is, of course, a possibility that most of these bacteria were saprovores. Sometimes, however, fossil remains show abnormalities which must have been caused by pathogens. An interesting aspect of this, related to early man, can be found in Calvin Wells' book *Bones, Bodies and Disease* (London: Thames and Hudson, 1964).

p. 125, Fig. 4.20. This is probably very similar to the Geisel Valley ecosystem described on pp. 122–4.

INVESTIGATION 4.1

PALEONTOLOGICAL COMPARISON (pp. 126–8)

This investigation should give pupils an appreciation of how a paleontologist organizes and analyses his data. It may, too, dispel the idea that a paleontologist's work only involves the digging up of fossils in some far-distant, romantic spot.

It would be ideal, of course, if pupils could work with actual fossil material, as a paleontologist does. The skull or jawbone of a modern horse makes the exercise more 'real', and pictures of restorations of the extinct equids also help. Sample skeletal material proves quite useful for illustrating some characteristics—in addition to span of cheek teeth—that have been used in reconstructing the phylogeny of the Equidae.

Sketches of some equids, with forelegs and molar teeth, are included in Fig. 3.13, *MHE*. *Hipparion* in this figure is a close relative of *Neohipparion*. Whether illustrative materials are available or not, with most pupils it is necessary to go over the 'Background Information' and 'Procedure' rather carefully. After this the actual construction of the chart can be done by pupils outside of class. *Note:* the name *Hyracotherium* has priority over the more familiar *Eohippus*. It is, however, perfectly permissible to use 'eohippus' as a common name.

Procedure

The following matters should be emphasized: (*a*) In the chart, p. 127, the points are to be plotted just as they would be plotted on an ordinary line graph, but they should be placed midway *between* the lines dividing the time intervals rather than on the lines. And, of course, the line connecting the points will branch. (*b*) The directions in 3 and 5 on p. 128 must be followed carefully; the aim is to show how data on a single characteristic—the span of cheek teeth—fit the scheme of phylogeny derived from the study of many characteristics. (*c*) *Miohippus*, like several other genera, is represented by two dots, which indicates existence of the genus at two time levels. It probably will not be obvious to the pupil that the genera *Anchitherium* and *Parahippus* could not very well have evolved from species of *Miohippus* that existed contemporaneously with them (in the early Miocene). Explain that *Anchitherium* and *Parahippus* were more likely to have arisen from species of *Miohippus* living in the late Oligocene.

You can facilitate the work of the pupils by providing them with copies of a chart similar to that on p. 127.

Conclusions

B. (*a*) 0.15 cm/million years. (*b*) 0.57 cm/million years. Some variation is to be expected in answers to this and the following two items. The difference between *Hyracotherium* and the first *Miohippus* is 8.4—4.3 = 4.1 cm. The time difference from the beginning of the Eocene to the beginning of the late Oligocene is about twenty-seven million years: 4.1÷27 = 0.15 cm/million years. (*c*) 0.35 cm/million years, using the earliest *Miohippus* and earliest *Equus*.

C. The principal generalization from these figures is that the rate of change in cheek-teeth span varied during the evolutionary history of the Equidae.

D. Although the general trend through the history of the Equidae was towards the lengthening of the cheek-teeth span, the change from *Merychippus* to *Calippus* was a shortening of the span, and in a few other instances (shown by the negative slope of the graph line) a similar reversal of the general trend occurred.

The Work of the Paleontologist (pp. 125–38)

p. 128, ¶ 11. A restoration of an extinct Irish elk, based upon skeletal material, is illustrated in Fig. 3.9, *MHE.*

p. 129, Fig. 4.25. The picture itself rather clearly suggests that the exact position of each bone in relation to the positions of all others must be carefully recorded. Less obvious is the need for recording the rock stratum in which the fossil occurs and the relation to other strata.

p. 129, Fig. 4.26. This is *Varanosaurus*, one of the earliest genera of reptiles. The caption question merely calls for an echo of the discussion in paragraph 3, p. 129, to 2, p. 130. However, in the case of this particular fossil, the reptilian giveaway is the hole—temporal opening—behind the eye socket (orbit). Although some early reptiles lacked temporal openings, any skull possessing one or more such openings behind each orbit is reptilian. Other reptilian characteristics that may be detected in this sketch are the somewhat arched skull (amphibians tend to have a flat-topped skull), the apparent single articulation (single occipital condyle) between the vertebral column and skull, and well developed pectoral and pelvic girdles and appendages.

p. 130, heading 2. This contrast in ideas will recur in Chapter 3, *MHE.* Here it is principally a historical matter; there it becomes a part of the concept of evolution.

p. 130, ¶ 6. There is a new hypothesis that man, rather than climatic change or other natural force, may have caused the extinction of some of the large Pleistocene mammals.

p. 131, Fig. 4.27. This figure is intended to emphasize the need for caution in accepting restorations. The evidence found in the rocks (top photograph) clearly requires a great deal of interpretation even before it undergoes transformation to the mounted skeleton (middle photograph). Still further interpretation is required before it arrives at a three-dimensional model—and interpretations may differ (bottom photographs). When only a few fragments are preserved, how much greater must be the distance between fossil and restoration!

p. 133, Fig. 4.29. The term 'pterosaur' in this figure is equivalent to the more familiar 'pterodactyl'.

p. 134, ¶ 3. One of the best examples of adaptive radiation involves the mammalian order of marsupials, whose primitive members within the isolated continent of Australia have occupied almost every niche represented by all other orders of the class Mammalia elsewhere. A good presentation on marsupials is to be found in David Bergamini's book *The Land and Wildlife of Australia* (New York: Time, Inc., 1964).

p. 135, Fig. 4.31. Both pupils and teachers may have used 'carnivores' in a popular ecological sense, usually as an equivalent to 'predator'. It may be well, therefore, to note that in this figure it is used in a taxonomic sense, as the English form of 'Carnivora' (used in Chapter 1, *DLT*).

p. 137, ¶ 2. Refer to Fig. 2.26 for illustrations of modern tapirs.

p. 137, Fig. 4.35. Determining the site of origin for taxonomic groups is an intriguing pursuit. The location of the oldest fossil representative of a genus provides some evidence. Since evolution proceeds most rapidly in areas of great variability, where many niches are available, such paleoecological situations are considered favourable centres of origin. Also, sites where there are many species of a genus might represent the origin centre of that genus.

GUIDE QUESTIONS (pp. 138–9)

8. By the time this question is encountered, its answer may very well differ from an answer obtained by reading Chapter 4, written in 1970. Pupils should be encouraged to present information superseding that in the textbook.

15. It is possible to give non-scientific reasons for differences; for example, differences in the skill or technique of artists. These, of course, are valid reasons but contribute nothing to the point in hand.

PROBLEMS (pp. 139–40)

2. There have been several explanations, but in very recent years the idea of continental drift has become more and more seriously considered by geologists, especially with respect to Antarctica. There are innumerable contemporary articles dealing with this topic. One of the best summaries of current thinking is Patrick M. Hurley's 'The Confirmation of Continental Drift', *Scientific American*, April 1968, pp. 52–64. The changing view with respect to continental drift and associated biological phenomena makes a good discussion topic, showing pupils why scientists change their minds and that scientific 'eternal truths' (such as the permanence of the continents) may not be very 'eternal' after all.

3. *Webster's New International Dictionary* (unabridged) is a good source of information on these names.

6. While pupils may jump to the conclusion that the large ichthyosaurs had eaten the smaller ones, some may remember that certain reptiles give birth to 'live' young, which have developed within eggs inside the female (ovoviviparous). Kinds of evidence that might be sought: where the small skeletons are located within the large skeleton; whether they are disarticulated or otherwise damaged as a possible result of having been eaten; whether they have embryonic characteristics.

7. Pupils will discover many hypotheses but little agreement. There is a pertinent chapter in Colbert, Edwin H., *The Age of Reptiles*, pp. 191–207 (New York: W. W. Norton & Co., Inc., 1965) about 'The Great Extinction', in which Colbert aptly notes that 'It was an event that has defied all attempts at a satisfactory explanation, for which reason, among others, it has fascinated paleontologists for decades.'

10. The situation in South America can be compared with that in Australia, which still is isolated from other major land masses and still contains an array of bizarre, primitive animals that have survived because of long isolation from more advanced animals. The primitive South American mammals did not fare so well because of the geologically recent re-establishment of the Panama land bridge and consequent contact with more 'efficient' predators and competitors from the north.

AUDIOVISUAL MATERIALS

Filmstrips

Reptiles Inherit the Earth. 'The World We Live In', Part V. New York: *Life* Filmstrips, 1955. Colour. (A wonderful gallery of paintings, with all the familiar and many unfamiliar Permian and Mesozoic amphibians and reptiles represented. But there is no mention at all of the fossil evidence on which the magnificent imagery is based.)

The Age of Mammals. 'The World We Live In', Part VI. New York: *Life* Filmstrips, 1955. Colour. (Another fine collection of paintings. Includes some introductory matter on stratigraphy, which links the pictures to the fossil evidence.)

Fossils. Encyclopaedia Britannica Educational Corp., EB 6413: five colour filmstrips and five phono-discs. Contents: *How Fossils Are Formed*; *Collecting and Interpreting Fossils*; *Fossils and the Relative Ages of Rocks*; *Fossils and Prehistoric Environments*; *Fossils and Organic Change*. (One of the best audiovisual treatments of this subject available.)

South American Fossils. 'Darwin's World of Nature', Part IV. New York: *Life* Filmstrips, 1960. Colour. (Deals with the development of the South American vertebrate land fauna after its isolation in the Cretaceous and with the results of the rejoining of South and North America in the Pleistocene. The pictures are good, and the zoogeographical principles have been well presented.)

Films

Australian Marsupials. BSCS Inquiry Film Loop, available from John Murray. (An inquiry reflecting adaptive radiation and the past as explanation of the present.)

Story in the Rocks. Shell Films, available through Petroleum Films Bureau. 16 mm, colour, 17½ min. (Deals interestingly with the more glamorous activities of paleontologists, emphasizing their skill in interpreting bits and pieces of evidence.)

Biochemical Origin of Terrestrial Life. McGraw-Hill Film Rental Library.

TEACHER'S REFERENCE BOOKS

AGER, D. V. *Principles of Paleoecology*. McGraw-Hill, 1963. (An 'introduction to the study of how and where animals and plants lived in the past'. Provides an excellent background for the point of view that dominates Chapter 4 of *PLW*.)

BRITISH MUSEUM (NATURAL HISTORY). *British Palaeozoic Fossils*. 3rd edition, 1969. *British Caenozoic Fossils*. 3rd edition, 1968. *British Mesozoic Fossils*. 3rd edition, 1967. (These three British Museum publications were written so as to provide a cheap simple text which would enable the young, or those without experience, to identify the fossils they find.)

CARRINGTON, R. *A Guide to Earth History*. Chatto and Windus, 1956. (Tries to answer simply some of the questions posed on how life has evolved on Earth.)

MOORE, R. *Man, Time and Fossils*. Cape, 1966. (This story of evolution focuses especially on the search for remains of fossil man.)

OAKLEY, K. P., H. M. MUIRWOOD. *The Succession of Life through Geological Time*. The British Museum (Natural History), 1958. (A concise paleontological review by geological periods, well illustrated and with maps.)

OPARIN, A. I. *The Origin of Life*. 2nd edition. New York: Dover Publications, Inc., 1953. (The source of much current thinking on the origins of life; originally published several decades ago.)

ROMER, A. S. *Vertebrate Palaeontology*. 3rd edition. University of Chicago Press, 1966.

LOOKING INTO ORGANISMS

Chapter 1 begins the second half of the twenty chapters in this series of books. But the half-way point has not yet been reached with respect to the learning load in the course. The density of ideas now becomes much greater than it was in the previous books; thus, increased pupil effort is required. The heavier burden in the second half of the course may, in part, be balanced by the pupils' increasing familiarity with techniques of studying biological science.

Heavy emphasis on the morphology and gross internal physiology of organisms has persisted in school biology teaching long after its decline in biological research. Much of this 'traditional' school biology is concentrated in *Looking into Organisms*. In contrast, also encountered in *LO* is material from the burgeoning field of biochemistry—a branch of science changing with such rapidity that the factual knowledge of even last year's graduate may be obsolete before he is established in his first year of teaching. The same can be said concerning the field of animal behaviour, except that even recent graduates may have had little preparation in this field.

The blend of the familiar and the unfamiliar in this book may prove challenging to the teacher. To achieve the purposes of the *Environmental Approach*, the writers found it necessary to reduce what has frequently been the substance of an entire course to less than one part.

1
THE CELL

MAJOR IDEAS

1. Observations and reasoning by men of many nations contributed to the development of the cell theory. This theory, though imperfect in details, is an important guide in the investigation of biological structure and function.
2. Cell structure varies and therefore generalizations are inevitably misleading. This applies both to the chemical composition of the cell and to the organelles visible within it.
3. Biologists aim to explain the functioning of cells in terms of physical principles. This is illustrated by a discussion of the transport of substances into, out of, and within cells.
4. Among cells that have recognizable nuclei, the process of cell division is remarkably uniform, involving a definite sequence of nuclear events—mitosis.
5. In species that are unicellular, cell division results in the production of new individuals; in multicellular species it results in the additions of cells, usually leading to the growth of the individual.
6. The process by which successive generations of cells within a multicellular organism come to differ from each other—the process of cell differentiation—remains one of the major puzzles of biology.
7. Unless they periodically divide, most cells appear to age. The causes of ageing in cells, tissues, and whole organisms represent another major problem in biology.

PLANNING AHEAD

If you have been keeping up with the planning advice given in the chapters of previous books, you have now reached a temporary rest point. Check the following: (*a*) Are you completely ready for the first chapter of *LO*? (*b*) Have you obtained pea seeds, glass beads, volumeters, fresh hydrogen peroxide, vacuum bottles, propanone (acetone), and petroleum ether (as well as the more usual materials—for the investigations in Chapter 2? (Try out the operation of a volumeter if you are unfamiliar with it, and prepare fine-pointed pipettes for Investigation 2.4.) (*c*) Have you arranged for the plant materials required in Investigations 3.1 and 3.2?

The only long-term preparation that might be undertaken at this time is to start growth of *Zebrina* (Investigation 5.1) and *Coleus* (Investigation 1.1, *MHE*) if you have facilities for producing the plants yourself.

GUIDELINES

As a term, 'cell' is probably not new to your pupils. But since experience with cells is likely to have been vicarious, arranging for some observational foundations is of first importance. You may therefore start work with Investigation 1.1. First 'seeing', then 'explaining' can lead to lasting, accurate understanding. This principle can be applied throughout the chapter.

The cell-physiology portion of the chapter (pp. 9–15) depends heavily on (*a*) an understanding of cell structure, and (*b*) acquaintance with the rudiments of the molecular theory. An adequate understanding of cell structure may be gained from the preceding part of the chapter. Some acquaintance with the molecular theory must, however, be expected in the pupil's background. The amount of understanding expected in this series of books is not as great as many persons might believe. The pupil who works through pp. 9–14 thoughtfully will have a sufficient (though certainly neither broad nor profound) view of diffusion.

The best presentation of mitosis is, of course, by means of a cine film. Avoid films combining mitosis and meiosis (or stop the film before meiosis appears). Meiosis is not relevant to Chapter 1 in this book and can only confuse. If mitosis is thoroughly understood now, meiosis will be relatively easy when it is encountered in the first Chapter of *MHE*. But viewing a film does not assure understanding. Need for Investigation 1.3 remains.

Chapter 1 is difficult. The writers have undertaken to make it as clear as possible and have ruthlessly pruned away what they considered inessential detail. They have used an unusually spare running vocabulary. But the material remains difficult. The difficulty must be surmounted, for some understanding of cellular structure and function is essential to the pupil's appreciation of modern biology. And many things in the remainder of the course depend upon that understanding.

Your first response to the difficulty of Chapter 1 should be to reduce the length of assignments; the second should be to increase the time allowed for discussion of each assignment. Five assignments are suggested: pp. 1–3 (guide questions 1–2), pp. 3–7 (guide questions 3–4), pp. 9–10 and 12–15 (guide questions 5–11), pp. 15–19 (guide questions 12–14), and pp. 21–25 (guide questions 15–18).

TEACHING NOTES

Some History (pp. 1–3)

This history illuminates the development of a biological concept and emphasizes the international character of science.

p. 2, ¶ 7. The development of the cell theory illustrates how more accurate understanding frequently depends on further technological developments. The use of dyes and of phase-contrast and electron microscopes has revealed increasingly detailed information about cells.

Cell Structure (pp. 3–7)

p. 4, Fig. 1.3. Compare this diagram with the photograph of a cell model which was constructed according to information obtained from the study of numerous electron-micrographs (book cover). After pupils have finished Investigation 1.1, ask them whether this diagram more closely resembles an animal or a plant cell.

p. 4, ¶ 2. The term 'cytoplast' for the structural region of a cell does not have the objectional connotations of 'cytoplasm', a term for a supposed substance. 'Cytoplasm' and 'protoplasm' are both avoided in this series of books.

INVESTIGATION 1.1

DIVERSITY IN CELL STRUCTURE (pp. 7–8)

This investigation is purely observational. The emphasis is on *living* cells that the pupils can directly associate with whole organisms. Of course, only a limited number of cell structures can be seen. If you want pupils to see other structures—mitochondria, for example—you may wish to use some commercial slides to supplement the investigation. A few microscopes equipped with oil-immersion lenses are desirable for such observations.

Materials

Onions can be cut into pieces at the beginning of the day; keep the pieces under water in crystallizing dishes.

If an onion is cut into orange-section-like pieces, a pupil can remove a piece of the fleshy leaf from a section. Bending this backwards until it snaps usually leaves a ragged piece of epidermis.

See p. T22 (Investigation 1.4, *WL*) for preparation of I_2KI.

Elodea is easy to maintain in an aquarium from which direct sunlight is excluded. To increase the likelihood of observing cyclosis in cells, place some of the material under a bell jar and illuminate it for at least twelve hours before using.

Make up packets of five to ten toothpicks in aluminium foil and sterilize in a pressure cooker or autoclave. Ask each pupil to break his toothpick after use to prevent its re-use.

The frog materials should be as fresh as possible. One frog should be used to provide material for morning classes; another should be used for afternoon classes. Etherize and pith the frog (see *UFAW Handbook*, reference in Note on p. T149, for pithing instructions). With a medicine dropper obtain blood from a large vessel, or flush an area of bleeding with Ringer's solution. Place the blood in a small beaker of Ringer's solution.

You may use skin that is shed in the water in which the frogs are kept. Or the skin of the freshly killed frog may be scraped with a sharp scalpel. The materials thus obtained may be kept in a small beaker of Ringer's solution. Whole skin, of course, is much too thick for use.

Methylene-blue solution: Add 1.48 g of the dye to 100 cm^3 of 95 per cent ethanol; let stand for about two days, stirring frequently; filter and store as stock solution. Before using, add 10 cm^3 of stock solution to 90 cm^3 of distilled water.

Saline solutions: Physiological saline for this work is 7 g of NaCl in 1000 cm^3 of water. Or use amphibian Ringer's solution:

KCl,	0.14 g
NaCl,	6.50 g
$CaCl_2$,	0.12 g
$NaHCO_3$,	0.20 g
H_2O (distilled),	1000.00 cm^3

Procedure

Two full periods are required for this investigation. The observations can be accomplished in one period; but sketching the cells, completing the chart, and answering the questions in the labbook make two periods necessary.

If pupils work in pairs, each pair may be provided with slides, cover slips, and dissecting instruments; but to save time and trouble, place stains and the materials to be observed in one or a few centrally located places where pupils can go to prepare their mounts. Or you can conveniently arrange this investigation by setting up five stations, one for each observation. The class may then be divided into five groups of pupils, each to begin at a different station.

Demonstrate the techniques of removing the onion epidermis and transferring cheek cells to a slide. Even at this point in the course, pupils need to be cautioned to use *small* pieces of material. Ask pupils to avoid using circles to frame their sketches. Such figures suggest the whole field of view and would require the drawing of everything seen in that field. Only a *small* section of the field of view should be drawn to show how cells are arranged. The cell drawings should be large enough to clearly indicate details of structure. An example drawn on the board helps to get this idea across; but use as an example a type of cell not to be observed by the pupils.

During this investigation a good demonstration to show ciliary action may be set up. Cut away the lower jaw of a pithed frog; then cut out small pieces from the lining of the mouth cavity in the region between the eye bulges and the throat. Mount a piece in Ringer's solution.

p. 8, A. *Elodea* leaf: Cyclosis (if seen) is usually considered by pupils as evidence of life.

Summary

Use pictures or prepared slides of cells to help increase the pupils' ideas of cell variability. Invite pupils to identify the cells pictured as plant or animal before you identify them.

A. Remind pupils of the limitations in magnification and resolving power of their microscopes. Moreover, adjustment of light intensity as well as special staining procedure may be involved in revealing cell structures.

D. The rigid cellulose cell walls of most plant cells are usually associated with definite cell boundaries and angular shapes.

Some Cell Physiology (in part) (pp. 9–10)

p. 9, ¶ 3. Advise pupils to differentiate between the steady state of a cell's composition and chemical equilibrium, in which no material is entering or leaving the system.

p. 10, ¶ 2. For a simple demonstration of diffusion, drop a large crystal of potassium permanganate into a small beaker of water and allow the beaker to sit undisturbed. A white card behind the beaker will make observation easier.

A somewhat more elaborate method: Lightly rub very small amounts of dry crystal violet, eosin, and methylene blue into the surface of writing paper. (Use a second, small piece of paper to do this—it will keep stains off your fingers.) Shake lightly to get rid of large particles. Hold paper with dye side down over a plate of 1.5 per cent agar. Tap hard with your finger to dislodge stain particles and to permit them to fall on the agar surface. Diffusion takes place very quickly, and different dyes diffuse at different rates.

To demonstrate quickly and effectively the diffusion of gas molecules, crumple a paper tissue in a dish and saturate it either with a fragrant cologne or with ammonium hydroxide.

p. 10, Fig. 1.8. Note that this does not illustrate the text discussion since only a single kind of molecule is involved.

p. 10, ¶ 5. Cell activity results in energy loss as heat. Since the heat must be dispelled anyway, that which is used to move molecules is not really a drain on the cell's energy supply.

INVESTIGATION 1.2

DIFFUSION THROUGH A MEMBRANE (pp. 11–12)

The equipment is simple and inexpensive, and when the investigation is done by groups of four pupils, everyone is close enough to the material to see the results. Presenting this investigation as a demonstration, though possible, is inadvisable.

Materials

Cellulose tubing with a diameter of 14 mm is convenient, but larger tubing may be used.

Prepare a soluble-starch solution by adding about 10 g of soluble starch to about 500 cm^3 of water. Stir or shake, and then filter. If you cannot obtain soluble starch, try laundry starch. Filter it through cloth and then through filter paper. It is desirable, but not essential, that the starch solution be clear. Some brands of soluble starch are reported to diffuse through cellulose membranes, so the starch should be tested before being used by pupils.

The glucose solution should be strong—close to saturation—but the exact concentration is not critical.

Iodine–potassium-iodide solution may be used for the iodine reaction with starch (see p. T22).

Wide-mouth jars may be used in place of the beakers. Baby-food jars are suitable.

Clinistix or Clinitest tablets, used by diabetics to test for sugar in urine, can usually be purchased at chemists. They do not require heating.

Two set-ups are employed because iodine in the water sometimes interferes with the use of Clinistix or Clinitest tablets. If burners are available, Benedict's solution or Fehling's solution may be used to test for glucose. In this case only one set-up is necessary, and both starch and glucose solutions can be placed in one tube.

Procedure

In order to draw any conclusions from this investigation, pupils must know the reactions between: (*a*) starch and iodine, (*b*) glucose and Clinistix, (*c*) starch and Clinistix, (*d*) water and Clinistix, (*e*) glucose and iodine. Before pupils begin their work, perform a quick, silent demonstration of these combinations. In the pupil's text Tes-tape is suggested. Clinistix are an alternative which is more easily available (from local chemists).

Cut the cellulose tubing into pieces of the proper length and soak in water. Have all solutions and other materials conveniently available when pupils arrive in the classroom at the beginning of the first period. The set-up should be completed as quickly as possible. About twenty minutes later, the glucose test can be made. The reaction of iodine with starch should be visible by the end of the period, but it will be more striking on the following day.

Discussion

It is important that the results of this investigation, which involves purely physical systems, be related to living things. Therefore the questions on page 11 must be given special attention. Glucose and starch are both common materials in living things; the free diffusion of the former and the lack of diffusion of the latter can be linked to storage of starch in plant cells, the need to digest starch, the possibility of feeding glucose by direct injection, and many other biological matters.

Studying the Data (p. 11)

C. The turgidity of the glucose tube after twenty-four hours not only indicates the diffusion of water but also demonstrates diffusion pressure. For further demonstration, place a lettuce leaf or a piece of fresh carrot in a 10 per cent salt solution for fifteen minutes. Then have pupils feel it for comparison with one soaked in plain water for an equal time. How is this related to the observation in the investigation? (Refer to results of Investigation 3.2, *PLW*.)

Turgor is perhaps the most important biological effect of diffusion pressure, but you may wish to demonstrate another effect by showing how diffusion pressure can support a column of liquid against the force of gravity. The apparatus shown in Fig. T-12 can be used for this. A more difficult set-up, but one that has a close connection with biology, can be provided as follows: Using the technique shown in Fig. 1.14, suspend a raw carrot that has been hollowed out to about half its length, the hollow having been nearly filled with syrup. Shape the opening in the carrot with care, so that a single-hole stopper with glass tubing may be sealed into its top with paraffin wax or an adhesive compound. Compare the results in this set-up with that in the purely physical set-up.

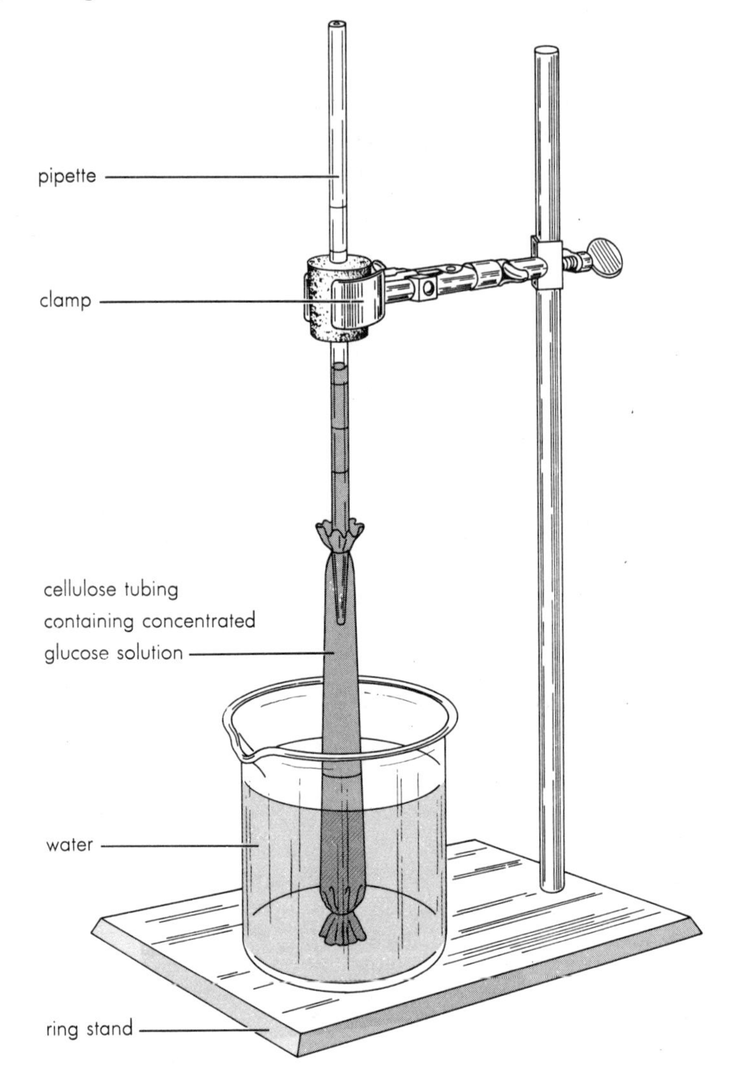

Fig. T-12

D. If the starch passed through the membrane, then the water in Beaker A should be blue. Actually, after twenty-four hours the water in Beaker A is usually clear, because all the iodine has been bound to the starch molecules in the tube.

Conclusions (p. 12)

A, B. The simplest hypothesis is that the tubing contains submicroscopic pores of such size that molecules of iodine, water, and glucose can pass through while molecules of starch cannot. This hypothesis merely assumes one of the points of the molecular theory: that even in solids there are molecular-size spaces between adjacent molecules.

Some Cell Physiology (contd.) (pp. 12–15)

p. 12, ¶ 2. The term 'differentially permeable' is preferred to 'semipermeable'; the latter is logically faulty, and the former is more descriptive. The introduction of 'differentially' here makes the later usage of 'differentiation' (p. 21) less strange.

p. 13, Fig. 1.10. In B, note there is the same number of molecules of the dissolved substance (coloured dots) within the membrane in both sketches, although in the second the membrane has expanded. Ask the pupils to explain why the membrane has expanded.

p. 13, ¶ 3. Ask the pupils to locate the *Paramecium's* 'pumps', shown in Fig. 1.2. Note the radial canals and discuss their function and that of the contractile vacuoles.

p. 14, ¶ 4. In discussing active transport, you may suggest the suspected role of enzymes in the membrane, or some other property of the membrane that forces materials to move against the diffusion gradient. These reactions may change a substance so that it can more readily enter a cell, and once inside, restore it to its original form. An informative reference on the role of certain fatty substances in transport of materials through cell membranes is an article by L. E. Hokin and M. R. Hokin, 'The Chemistry of Cell Membranes', *Scientific American*, October 1965, p. 78.

The importance of stressing the energy expenditure of active transport, pinocytosis, and cyclosis is evident when one looks ahead to the emphasis on bioenergetics in Chapter 2. The significance of that chapter can be strengthened by developing an understanding now of some of the cell's energy needs.

p. 15, ¶ 3. If pupils have not observed cyclosis in *Elodea*, you may wish to demonstrate it in slime moulds (pp. T 117–18).

Cell Duplication (pp. 15–19)

p. 15, ¶ 4. When contemplating why cells divide, pupils must remember that everything to maintain the volume of living matter in a given cell must pass through that cell's membrane. Refer to p. T128 for a discussion concerning problem 3, p. 61.

p. 15, ¶ 5. *Rhizopus nigricans* may be considered an exception to the formal cell theory in that it exists with a multiplicity of nuclei and an absence of separating membranes, an arrangement referred to as coenocytic. However, this may be a derived condition.

p. 17, Fig. 1.12. Clearly distinguishable are cell membrane, cytoplast, centrosome, chromosomes, and spindle fibres.

p. 18, ¶ 1. The names of the phases of mitosis are omitted in the text because their use tends to belie the uninterrupted continuity of the mitotic process. Though useful to the cytologist, they do nothing for the O-grade biology pupil but add terms to an already heavy vocabulary load. For the more advanced pupils, the teacher can add the names of the phases of mitosis. For an explanation of the various names see *The Chromosomes* by M. J. White, 5th edition (Methuen, 1963).

INVESTIGATION 1.3

PROCESS OF CELL DIVISION: ONION CELLS (pp. 19–21)

For many pupils, the observation of prepared slides is little more than a difficult way to look at pictures. This investigation allows pupils to associate a familiar macroscopic organism directly with the somewhat unconvincing story of mitosis told in the text.

Materials

Onions obtained from a grocery store can usually be used. In some cases, however, you will find that onions have been treated to inhibit root formation. In this case, onion sets will provide good material.

Root tips (1 to 2 cm) should be cut between 11.30 a.m. and noon, or just before midnight. Tips collected at this time show a maximum number of mitotic figures. Very poor results may be obtained from roots harvested at other times.

Fixative: Fix the root tips for twenty-four hours in the following mixture: Absolute ethanol (alcohol), 3 parts; glacial ethanoic (acetic) acid, 1 part.

If absolute ethanol is not available, denatured alcohol containing methanol and propan-2-ol (isopropyl alcohol) as denaturants will work satisfactorily. The following has been used successfully: Specially denatured alcohol, 95 parts by volume; propan-2-ol, 5 parts by volume. Fix for twenty-four hours, then store (indefinitely) in 70 per cent ethanol.

Carnoy's fluid: absolute ethanol, 6 parts by volume; trichloromethane (chloroform), 3 parts by volume; glacial ethanoic acid, 1 part by volume.

Aceto-orcein stain: orcein (synthetic), 1 g; glacial ethanoic acid, 45 cm^3; distilled water, 55 cm^3.

Dissolve orcein in hot glacial ethanoic acid. Use a large beaker, as foaming may result when dry stain is added to acid. Cool and then add distilled water. Filter just before use.

Procedure

Pupils should observe the set-up of the onion bulbs and the growth of the roots; this will allow them to clearly associate the small pieces of root (used later) with a real, living organism.

Emphasize that drawings should be large enough to show all detail observed.

If you place dishes containing water and detergent at strategic locations in the laboratory, pupils may conveniently wash their slides and leave them clean for the next group.

Discussion

By comparing cell size in embryonic tissue with size of cells observed in older tissue farther away from the root tip, develop the idea that growth of root depends upon both the increase of number of cells and their enlargement after cell division.

Observations (p. 21)

B. Further discussion, concerning changes in shape as well as in size, will lead effectively into the text's next consideration—differentiation.

If a film on mitosis has been shown, discuss the dynamic activity within the cell during mitosis—activity which clearly evidences the need for energy to accomplish this work of the cell.

D. Emphasize similarities, but note the formation of the new cell wall and the absence of centrosomes in plant cells. Ask pupils why plant cells do not 'pinch in' during division, as do animal cells.

Differentiation (pp. 21–24)

p. 22, ¶ 1. Pupils have no difficulty in recognizing the distinct differences of cells in various tissues, and in concluding that differentiation is a reality. They are less likely to see that the characteristics of a cell are to be expected in its daughter cells. Thus pupils often miss the point of the paradox. Questioning the obvious is not a frequent practice among pupils; trying to increase its frequency is one of the aims of this course.

Ageing (pp. 24–25)

It is difficult to interest school pupils in this topic. Yet because it is a fundamental biological problem and one that is receiving increasing attention from researchers, some consideration should be given it. Note that the discussion is primarily concerned with the ageing of cells; the senescence of multicellular organisms—particularly animals—involves additional complications.

PROBLEMS (p. 27)

1. Recently developed phase-contrast microscopy has made possible more effective examination of living, unstained cells.

3. Seen with a microscope, the presence or absence of cell structures such as chloroplast, cell wall, the centrosome, would be evidence that the material is either plant or animal. Chemical tests for cellulose would confirm the identification.

4. The jellyfish would swell up if placed in fresh water, and the frog would dehydrate if placed in sea water. Fish that live in both fresh and salt water maintain water balance in their tissues by means of various mechanisms. In salt water, they continuously drink, void very little urine, and excrete salts actively through their gills. In fresh water, they take in only a little water by drinking, absorb salts through their gills, and excrete much water as urine.

5. (*a*) In some 'unicellular organisms'—such as *Paramecium*—the macronucleus merely pulls apart, but the micronucleus divides by mitosis. (*b*) In division of a blue-green alga, presumably equal portions of chromatin material find their way to each of the two daughter cells.

6. In many ways cancer cells act as rejuvenated, undifferentiated cells. If it can be learned why and how cells differentiate, a way may be determined to control the reversal of the process in cancer cells.

7. In multicellular animals connective tissues in wide variety hold cells together. All these tissues have an extensive matrix of extracellular material containing fibres. Such extracellular material is manufactured by the connective tissue cells and serves to hold these cells together; and in turn, the connective tissue holds other tissues, such as muscle and nerve. Plant cells are held together primarily by the adhesive qualities of cellulose cell walls.

8. The term 'protoplasm' suggests that there is a fundamental substance in living cells. This is anything but accurate. Not only does the matter within living things differ among different kinds of organisms, but within a particular organism it is of very varying characteristics and composition. Further, within a given cell, substances are constantly undergoing physical and chemical changes. The matter within a living being is a dynamic condition, not a single substance.

9. Electron microscopes have provided information on intracellular fine structure. Phase-contrast microscopes make possible the study of living, unstained material, thus eliminating the physiological effects of the usual histological fixing and staining procedures.

SUPPLEMENTARY MATERIALS

Slime Mould Demonstration

Not only is a slime mould culture useful for demonstrating cyclosis; it is also convenient for reviewing the problems associated with a two-kingdom system of classification and for illustrating some difficulties in the cell theory.

Preparation for this demonstration must be started fourteen days before its use.

Cultures of *Physarum polycephalum* should be started from sclerotia, which may be purchased from most biological supply companies.

Sterilize a large glass dish or pneumatic trough (about 20 cm diameter) with sodium chlorate(I) and rinse thoroughly. Cut two circular pieces of muslin; the diameter of each piece should be about 5 cm greater than the diameter of a petri dish. Boil them. Cover the outer surface of one half of a petri dish with the double thickness of muslin, and fold the edges tightly under the rim of the dish (see Fig. T-13). Place this muslin 'table' in the culture dish, and add water to a depth of about 5 mm. Place paper bearing the *Physarum* sclerotium in the middle of this 'table', coloured side (the sclerotium side) up. Cover the culture dish with a sheet of glass or a larger dish. A paper towel helps maintain a humidity gradient in the chamber and, in addition, prevents condensed moisture from dropping on to the slime mould.

Store the set-up in *indirect* light at a temperature of about 20 °C. When the slime mould has 'crawled' on to the muslin, remove the paper. Then feed daily by sprinkling on the surface of the plasmodium a small quantity of oatmeal that has been pushed through a wire kitchen strainer. Avoid dropping oatmeal in the water. Change the water daily. If contaminating moulds appear in the culture, they should be cut out with a sterile scalpel. Always cut outside the apparent margin of the mould.

When the slime mould has spread over the entire muslin 'table', transfer small bits of the plasmodium to a petri dish containing a *thin* layer of non-nutrient agar. (Prepare by dissolving 1.5 g of agar in 100 cm^3 of hot water.) It is not necessary to autoclave the dishes or the agar, since slime moulds live on bacteria and mould spores. Maintain these cultures for *two days* under the same light and temperature conditions as above. If necessary, the petri dishes containing agar and bits of plasmodium may then be stored for a few days in the

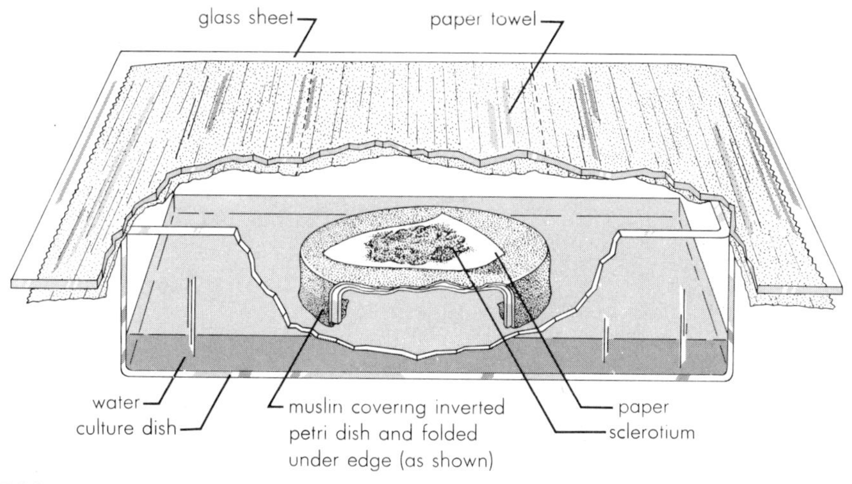

Fig. T-13

refrigerator. Several hours before use, they should be removed from the refrigerator and allowed to warm to room temperature.

The remainder of the plasmodium in the humidity chamber may be sclerotized and stored for later use, as follows: Spread a piece of moistened absorbent paper on the walls of the culture chamber. Move the muslin 'table' to the edge of the culture chamber so that it is in contact with the vertical layer of paper. Discontinue feeding. In a short time the plasmodium will 'crawl' on to the paper. Remove the paper, place it in a loosely covered container, and allow it to dry slowly. After several days the paper containing the sclerotium may be removed, cut into pieces of convenient size, and stored in an envelope for later use. The dormant organism will remain viable for one to two years.

A Mitosis Model

In Investigation 1.2, *MHE*, meiosis is illustrated by means of a model. A similar model illustrating mitosis may be used in Chapter 1 of this book if time permits, especially if a good film showing mitosis in living cells is not available. In the meiosis model, strands of Poppit beads represent chromatids; for mitosis, pipe cleaners will serve just as well, since it is not necessary to show crossing-over and, therefore, the 'chromosomes' do not need to be broken. Pupils can make the 'chromosomes' by threading two pipe cleaners through a small bead representing a centromere. If they use Poppit beads, they can make a 'chromosome' by fastening two strands together with a piece of pipe cleaner representing a centromere.

In the procedure given below, maternal and paternal 'chromosomes' are differentiated by colour. At this point the distinction is not necessary, and therefore a single colour can be used. But there is something to be said for the pedagogical value of anticipating future developments (if such action does not obscure the present point), so you may wish to use two colours and briefly answer the inevitable questions. Be sure, however, to emphasize that the colours are symbolic rather than actual.

Make up 6 'chromosomes': 2 of them long, with the centromere in the middle; 2 medium-sized, with the centromere a quarter of the way from the end; and 2 short, with the centromere in the middle. In each pair, make one 'chromosome' of one colour, one of another (see Fig. T-14).

Using a crayon, draw a spindle on a large piece of wrapping paper spread out on a table. The spindle should be large enough to accommodate the six 'chromosomes' when they are arrayed on its equator.

Begin with the 'chromosomes' lying at random on the spindle. Bring them from their position into line at the equator. To represent the splitting of the centromere, uncouple the 'chromatids' and provide each pipe cleaner with a separate bead (or provide each strand of beads with a piece of pipe cleaner).

Now move one pipe cleaner of each pair

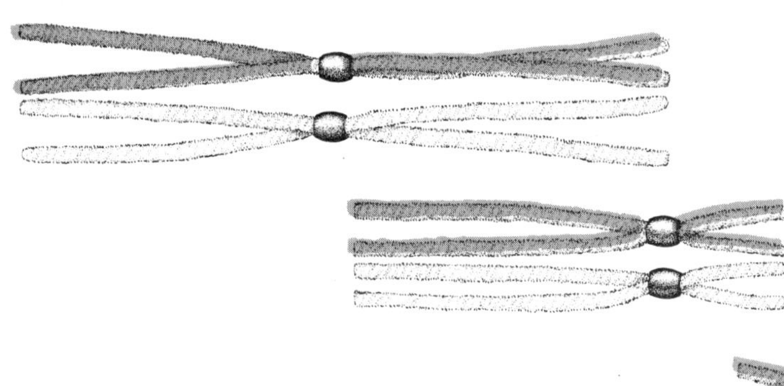

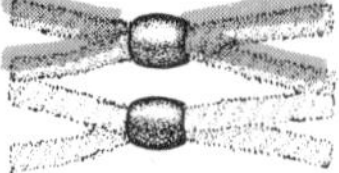

Fig. T-14

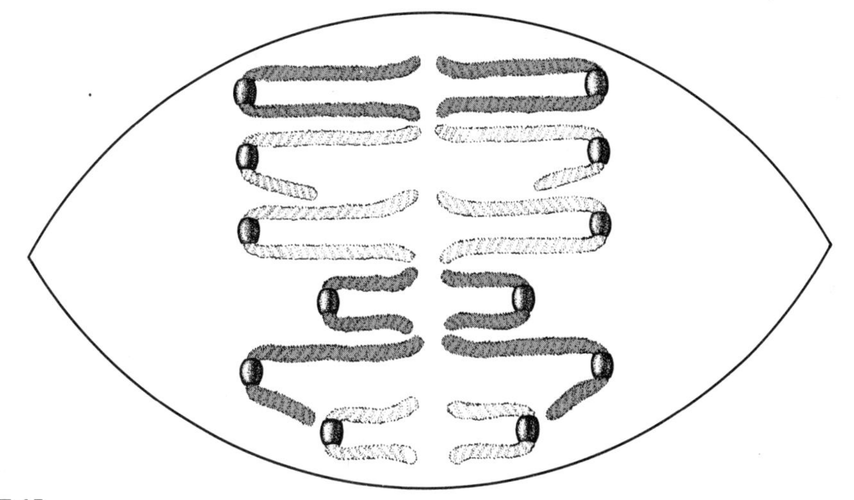

Fig. T-15

towards one pole of the spindle and the other towards the other pole of the spindle. If six pupils are lined up on one side of the table and six on the other side, all the pipe cleaners can be moved simultaneously, just as the chromosomes move in mitosis. The pipe cleaners should be bent at the bead to represent the shape of chromosomes as they migrate to the poles of the spindle (Fig. T-15). Finally, compare the makeup of the set of 'chromosomes' at each pole with that of the original set of 'chromosomes'.

AUDIOVISUAL MATERIALS

Films

The Cell, Structural Unit of Life. Coronet Films, available from Gateway Educational Films Ltd. 16 mm, colour, 10 min. (A short and simple film, useful as an introduction.)

Mitosis. Encyclopaedia Britannica Films. 16 mm, colour, 24 min. (Photomicrography shows the process of cell division in a living cell. Includes the effects of chemicals and radiation on mitosis.)

Mitosis: In Endosperm of Haemanthus katherinae. Ealing. Film Loop.

Mitosis. Commonwealth Scientific and Industrial Research Organization, distributed by Boulton-Hawker Films, on hire from EFVA. Colour, 9 min.

Cytoplasmic Streaming in Plant Cells. Ealing (R. Allen and M. Allen, Princeton University). Film Loop, colour, 4 min. 5 s.

An Inquiry: The Importance of the Nucleus. BSCS Inquiry Film Loop, available from John Murray. (What is the role of the nucleus? How would you test your ideas?)

Mitosis. BSCS Inquiry Film Loop, available from John Murray. (Attempts to lead the pupil to see the significance of mitotic events.)

TEACHER'S REFERENCE BOOKS

DARLINGTON, C. D., L. F. LA COUR. *The Handling of Chromosomes*. 4th edition. Allen & Unwin, 1962.

DODGE, J. D. *An Atlas of Biological Ultrastructure*. Arnold, 1968. (Illustrations of plant and animal cells, bacteria and viruses.)

KEMP, R. *Cell Division and Heredity*. Arnold, 1970.

LOCKWOOD, A. P. M. *The Membranes of Animal Cells*. Arnold, 1971.

SHAW, G. W. *Modern cytological techniques*. Modern Science Memoirs No. 40. Murray, 1959.

2
BIOENERGETICS

MAJOR IDEAS

1. Bioenergetics is the study of the chemistry of energy capture by organisms and energy transfer through the biosphere.
2. Chemical reactions in living cells are catalysed by enzymes, which are proteins synthesized by the cells. Enzymes are highly specific in function.
3. Cellular respiration is a chemical process in which large multicarbon molecules release their energy as they are converted to smaller molecules. Oxygen is required in this process, and carbon dioxide and water are products.
4. Fermentation processes release energy from multicarbon molecules without the need for oxygen but result in products that still contain usable energy.
5. Energy released by respiration or fermentation is temporarily stored in adenosine triphosphate, from which it is released as needed for the activities of cells.
6. One such activity is the synthesis of the complex organic substances that account for almost all the dry weight of cells. For the most part, these cell substances turn out to be the familiar food substances: carbohydrates, fats, and proteins, plus nucleic acids.
7. Man's understanding of photosynthesis began with the chemical revolution that brought about the downfall of the phlogiston theory and reached a basic formulation in the first half of the nineteenth century. In the twentieth century new methods, especially chromatography and isotope techniques. have revealed the enormous physicochemical complexity behind the simple term 'photosynthesis'.
8. The present understanding of photosynthesis may be summarized in three sets of reactions. Two involve the transfer of light energy into chemical energy; the third, not requiring light, incorporates carbon dioxide into organic compounds. It appears that only 'light capture' is unique to chlorophyll-bearing organisms.

PLANNING AHEAD

If you have no experience with the apparatus to be used in Investigation 3.1, it would be wise to set it up and test its operation. Collect some dormant twigs and if necessary store in a refrigerator for use in illustrating the structure of stems.

Obtain frogs for Investigation 4.1 and *Daphnia* for Investigation 4.3. If you do not have facilities for maintaining these animals in your laboratory, you will have to estimate closely the time at which you will require them and either collect at a suitable time or order from a supplier specifying a posting date on your order. *Daphnia* are more difficult than frogs to maintain after arrival from the supplier.

Have you obtained the *Zebrina* that you will require for Investigation 5.1?

GUIDELINES

You may feel that an understanding of biochemistry on the level treated in this book is insufficient. The pertinent arguments in rebuttal are given on pp. T5 and T6. Nevertheless, although the writers see no reason for leading the majority of school pupils deep into the terminological jungles of biochemistry, they certainly believe that the teacher should be well acquainted with the scheme of modern biochemistry and thus prepared to expand on the text with the occasional interested pupil or with science-oriented classes.

Though energy flow is continuous, it must be reduced to sequential steps for study. You must do what you can to clarify the separate steps for your pupils, but convey the final impression that energy *flows* through the biosphere.

Work on bioenergetics can be conveniently begun by setting up Investigation 2.1. This should stimulate thinking about biochemical processes. While work on the investigation proceeds, a first study assignment might be p. 29 (guide question 1) and 32–33 (guide question 3).

Following the discussion of results from Investigations 2.1 and 2.2 there is a rather long assignment (considering its complexity), pp. 34–41 (guide questions 4–9). Here considerable teacher explanation may be required, but only as much as you think suitable for the particular pupils in each class. A rough understanding of Fig. 2.10 may be all that can be achieved in many cases, and Investigation 2.3 may be assigned to a group of pupils for demonstration to the class.

Pp. 43–47 (guide questions 10–15) form another difficult assignment. Help will be required on structural formulae; they look quite formidable to beginners. You will find it useful to have molecular-model kits available; you may be able to borrow these from the chemistry department. Otherwise, quite serviceable 'exploded' models can be constructed from polystyrene balls and toothpicks. Avoid getting involved at this time in the genetic significance of nucleic acids.

Pp. 48–50 (guide questions 16–17) form a relatively simple prelude to Investigation 2.4. Finally, pp. 52–58 (guide questions 19–23) can be skimmed rather lightly if your classes begin to show signs of restlessness with biochemistry. However, some attention should be given to Investigation 2.5.

TEACHING NOTES

Life, Energy and Cells (p. 29)

Stress that the bulk of living things are alike in the way they store and utilize chemical energy. Plants and animals (as well as most protists) utilize the same chemicals and processes to obtain energy from their stored fuel.

INVESTIGATION 2.1

BIOENERGETICS: AN INTRODUCTORY VIEW
(pp. 30–32)

Materials

Any variety of pea will do. Beans may be substituted.

The jar should be filled with tap water and allowed to stand overnight so that the water will come to room temperature. Some volumeters come equipped with syringes on the adjustment tube. If you do not have syringes, you may insert dropper pipettes in the tubing above the spring clips. The syringes or dropper pipettes will allow you to adjust the position of the indicator drops in the pipettes after the stopper has been placed in the test-tube. Inexpensive plastic syringes are handy for placing the indicator drop in the pipette. Glass tubing drawn out to a fine point will also aid in doing this tricky job. The indicator solution should have a little detergent added to it so that the drop will hold together in the pipette. Food colouring can be used in place of eosin. Attach the pipettes so that the lower numbers are towards the volumeter. This gives negative numbers when zero readings are subtracted if the indicator drop moves towards the volumeter (as it should in the pipettes attached to respiring peas). Small-bore glass tubing may be attached to rulers if pipettes are not available.

Glass beads may be of any size, but the nearer to pea size the better.

When inserting glass tubing or thermometers into rubber stoppers, use a cork-boring set. (See p. T68.)

Procedure

A schedule is presented here to assist you in organizing this investigation:

Day 1: Special group assembles vacuum bottles and obtains two batches of peas, each equal to 200 cm^3 in volume. One batch of peas is covered with water, the other left dry. Other teams assemble volumeters and count out peas. One set of peas is soaked overnight, the other left dry. Volumeter jars are to be filled with tap water.

Day 2: Special group places peas in Vacuum Bottles A and B as directed and takes initial temperature readings. Other groups place peas and beads in volumeter tubes and make readings of the positions of indicator drops.

Day 3: Special group dismantles apparatus after taking final readings. Ask this group to record any odour noted in the vacuum bottles. (The lack of ethanol will be useful in comparing this investigation with 2.3 later.) Special group reports results to class. Other groups dismantle apparatus and plot data.

Great care must be taken in placing the potassium hydroxide on the cotton. The inside walls of the test-tubes must be thoroughly dry. Moisture coming in contact with potassium hydroxide liberates heat, causing an expansion of the air in the closed system. This upsets readings. The ability of potassium hydroxide to absorb CO_2 should be demonstrated.

Working with volumeters is sometimes discouraging. A number of factors have noticeable effects on confined gases (changes in temperature, changes in atmospheric pressure), and gravity can affect the indicator drop. But perhaps the greatest source of error is a system that is not airtight.

If possible, provide time for two or more sets of readings of the volumeter. You may want pupils to graph class averages rather than individual group results.

Discussion

The discussion should be based on the graphs produced by the groups. The chemistry involved here is simple: germinating seeds require oxygen to release the energy of their stored food. As oxygen is used by the germinating peas (dry peas use so little that, at most, only a slight change in the volume of the air will be noted), atmospheric pressure exceeds the pressure of the confined air, and the indicator drop is forced farther into the pipette.

Procedure

B. The glass beads are used to equalize the volume of solid matter in the three tubes.

C. The third tube is an overall control.

Studying the Data (p. 31)

A, B. If the pipettes are attached so that the zero is at the end next to the rubber tubing, the numbers recorded as the drop moves towards the volumeter are smaller than the initial rest-position number. Then subtracting the rest position from the readings produces negative numbers; these correspond to decreases in the volume of confined air. Positive numbers indicate that the confined air is expanding. Of course, if the zero on the pipette is distal to the volumeter (as seems to be the case in Fig. 2.1), the reverse is true: a positive number indicates volume decrease and a negative number indicates volume increase.

D. The tube containing dry peas and glass beads should perform much like the tube containing only glass beads.

E. Soaked peas use up oxygen, thus decreasing the volume of air in that chamber. Dry peas do not use enough oxygen to produce a similar change.

F. There is, of course, no evidence in this investigation to rule out nitrogen. See that the cocksure pupil is corrected in this regard.

G. Use of potassium hydroxide implies that something is likely to happen that would increase the volume of CO_2, thus masking volume change due to other air gases. Could it be that germinating seeds produce CO_2?

Conclusions (p. 32)

A. The amounts of substances are changing. Oxygen or nitrogen from the air is apparently being used by the germinating seeds, and it seems reasonable to suppose that CO_2 is produced.

B. If all goes well, the temperature of the germinating peas is higher than that of the dry peas. Thus, the germinating peas must liberate heat, a form of energy, while the dormant peas do not.

Energy-releasing Processes (in part) (pp. 32–33)

p. 32, ¶ 4. A little practice with the calculation of joules will help establish the concept. How many joules are absorbed by 20 g (20 cm^3) of water if the temperature of the water rises 3 degrees Celsius? A medium-sized apple contains about 200 kJ. By how many degrees would this much energy raise the temperature of one litre (1000 g) of water? Pupils, especially those interested in slimming and its attendant problems may not have seen joules in the literature. The energy content of foods is still referred to in calories in many books and magazines; thus some explanation of the relationship between calories and joules should be made at this point to avoid confusion.

p. 33, ¶ 3. Pupils who have encountered enzymes in a previous science course may have the idea that enzymes act only in digestion. This is a difficult idea to eradicate; Investigation 2.2 should help.

INVESTIGATION 2.2

CELLS: BIOCHEMICAL REACTIONS (pp. 33–34)

Materials

20 volumes (or 6%) hydrogen peroxide is available at the chemist. For use in this experiment, dilute to 10 volumes (or 3%) by adding an equal volume of water. It is better not to buy it until you are ready to use it—the fresher the better.

Use only fresh liver; it is important to be sure that it has not been frozen.

Fine sand intended for grinding is available from supply companies, but washed builder's sand will do.

It is not necessary for each group to have its own boiling water. Have a couple of litres of hot water ready at two or three points in the lab, and as the work begins bring these to boiling point.

Procedure

Directions to the pupil are quite simple, and you should not need to do much more than provide the equipment and materials. If everything is ready, the procedure can easily be carried out in only one period.

Discussion and Conclusions (p. 34)

No investigation in the course has tighter reasoning from observations to conclusion than this one. Be sure that you guide the discussion along lines of disciplined thought in order to take advantage of this construction.

A. Tube 1 establishes the lack of reactivity between MnO_2 and H_2O. This is necessary because the remaining tubes contain 97 per cent H_2O and only 3 per cent H_2O_2.

B. The only evidence is that there appears to be as much MnO_2 following the reaction as there was before.

C. If the pupil weighed the amount of MnO_2 added, and after the reaction reclaimed an equal amount of MnO_2, he might conclude that the MnO_2 was a catalyst rather than a reactant.

D. Both $H_2O_2 \rightarrow H_2 + O_2$, and $2H_2O_2 \rightarrow 2H_2O + O_2$ are logical possibilities for production of the gas evident in the reaction.

E. A glowing splint bursts into flame when placed in the mouth of a test-tube where oxygen is being produced, but if a mixture of H_2 and O_2 is produced, a small explosion occurs on ignition. You may have to supply some background at this point and repeat the operation of Tube 2 during the discussion so that the splint test may be applied.

F. If liver contains a substance that catalyses decomposition of H_2O_2, grinding might liberate more of the substance.

G. To demonstrate that sand has no effect upon H_2O_2.

H. The substance in liver that catalyses decomposition of H_2O_2 is destroyed by heat.

I. The catalyst in liver is made inactive by heating, but MnO_2 continues to act as a catalyst after being heated. In addition, of course, no black powder is visible in liver, but most pupils recognize this as weak evidence.

K. The H_2O_2-decomposing substance is found in plants as well as animals.

By this time most pupils will recognize that the substance decomposing H_2O_2 is probably an enzyme. The teacher may supply the information that enzymes are proteins and that proteins are sensitive to high temperatures.

Energy-releasing Processes (contd.) (pp. 34–41)

p. 34, ¶ 1. A distinction between breathing and respiration can be made. Breathing is the process that moves respiratory gases to and from respiratory organs such as lungs or gills; respiration is the exchange of O_2 and CO_2 and, by extension, the chemical process that involves this exchange in cells.

p. 35, Fig. 2.2. You will find molecular-model kits effective teaching aids in introducing pupils to structural formulae. Point out that the three-dimensional character of molecules such as glucose allows points of attachment for enzymes. The structural formulae given here and on following pages are, of course, not to be memorized. They are intended to show the complexity of even relatively simple organic compounds and to provide a mental image for differentiating large classes of such compounds.

p. 36, ¶ 2. ATP is a large molecule. When a phosphate group is hydrolysed from the parent molecule by enzymes, electrons throughout the molecule are allowed to assume lower energy states. Thus the electrons in ATP as a whole—not only at the position indicated by the wavy line—are of higher energy than are the electrons in ADP and the released phosphate group. It is the stabilizing of the entire ADP molecule by lowering the energy states of the electrons that is believed to account for the release of energy in the reaction:

$$ATP \rightarrow ADP + \text{(P)} + \text{energy}$$

The energy state of ATP is lowered by at least 29 000 joules per mole when ADP is produced from it. It has been calculated that a minimum of 2 500 000 molecules of ATP are broken down to ADP and phosphate per second during synthesis in a bacterial cell.

p. 37, Fig. 2.4. If your pupils are encountering chemical formulae for the first time, they may easily confuse the biochemist's abbreviations, such as ATP and ADP, for formulae. Even if pupils have some familiarity with formulae, it is as well to emphasize the distinction.

p. 37, Fig. 2.5. The diagram will have to be discussed in class. Help the pupil to trace the reactions, but do not dwell on specific steps. The point to make is that during chemical breakdown of a fairly large molecule, such as glucose, to small molecules of H_2O and CO_2, the energy of the larger molecule is transferred (as high-energy electrons) to $ADP + \text{(P)} \rightarrow ATP$. The coloured lines represent principal energy transfers within the system of reactions.

pp. 37–39. These pages explain Fig. 2.5. Electron carriers are the large organic molecules flavin adenine dinucleotide, FAD, and nicotinamide adenine dinucleotide, NAD. (In older works NAD is referred to as diphosphopyridine nucleotide, DPN.) Both FAD and NAD are derived from vitamins of the B group. Electrons are transferred from NAD and FAD to cytochromes on their way to the H^+ and O_2. Cytochrome actions can be blocked by minute amounts of cyanide, explaining the toxicity of that compound.

p. 40, Fig. 2.10. In reviewing this diagram, stress the one-way flow of energy in the biosphere. Refer pupils to Fig. 1.10 and 1.18, *WL*. Pupils do not automatically link their understanding of biochemistry to their understanding of ecology.

p. 41, Fig. 2.11. Fermentation begins with glycolysis, as does cellular respiration (upper left of Fig. 2.5). But in mammals lactic acid resulting from intensive work by skeletal muscle may be carried from muscle to liver and there oxidized aerobically or even regenerated into glucose.

p. 41, ¶ 2. Note that *Clostridium tetani* cells, like those of other strictly anaerobic organisms, do not possess the enzymes of the electron carrier system, and so they lack mitochondria (p. 40, paragraph 2).

INVESTIGATION 2.3

FERMENTATION (pp. 42–43)

The objective in this investigation is to provide the pupil with an opportunity to obtain some qualitative data on the most familiar kind of fermentation.

Materials

The most convenient size for the vacuum bottle is 1 pint. You will need to make the limewater a day or two before use. Mix 10 g of calcium hydroxide, $Ca(OH)_2$, in a litre of water at room temperature. Decant and use the saturated solution formed, but do not disturb the $Ca(OH)_2$ that settles out.

Make up the treacle solution to a relative density of approximately 1.30. Undiluted grape juice to which about 5 g of sugar per 100 cm^3 has been added may be used in place of the treacle solution.

Procedure

This investigation takes one period to set up, and portions of two other days for observations.

See Investigation 1.4, p. T68, for a method of inserting the thermometer into the rubber stopper.

Studying the Data (p. 42)

A, B. In addition to the temperature change, the odour of ethanol, if detected, is evidence of a chemical reaction, as is any change noted in the limewater. An abundance of foam is evidence of the formation of a gas. If yeast or bacteria are accidentally introduced into Bottle B, some reactions may be noted in it, but they are not likely to be as great as in Bottle A. Compare odour with that detected in the vacuum bottle containing germinating peas in Investigation 2.1.

C. Conditions are not strictly anaerobic, since air is trapped above the solution. Yeasts are facultative anaerobes, using oxygen if it is available and reverting to fermentation when necessary.

Conclusions (p. 43)

A. The variable is the yeast introduced into Bottle A.

B. Heat is produced as in Bottle A of Investigation 2.1.

C. There was no ethanol produced by the peas.

Syntheses (pp. 43–47)

Molecular models are useful throughout the teaching of this section.

p. 44, ¶ 1. If pupils use models of glucose and fructose molecules and attempt to combine them into a model of sucrose, they will eventually note that the equivalent of water—two H's and an O—has to be removed. This illustrates dehydration synthesis. But emphasize that enzyme mediation is necessary for the process to occur in cells.

p. 44, ¶ 4. Glycerol is an alcohol sold under the commercial name glycerine.

p. 44, ¶ 5. Working with models again demonstrates dehydration synthesis. If you call attention to the energy your pupils expend in making fat-molecule models, you may consolidate the idea that all syntheses require energy available from ATP. Keep the thought of the expenditure of energy in biosynthesis before the pupils all the time. This is the real point of taking up synthesis in a chapter dealing with bioenergetics.

p. 47, ¶ 2. Deoxyribose differs from ribose merely in having one less oxygen atom. Unfortunately the term 'base' appears here in association with nucleic *acids*. Pupils must be cautioned that in this context 'base' does not mean 'alkali'. It simply means 'that to which something else is attached'.

p. 47, ¶ 4. This is not the time to enlarge on DNA and its structure; this is done later when it has meaningful application. Here nucleic acids are merely a class of organic constitutents of living matter. At this point the most meaningful relationship is that between nucleic acids and ATP and ADP.

Photosynthesis (in part) (pp. 48–50)

From here on, the discussion deals with the cellular capture of external energy instead of the release and transformation of energy within organisms.

p. 48. This is another opportunity to emphasize the international character of science.

p. 49, Fig. 2.19. Pupils may be struck by the similarity in structure between chloroplasts and mitochondria. Though the physiology is different, the stepwise changing of one compound into another requires an extensive and intimate interaction of enzymes. A layered construction appears to suit this purpose.

A single granum is made up of alternating protein and lipid layers. In one protein layer are the enzymes needed for the 'light reaction' in photosynthesis. Below this, like the filling in a sandwich, is a lipid layer containing chlorophylls, carotenoids, and phospholipid molecules. The lower protein layer contains the enzymes of the 'dark reactions'.

INVESTIGATION 2.4

SEPARATION OF LEAF PIGMENTS (pp. 50–52)

Though more sophisticated separation methods are now widely employed by biochemists, paper chromatography remains useful. And as a demonstration of the basic principle of chromatography, the separation of leaf pigments is highly effective and meaningful to school pupils. The materials and equipment required for this investigation are easily obtained in quantity, the manipulations are simple, and the pupils' satisfaction with their results is great; therefore it is desirable to have very small groups, preferably pairs.

Materials

Use strips cut from a roll or large sheet of Whatman No. 1 filter paper. Care should be exercised in handling the paper so that fingerprints are kept to a minimum. The discs of filter paper ordinarily used in chemistry classes are too small to yield a strip of the required length.

Ordinary ether (ethoxyethane, or diethyl ether) *cannot be used.*

Theoretically, any leaves containing chlorophyll are usable, but some kinds are better than others. *Pelargonium* (household geranium) is good; but the pigments in spinach leaves are especially rich and easy to extract (frozen spinach is easily obtained).

Pipettes with suitably fine tips may be made by heating a piece of tubing, pulling it out to a small diameter, allowing it to cool, and cutting it in two at the middle of the constriction. The large end of each piece can be made to fit the rubber bulb from an ordinary medicine dropper by heating to softness and enlarging with the shank of a triangular file.

Procedure

One period is required to set up the chromatographic chamber and to prepare the leaf extract. An additional period is needed for developing the chromatogram, making observations, and discussing the results. If you do not have a double period, the assembly of the apparatus and extraction of the pigments may be done one day, and the development of the chromatogram the next. Store the pigment extract in a refrigerator. Keep the amount of propanone (acetone) very small to secure as concentrated a solution as possible.

Studying the Data (p. 52)

A. If good separation has been obtained, several bands of pigment can be seen. A band of yellow pigment is found close to the leading edge of the solvent (see Plate III, p. 57). It is likely to be the palest of the bands. It contains carotenes and has an R_f of almost 1.00. (R_f = distance of band from origin divided by distance of leading edge of solvent from origin.) Much lower on the paper, starting at about R_f 0.4, are, first, another yellow band (sometimes distinguishable as two separate bands) that contains xanthophylls; second, a band of bluish-green chlorophyll *a*; and a third, a band of yellowish-green chlorophyll *b*. (If the pigment extract has been allowed to stand for some time, a grey band may be seen above the xanthophylls; this is phaeophytin, a decomposition product of chlorophylls.)

D. Usually the chlorophylls are so abundant that other pigments are masked. But in some leaves, especially those of ornamental varieties developed by horticulturists (e.g. varieties of Japanese maples), red anthocyanins may mask the chlorophylls. Anthocyanins are soluble in hot water and insoluble in propanone (acetone), so they may be separated by differential solution.

E. Spinach leaves do not normally contain anthocyanins. However, since the mass of ground leaves is not likely to be left white by the process used in this investigation, pupils have little basis for answering this question except for a healthy scepticism.

F. Chlorophylls are rather unstable pigments. As they decompose, the other, more stable pigments in the leaf are unmasked. In addition, glucose in the leaf may be converted to anthocyanins.

G. They all started from the same point.

H. They all started at the same time (when the developer reached the pigment spot), and they all stopped at the same time (when the paper was removed from the tube and the developer dried).

I. Since they all moved for the same time but travelled different distances, they must have moved at different speeds. These differing speeds depend upon differences in adsorption to the molecules of the paper.

For well-known mixtures, such as leaf extract, identification of the major groups of substances (carotenes and xanthophylls are still mixtures) is sufficient on the basis of colour and R_f. Positive identification is made by cutting the bands apart, separately redissolving the substances, running absorption spectra in a spectrophotometer, and then comparing the curves with curves obtained from the pure substances.

Photosynthesis (contd.) (pp. 52–58)

Do not stress the terminology but emphasize the flow of energy—the bioenergetics. It is *not* intended that the pupil should memorize the details of biochemistry involved in the description of photosynthesis. At the least, however, presenting the whole sequence of chemicals in diagram form (Figs. 2.21, 2.22, and 2.23) should lead to some appreciation of the complexity of an apparently (p. 52) simple process, and of the work of biochemists who have used great ingenuity in unravelling the tangled threads of evidence. In addition, most pupils should retain the knowledge: that photosynthesis, as usually defined, consists of an initial light-trapping stage and a subsequent light-independent synthesis; that research in photosynthesis uses chromatography and analytical methods involving isotopes; and that knowledge has not yet reached the point where man is independent of plants. Even to accomplish these minimum aims, however, some details must be given; it is not enough merely to *state* that photosynthesis is complex.

p. 52, ¶ 1. You may find pupils saying that animals use respiration to obtain energy, but green plants use photosynthesis. This is true, but a misunderstanding often exists. Photosynthesis *is not* a green plant's way of obtaining energy from its stored food. Point out the difference in the source of energy in photosynthesis and respiration.

p. 57, Plate III. Point out that the horizontal axis of this graph represents a segment of Fig. 1.9, *WL*. Plate III, p. 57, gives a partial answer to the question. Why are plants green? Since chlorophylls absorb most light in the blue and the red ends of the spectrum, the green (mixed with some yellow), which chlorophylls absorb least, is reflected to our eyes. The figure does not, of course, give any clue to the reason chlorophyll fails to utilize energy in the green portion of the spectrum.

p. 53, ¶ 6. Remind pupils that ATP, ADP, PGAL, NADP and RuDP, encountered here and later, are abbreviations rather than chemical formulae. Unfortunately, the situation is somewhat complicated by $NADPH_2$, where the H_2 does indicate two hydrogen atoms.

p. 54, Fig. 2.21. The electron carriers in photosynthesis are cytochromes (one is the same as used in respiration) and ferredoxin. Note that this diagram shows only one of the ways in which energy of sunlight is transferred to ATP in the 'light reactions' of photosynthesis.

p. 56, Fig. 2.23. This sequence transfers the energy stored in ATP and $NADPH_2$ to phosphoglyceraldehyde, a three-carbon compound from which glucose (by now a familiar molecule to your pupils) can be made.

p. 56, ¶ 2. Literally speaking, the 'dark reactions' are not a part of *photo*synthesis. They might occur in organisms that do not contain chlorophyll, and some biochemists think they do. The end point of photosynthesis is a matter of convention. We end with PGAL; often the production of glucose is considered the end, and for some purposes, the production of starch is considered a test for photosynthesis (Investigation 3.2).

INVESTIGATION 2.5

PHOTOSYNTHETIC RATE (pp. 58–59)

Materials

Large, healthy stalks of *Elodea* are necessary for this investigation.

The reflectors on many flexible lamps do not accommodate 150 watt bulbs. You may need to experiment with improvised reflectors. Heavy aluminium foil is a good material for this.

Procedure

If the water bath is set up a day ahead, most of the air and chlorine will escape. Otherwise, gases escaping from the water may obscure O_2 bubbles.

Discussion (p. 59)

There are only three points to be graphed. Pupils have a tendency to bunch these points. Be sure the entire available width of the grid is used for the horizontal scale.

D. The age and condition of the plant are important considerations. Though the pupils have not yet studied leaf structure, some may know that pollutants in water and air can clog the stomata through which gases are exchanged. CO_2 must be present. The temperature of the environment is also important.

E. CO_2 is supplied by the $NaHCO_3$ in a sequence of reactions that may be summarized:

$$H^+ + HCO_3^- \rightleftharpoons H_2CO_3 \rightleftharpoons H_2O + CO_2$$

Temperature is controlled and healthy *Elodea* is used.

F. Organisms in the water capable of producing O_2 by photosynthesis, and organisms capable of using the O_2 produced, are uncontrolled factors. Filtering the aquarium- or pond-water would reduce this variable.

GUIDE QUESTIONS (p. 60)

4. You may want a pupil to report on bomb calorimeters, in which total combustion of a substance occurs, permitting accurate measurement of the heat energy produced.

15. Ribose, phosphate groups, and a base (adenine in the case of ATP) are common to both.

PROBLEMS (p. 61)

2. If nitrate reductase can provide a ready source of nitrogen for the amine group of amino acids, then conceivably artificially produced proteins could be made to augment the diet of man and his domestic animals. This source of food would circumvent the slow nitrogen cycle.

3. The point here is that volume increases much faster than surface area. All the materials utilized and discarded by the volume of a cell must enter via the surface; as a result, the surface/volume ratio sets a practical limit to the size of a physiologically active cell. A sphere is the extreme demonstration of surface/volume ratio; flattening and elongation reduce the disparity. See C. P. Swanson, *The Cell*, 2nd edition (Prentice-Hall, 1964).

4. The oxygen debt arises when ATP is used faster than it can be restored through aerobic respiration. The limiting factor is the rate at which oxygen can be delivered to the cell to act as electron acceptor in the last part of respiration. When oxygen is not available, anaerobic energy release occurs, leaving lactic acid. Following exertion, heavy breathing is required to supply oxygen needed for final oxidation of lactic acid.

5. Enzymes put amino acid building blocks in sequences characteristic of wheat physiology. Through digestion, wheat proteins are reduced to amino acids and these are then synthesized into new proteins characteristic of human physiology. Unfortunately, wheat proteins do not contain the various amino acids in proper quantities to synthesize human proteins adequately, and human physiology is deficient in ability to synthesize amino acids; therefore wheat is not a sufficient source of protein in human diet (see p. 46, paragraph 2).

6. By rapid photosynthesis the by-product oxygen can be supplied in quantities sufficient to meet a plant's needs. The increased photosynthesis is a homeostatic mechanism that would tend to maintain the equilibrium that normally exists between CO_2 and O_2 in our atmosphere. This tends to confirm Oparin's assumption that the primitive atmosphere was deficient in O_2 and the consequent idea that photosynthetic organisms changed the composition of the atmosphere as they became numerous.

7. CO_2 in the atmosphere produces a so-called 'greenhouse' effect: radiant energy from the sun penetrates the atmosphere to reach the hydrosphere and lithosphere, but the radiation of heat from the Earth is hindered. If there were higher concentrations of CO_2 during the Carboniferous, this might explain the tropic-like conditions then prevailing in vast areas of middle and even high latitudes. An abundance of CO_2 would also have promoted the rapid photosynthesis that evidently prevailed in coal age forests. The Industrial Revolution, dating from the late eighteenth century, has returned much of the carbon in coal beds to the atmosphere in the form of CO_2. Thus there has been a steady increase in the CO_2 concentration in our atmosphere. (See p. 171, paragraph 1, *MHE*.)

8. See note to p. 56, paragraph 2 (p. T127).

9. Once amino acids were available, enzymes could be produced by dehydration synthesis. Any such enzyme that could catalyse reactions using radiant energy would have given proto-organisms that possessed it a distinct advantage since radiant energy was much more abundant than the declining heat and electrical energy resources of the ageing Earth.

SUPPLEMENTARY MATERIALS

ADDITIONAL PROBLEMS

1. Although they are built of the same amino acids, the proteins in the cells of a bean plant are different from those in the cells of a dog. How can you explain this? [*This involves the same reasoning as problem 5, p. 61.*]

2. Why does an enzyme catalyse only a specific reaction rather than many different reactions? [*This involves current theory of enzyme action. See G. G. Simpson and W. S. Beck*, Life: An Introduction to Biology, *2nd edition (Routledge & Kegan Paul, 1965), pp. 59–61.*]

3. Most consumers, including man, are unable to synthesize many of the substances needed for their metabolism. For example, man cannot synthesize many of the amino acids and some of the essential parts of the enzymes involved in respiration. How does this situation complicate the nutrition of such organisms? How can you explain an inability to synthesize a substance? [*Such organisms require more than an energy source and inorganic materials. They must also take in with their food the molecules they cannot synthesize. Some of these, precursors of enzymes, are called vitamins. Synthetic inabilities are due to lack of the necessary enzymes.*]

4. When isolated chloroplasts are used in research on photosynthesis, what assumption must be made? [*It must be assumed that all of the enzymes required are located in the chloroplasts.*]

5. ATP from the cells of plants has been used to energize the life activities of animal cells. What further question does this suggest to you? [*Perhaps the most obvious is: Can ATP derived*

from animal cells be used to energize the dark reactions of photosynthesis in plant cells?]

AUDIOVISUAL MATERIALS

Films

Biochemistry and Molecular Structure. A Chem-Study Film. University of California, available through Guild Sound & Vision Ltd. Colour, 13 min. (An excellent background film.)

Pattern of Energy Transfer. McGraw-Hill Publishing Co. Colour, 28 min. (Considers energy transfer from sunlight to utilization in cell metabolism.)

Cell Biology: Life Functions. Coronet Films, available through Gateway Educational Films. Colour, 19 min. (Considers both physical and chemical functions. Animation helps to explain some organic chemistry.)

Photosynthesis. Encyclopaedia Britannica Films. Colour, 21 min. (Laboratory demonstrations are used to show how the process of photosynthesis is studied.)

Measuring Rate of Photosynthesis. J. W. Kimball, Phillips Academy, Andover, Mass., available from Ealing Scientific Ltd. Colour, 4 min. (Discs of bean leaf are floated on $NaHCO_3$ solution. When oxygen is removed from the leaf, it sinks. As irradiated discs produce oxygen through photosynthesis, they float to the top.)

Photosynthetic Fixation of Carbon Dioxide, Parts I and II. Iowa State University, available from Ealing Scientific Ltd. Colour, 7 min. (Carbon-14 is used as a tracer in *Coleus* leaf photosynthesis.)

TEACHER'S REFERENCE BOOKS

BARON, W. M. M. *Organization in Plants.* 2nd edition. Arnold, 1967. (Plant pigments and other aspects of photosynthesis.)

BRYANT, T. C. *The Biology of Respiration.* Arnold, 1971.

CHANGEUX, J. P. 'The Control of Biochemical Reactions', *Scientific American,* April 1965, p. 36.

CLARK, B. F. C., K. A. MARCKER. 'How Proteins Start', *Scientific American.* January 1968, p. 36.

DAWKINS, M. J. R., D. HULL. 'The Production of Heat by Fat', *Scientific American,* August 1965. P. 62.

GREEN, D. E. 'The Mitochondrion', *Scientific American,* January 1964, p. 63.

MORRIS, J. G. *A Biologist's Physical Chemistry.* Arnold, 1968.

STONEMAN, C. F. *Nuffield Advanced Biological Science Topic Review—Photosynthesis.* Penguin, 1970.

WOOD, D. D. *Principles of Animal Physiology.* Arnold, 1968.

3

THE FUNCTIONING PLANT

MAJOR IDEAS

1. The growth and development of plants depend upon use of nutrients, transformation of energy, internal circulation of materials, synthesis of new molecules, and chemical controls, much as do the growth and development of animals.
2. A high degree of differentiation of tissues and organs is characteristic of the most abundant and conspicuous land plants.
3. As specialized organs of photosynthesis, leaves exhibit (both externally and internally) close correlation between form and function.
4. The complementing of form and function is as conspicuous in roots as in leaves and is only a little less so in stems.
5. In the majority of land plants, water is constantly being lost (principally through leaves) and is constantly being absorbed from soil (through roots). In addition to performing the function of absorption, roots usually furnish plants with anchorage and with a site for the storage of food.
6. The complex microscopic structure of stems is primarily associated with conduction of liquids in the plant body. As yet, biophysical explanations of conduction do not completely account for all observations.
7. In contrast to animal growth, the growth of multicellular plants is a result of mitotic activity in persistently undifferentiated tissues —meristems.
8. After plant cells have been formed by mitosis, their increase in size is, in many cases, under the control of auxins. Auxin distribution is influenced by gravity and light; differential distribution of auxin results in changes in the direction of stem elongation.
9. Because they are the most conspicuous plants and the most important ones to man, tracheophytes receive the major emphasis in studies of plant structure and function. But all plants have varying synthetic powers—even the fungi, which lack photosynthesis—and they all exchange substances with their environment.

PLANNING AHEAD

If you have delayed obtaining frogs for Investigation 4.1 and *Daphnia* for Investigation 4.3, this is the last possible moment to order them. To avoid a last-minute rush, prepare the hydrochloric acid dilutions and the sucrose, ethanoic (acetic) acid, quinine sulphate, and sodium chloride solutions required for Investigations 4.2 and 4.4. However, the solutions for Procedure C of Investigation 4.4 must be prepared just before use. Quinine sulphate may be difficult to obtain.

If you have a greenhouse, you should be propagating quantities of *Zebrina* for Investigation 5.1 and *Coleus* for Investigation 1.1, *MHE.* If you do not, you should have made arrangements for obtaining these materials at the time when they will be needed.

In many localities fertile chicken eggs are not easy to obtain. You might, therefore, begin to look for egg sources for Investigation 1.3, *MHE.*

GUIDELINES

By and large, secondary school pupils are not greatly enamoured of plants. This makes it desirable for Chapter 3 to have a beginning that will capture attention, if not engender enthusiasm. You undoubtedly know best how to do this with your pupils, but here are a few possibilities:

1. Begin by setting up Investigation 3.3. Many pupils develop special interest in plants they have themselves nurtured, and thereby become involved in plant study. Further, this approach allows you to finish this long investigation before you must shift attention to the more fascinating animals. Nothing in the earlier parts of the procedure requires previous study of the text materials.
2. Begin by displaying a variety of plants, including amongst the usual kinds—such as *Pelargonium* and *Coleus*—some that exhibit special adaptations. Some suggestions for the latter: *Aloe* or *Agave* (with thick fleshy leaves that resist desiccation), *Kalanchoe* (plantlets on leaf margins), *Monstera* (perforated leaves), *Maranta* (leaf enfolding at night), and, of course, various cacti. For information on looking after the plants listed and others consult *Pot Plants*, Bulletin 112, Ministry of Agriculture, Fisheries and Food (HMSO, 1969).

Parts of this chapter may be quite familiar to pupils and teachers. Insofar as material is familiar, no learning occurs, of course; you must not allow nostalgia to induce unprofitable lingering. If pupils are already well acquainted with macroscopic anatomy of vascular plants, attention should be focused on microscopic anatomy and, even more, on physiology. Although morphology and physiology are complementary, it is possible to approach the systems level of biology either by way of structure, which is then elucidated in terms of function, or by way of function, which is then explored in the light of observable structure. On the secondary school level the second approach seems preferable wherever a happy compromise between the two extremes cannot be arranged.

Depending upon the background of pupils and the extent of their need for concrete references for their thinking, the teacher may introduce some purely illustrative laboratory work. A session or two devoted to observing the microstructure of plant organs may be desirable, preferably beginning with hand-sectioned materials.

A reasonable plan of study assignments for the chapter is: first, pp. 63–68 (guide questions 1–4), to which Investigations 3.1 and 3.2, concerning leaf functions, properly belong; second, pp. 72–84 (guide questions 5–13), with which some laboratory investigation of microscopic anatomy may be included; third, pp. 84–89 and 91–93 (guide questions 14–18) taken together while work on Investigation 3.3 is being concluded.

TEACHING NOTES

Vascular Plants: Leaves (pp. 63–68)

p. 63, ¶ 2. Reviewing Chapter 3, *PLW*, p. 78 and pp. 93–94, may help pupils recall the importance of phytoplankton to productivity in the biosphere.

p. 64, Fig. 3.2. This is a wide open question. Examples: many ferns, clovers, vetches, ash, etc.

p. 65, Fig. 3.3. Examples of differences in lower compared to the upper leaf: thinner, one layer of palisade cells, palisade cells shorter. The term 'palisade' need not be used.

p. 66, Fig. 3.4. Have pupils compare the epidermis of the leaves shown here with the sketch of *Elodea* leaf epidermis drawn for Investigation 1.1. What explanation can be given for any difference noted?

p. 67, ¶ 2. Respiration and photosynthesis both play an important role in the CO_2 supply. When CO_2 accumulates, pH shows a tendency to fall. This favours formation of starch from soluble sugar, reducing turgor and resulting in narrowing of the aperture. When a leaf is illuminated, photosynthesis uses up CO_2, the pH rises, and hydrolysis of starch is favoured. Water then enters the guard cell, increasing turgor, and the stomatal aperture widens.

INVESTIGATION 3.1

TRANSPIRATION (pp. 68–69)

This study quantitatively investigates the validity of a pupil-formulated hypothesis. It *can* be done as a demonstration set up by a single group. But if the rather simple apparatus is obtainable in sufficient quantity, work by a number of groups is much to be preferred. You then have another opportunity to point out the effect of pooling results from replication. Groups of four pupils should be able to work efficiently. Some pupils are disappointed by negative data; try not to assign them to the group for the control set-up.

Materials

If a number of groups are to work, all materials and equipment must be conveniently arranged when pupils enter the classroom, so they can pick them up easily as needed. Ideally, the leafy stems should be woody so that they can be forced through the stopper holes without damage. If leafy woody plants are not available at the time this work is to be done, tomato or sunflower plants grown indoors can be used. Be sure the potted plants have been well watered and have not wilted. The stems must be selected to fit the holes in the rubber stoppers and inserted with care.

If shoots are to be taken from different kinds of plants, some comparisons can be made among species, but the value of replication is lost. The bore of glass *and* rubber tubing should be small.

The water in the flasks must be at room temperature; otherwise, expansion caused by warming will make the water column move upwards. The flasks may be filled before class time and allowed to come to room temperature, or containers of water at room temperature may be provided. Sinks filled with water that has come to room temperature will eliminate the need for large battery jars or bowls. When pupils are cutting the stems and placing them in the stoppers, they should keep the stems below water level all the time. When a flask is filled, the water level must come up to the top of the neck so that the insertion of the stopper will force the water out, leaving no air space.

If paraffin wax is to be used to seal connections, have it molten. Small brushes are handy to use as applicators. Vacuum-pump compound has been found to be a good sealer. Vaseline is unsatisfactory.

If pipettes are not available, lengths of small-bore glass tubing may be used. With a rubber band, fasten a metric rule behind each piece of tubing. You may determine the volume of the tubing per centimetre of length by filling a piece with water and emptying this water into a small graduated cylinder. (The last of the water may need to be blown out.) Then divide the volume of water (in cm^3) by the length of the tube (in cm) to find the cm^3 per centimetre.

Procedure

If pupils did not use pipettes in Investigation 2.1, practice in reading them should be given. Pupils may also have to be shown how to fill pipettes. Some teachers have found 5 cm^3 pipettes more satisfactory than the 1 cm^3 size, but 5 cm^3 pipettes require a longer time between successive readings.

Be sure pupils keep the cut end of the plant under water at all times. Also, the rubber tubing must be filled completely when the pipette (also filled with water) is inserted into it.

Before connections are sealed, each of the areas of connection should be dried with a piece of cotton wool or paper tissue.

Uniformity among groups can be enhanced if, when all set-ups are ready, an interval timer, set for two-minute intervals, is used as a signal for all groups to record their readings.

Studying the Data (p. 69)

Groups may record their data on a large chart on the blackboard. This will show distinctly the readings from the control group. Then the average for all the groups can be used for plotting the graph, reducing experimental error.

If the leafy shoots are taken from a shrub near the school or from a greenhouse, an estimate of water loss from the whole plant may be made. Estimate the number of leaves on the shrub, divide by the number of leaves on the experimental shoot, and multiply by the volume of water lost in the pipette during the first ten minutes. You can multiply this figure by 6 to obtain the water loss per hour. Using the data for the second ten minutes gives an estimate of water loss from the plant on a windy day. This procedure provides an experience that helps to explain how data such as those given on p. 68 are obtained.

In all discussions of transpiration, make certain that pupils speak only of movement of water through the plant system, not movement of solute molecules. Water and solutes move independently of each other; they may even move in opposite directions.

A. *If* all connections are tight, water can be lost from the open end of the pipette (a very small free surface and therefore a negligible source of loss) and from the plant. Previous study of leaf structure suggests that the major loss from the plant might be through the leaves. Again, *if* all connections are tight, the control set-up shows little loss of water. Therefore, greater loss in the experimental set-ups is linked to the factor by which these set-ups differed from the control—the plant.

B. The water 'disappeared into thin air'; that is, it evaporated.

C. It is possible that vapour may condense on the inside of the plastic bags.

D. The amount of water loss can be read directly from the pipettes. If plain glass tubing has been used, see p. T132, paragraph 5. Note that the question concerns loss from the *apparatus*; this is loss from the plant only if all connections are tight.

E. The variable is the movement of air around this plant.

Some additional questions that may be used in discussion:

1. What other environmental factors might influence the rate of transpiration? [*Chiefly humidity and wind; to some extent cloud cover. Soil water supply has little effect except when extremely deficient.*]
2. What variations in plant structure might influence the rate of transpiration? [*Size of leaves, thickness of cuticle, number of stomata—amongst others.*]
3. When a plant is transplanted, its ability to obtain water may be impaired by the destruction of roots. On the basis of the results from this investigation, what suggestions might be made for treating a transplant to reduce dangers of wilting through excessive loss of water? [*Reduce number of leaves by pruning; protect from wind; increase humidity. All are used by horticulturists.*]

INVESTIGATION 3.2

STOMATA AND PHOTOSYNTHESIS (pp. 70–71)

The procedures in this investigation, especially Part B, are frequently presented by teacher demonstration. This requires fewer materials and less equipment, but it robs the pupils of involvement in the study. Give them the chance to do the work, and it will be that much more meaningful to them.

The work on this investigation must extend over several days, e.g. if Procedure A and the original setting up of the plants for Procedure B can be done on a Friday, the first part of Procedure B can then be done on the following Monday. Then on Thursday the final part of Procedure B can be done.

Materials

Be sure leaves are kept from wilting. You need small potted plants, with abundant but not large leaves. *Coleus* or *Pelargonium* is suitable, but well-grown bean plants may also be used.

For the iodine solution, see instructions for Investigation 1.4, p. T22.

Procedure

You may need to assist some pupils in reviewing the computation of field-of-view area (Investigation 1.3, p. 13, *WL*). Be sure pupils do not forget to check for watering of plants kept in the dark.

Two or three beakers of boiling water at strategic places in the laboratory may suffice for all groups. Note well the caution on heating of ethanol.

Studying the Data

Procedure A (p. 70)

B. Many counts must be made and the counts averaged. Care should also be taken that counts are made on comparable leaves of the two plants: age of plant and leaf, position of leaves with respect to sun and shade, etc.

C. Explanations may be based on the discussion on p. 67, paragraph 2, but all must be inconclusive.

Procedure B (p. 71)

A. The first set of tests established a condition in plants exposed to light—photosynthesizing plants.

B. You assume that the presence of starch is an indication of photosynthesis. This is not necessarily true. Starch might be synthesized from glucose transported into a leaf from elsewhere in the plant. On the other hand, the synthesis of multicarbon compounds need not necessarily be carried as far as starch; it *might* stop with glucose or even phosphoglyceraldehyde. Nevertheless, starch in leaves is commonly regarded as an indication of photosynthesis and lack of starch as an indication of lack of photosynthesis.

Vascular Plants: Roots (pp. 72–76)

p. 72, ¶ 1. In addition to the primary root systems, there are roots that arise in many plants from aerial portions of the stem and from rhizomes, corms, and cuttings. These are adventitious roots.

p. 73, Fig. 3.8. In the photograph the shoot has still not emerged from the seed coat. Emergence of the root first is the rule; pupils may have observed this for themselves in Investigation 1.2, *WL*, and they will see it again in Investigation 2.1, *MHE*.

p. 74, Fig. 3.9. Have pupils suggest how the structure of the root hairs and the structure of other cells shown here (compare with root diagram in Fig. 3.10) fit the functions each performs.

p. 75, Fig. 3.10. Some of the labelled structures are mentioned briefly or not at all in the text. These may raise pupil questions that can be the basis for reference to more advanced books. You may want to employ 'pericycle' and 'endodermis' if you undertake to explain the origin of secondary roots.

p. 76, Fig. 3.11. Compare with Fig. 3.7.

Vascular Plants: Stems (pp. 76–84)

p. 77, ¶ 3. A variety of dormant woody stems are effective for comparison with the printed illustration (Fig. 3.14) when the pupils are studying macroscopic structure. If your season is early, collect the twigs before they begin developing and store them in a refrigerator. Bud packing can be examined without the aid of a lens if the end bud of horse chestnut is opened with a dissecting needle.

p. 79, Fig. 3.15. Note the asymmetrical growth of the top log; the tree may have had competitors on the right side. The bottom tree was the youngest. It was nineteen years old when felled. Sample slices of small tree trunks are not difficult to collect. They make a good interest-retaining device; pass out a selection and ask pupils to report what they can 'read' from them.

pp. 78–80. Watch for flagging pupil interest. Some microscope work with prepared slides may help. The only terminology essential for the physiology to come is 'cambium', 'xylem', and 'phloem'.

p. 83, ¶ 3. Xylem sometimes contains large amounts of dissolved food. This is so when material stored in roots or underground stems is being transported to the upper portions of the plant body; this is especially true after a period in which photosynthetic activity of the plant has been reduced or stopped completely—for example, during early spring. Sugar-rich maple sap comes from xylem.

p. 84, Fig. 3.19. This is another example of adaptive convergence (pp. 135–6, *PLW*): the stems store water, as do those of many cacti.

Vascular Plants: Growth (pp. 84–89)

p. 85, Fig. 3.22. The base below the girdle and the tree roots contained enough food for survival up to the present time. The girdle has not cut off the supply of water and minerals, which move through the xylem, and so the crown of the tree has continued growth. However, food can no longer reach the roots and base because the phloem has been destroyed by the girdle, and so they will ultimately die; this of course will be followed by death of the plant.

pp. 86–89. This is another example of a historical approach to a field of biological study. It is also an excellent opportunity for a selected group of pupils to demonstrate the effect of auxins and gibberellins on plant growth.

p. 86, Fig. 3.23. Greater diameter, additional growth ring, relatively smaller size of pith, more cracking of bark.

p. 89, Fig. 3.27. The picture really illustrates the answer. Holly cuttings were once extremely difficult to root, and so varieties that would not come true from seed were difficult to propagate. Growth hormones have changed this, and greatly increased supplies of marketable hollies have resulted.

INVESTIGATION 3.3

RATE OF GROWTH: LEAVES (pp. 90–91)

Materials

Bean seeds are especially susceptible to mould, so it is necessary to soak them in fungicide solution. See p. T20 for information on fungicides.

Almost any kind of wooden box 8 to 15 cm deep may be substituted for seed boxes. Germination trays can also be made from cardboard shirt, shoe, or sweater boxes. To make these waterproof, line them with polythene film. This film, 2 to 4 mm thick, should be shaped to fit the bottom and continue up and over the sides. After folding the edges as in wrapping a package, staple near the top edge of the sides so that the film is not punctured in the parts where moisture is to be retained.

Procedure

If this work is started on a Monday (with planting to be done on Tuesday), the days scheduled for measuring will be school days. Of course, adjustments can be made in the schedule, but the intervals between measurements should be about equal.

Discussion

Making the comparisons suggested ought to result in recognition of a certain similarity in these curves—the sigmoid shape characteristic of the generalized growth curve. Many factors may alter or obscure this curve. In Investigation 2.1, *WL*, the curve does not flatten out; in Investigation 2.2, *WL*, the curve usually declines; in Investigation 2.3, *WL* (field mouse) the curve approaches a fluctuating equilibrium. None of these patterns is likely to occur in the growth curve of an individual or in a part of an individual. However, each has the general characteristic of beginning with a gentle slope that steadily steepens.

Non-vascular Plants (pp. 91–93)

Of chief significance here are the physiological differences necessitated by the absence of xylem and phloem tissues in non-vascular plants. This section obviously dangles—almost appearing to be an afterthought. But under some circumstances, such as in a school situated near a coast studded with tide pools, you might want to make much more of it. At the least, you should provide plentiful examples: a variety of living mosses, some fresh-water algae in aquaria, good herbarium sheets of larger marine algae, and some prepared microscope slides of planktonic algae. And refer pupils to the Appendix of *DLT* again.

p. 92, ¶ 1. Ask pupils to avoid the use of the terms 'root', 'stem', and 'leaf' when referring to non-vascular plants. Botanists do not always do so, but terminological looseness that is permitted the initiated only confuses the learner.

PROBLEMS (p. 94)

1. Leaves of water lilies float because of their highly developed spongy tissue. Their stems and long leaf petioles are fragile and soft, lacking the abundance of fibrous tissue characteristic of terrestrial tracheophytes; this correlates with the fact that the weight of the plant is largely supported by water. Water being directly available to all plant parts, the roots are not as extended or possessed of as many root hairs as are the roots of land plants.

2. Most parasitic or saprophytic tracheophytes have no leaves or only very small ones. Tall stems from which leaves spread upwards and outwards to receive sunlight being of no advantage, these plants are in most cases small, though some are twiners. Roots function principally as organs that penetrate host organisms and absorb food from their tissues or (in saprophytes) from dead organic materials. Sometimes special absorptive organs are produced on twining stems.

3. The point of this question is that root cells are alive and carry on respiration at all times. Therefore, gas exchange between roots and soil air (or air dissolved in soil water) is primarily an intake of O_2 and an output of CO_2, just as in aerobic consumers—or in leaves during hours of darkness.

4. Cultivation breaks soil particles apart, increasing the size of the air spaces and thus reducing the movement of water upwards by capillarity. At the same time, it allows air to penetrate into the soil.

In some countries dry farming is practised in areas where annual precipitation is less than twenty inches and irrigation is impracticable. In a given strip of land, a crop is planted only every second or third year. Between crop years the soil is kept tilled to achieve maximum penetration of precipitation and minimum loss through capillarity and transpiration. Often contouring is used to reduce run-off of rain.

5. The attached end of the fence is the same height above the ground as it was ten years ago. The tree grew in height, but only the apical meristems moved upwards; all tissues to which the fence was attached remained where they were when formed.

6. (*a*) An annual ring consists of an inner layer of large cells formed by the cambium during rapid growth in the spring of the year, and an outer (not necessarily distinct) layer of smaller cells formed during slower growth in the summer. The thickness of the cell walls differs in these two layers also. The result is a light–dark pattern (see Fig. 3.15), the combination being the ring. (*b*) A wet year causes an annual ring to be wide; a dry year results in a much narrower ring. (*c*) Drought or defoliation by insects may temporarily slow the growth, producing a band of small, thick-walled cells, followed by a band of large, thin-walled cells produced during late summer rains or after appearance of a new crop of leaves. This pattern produces 'false' growth rings, suggesting two growing seasons. (*d*) Especially in arid and semi-arid climates, past climatic patterns are reflected in the patterns of growth rings in trees. This makes possible the determination of dates of events in former periods by comparative study of the sequence of growth rings in trees and aged wood. (*e*) Phloem tissue cells are short lived; the functional life of the sieve tubes is usually a single growing season. The cells are soon crushed and broken by the outward growth of other stem tissues and are eventually sloughed off as part of the old outer bark. (*f*) Indistinct rings are produced in the bark by cork cambium and by growth of phloem from the principal cambium. But these rings are compressed by the outward growth of the wood. Eventually outermost layers of dead cells, unable to increase in circumference, break, forming a scaly, ridged or roughened bark.

7. The plant might be non-photosynthetic. If it is a green plant, light energy or proper wavelengths may not be present. Or amounts of water or mineral nutrients may be insufficient. Or temperatures may be too low.

8. Comparatively large amounts of water are lost through the apple leaves, but this occurs only during the season when living leaves are on the trees. Comparatively small amounts of water are lost through the pine needles, but to some extent this occurs throughout the year.

SUPPLEMENTARY MATERIALS

ADDITIONAL INVESTIGATION

CHEMICAL ACTION IN A PLANT

Purpose

This is an investigation of chemical action produced by living plant tissues.

Materials and Equipment (for each team)

Maize grains, 6
Paper towels, 2
Beaker, 2
'Formalin–acetic alcohol' (FAA)
Glass-marking crayon
Petri dishes containing sterile starch agar, 3
Petri dishes containing sterile plain agar
Scalpel
Forceps
Iodine solution
Medicine dropper
Clinistix

Procedure

Day 1. Soak a paper towel in water. Wrap three maize grains in the towel. Place an inverted beaker over the towel.

Day 3. Remove the grains from the towel and place them in a beaker containing 'formalin–acetic alcohol', a fluid that kills and preserves the grains. Soak another paper towel in water. Wrap three more maize grains in the towel. Place an inverted beaker over the towel.

Day 5. Number three petri dishes containing sterile starch agar, 1, 2, and 3. Number a dish containing sterile plain agar 4.

Take two maize grains from the paper towel. Using a scalpel, cut the grains longitudinally and parallel to the flat surfaces (Fig. T-16). Using the forceps, carefully place each half grain (cut surface downwards) on the starch agar in Dish 1. (See Figs. 3.9 and 3.10, *PLW*, for technique of preserving sterile conditions in dish.) Avoid pressure that might break the surface of the agar. Using the same techniques, cut two of the

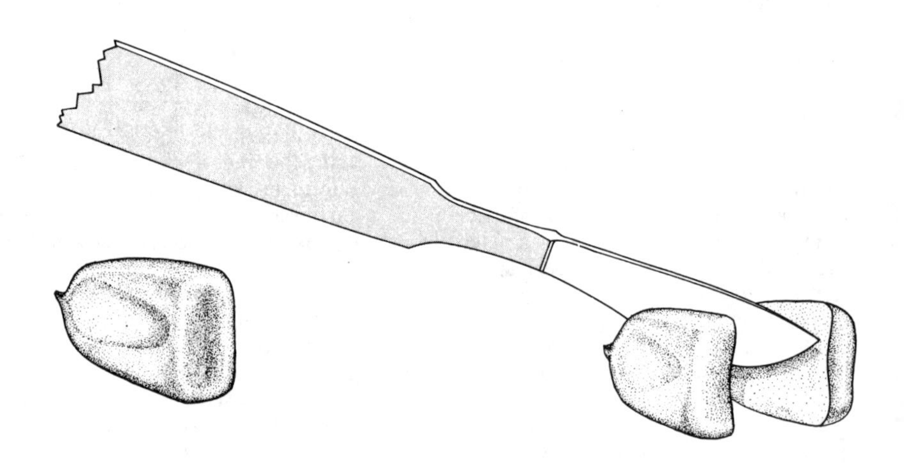

Fig. T-16

maize grains that were killed and preserved on Day 3, and place the four halves of these grains on the agar in Dish 2. Nothing is to be added to Dishes 3 and 4. Put all the dishes in a place designated by your teacher.

Cut the two remaining maize grains and test the cut surfaces with iodine solution. Note the result in your lab-book.

Day 7 or 8. Remove the lids from all the dishes. Using forceps, carefully remove the half grains from the surface of the agar. Place strips of Clinistix over the places where the grains lay in Dishes 1 and 2. Test the surfaces of the other dishes with Clinistix. Record the results in your lab-book.

Studying the Data

A. First, compare the results of the Clinistix test in Dish 1 with the results in Dish 3. What do these results indicate?

B. Compare the appearance of Dish 1, after you flood it with iodine, with the appearance of Dish 3. Is there any difference? If so, devise an explanation.

C. What chemical change (if any) has occurred?

D. What kind of substance may have been involved in any reaction that you think may have occurred?

E. What name can be given to it?

F. Compare the results of both Clinistix and iodine tests in Dish 1 with the results in Dish 2. Explain the differences, if any.

G. What is the purpose of Dish 4?

Conclusion

H. What do the results of this experiment reveal about the physiology of a germinating seed?

For the Teacher

To prepare FAA, mix 500 cm^3 of 95 per cent ethanol, 20 cm^3 of glacial ethanoic (acetic) acid, 100 cm^3 of 40 per cent methanal (formaldehyde) solution, and 400 cm^3 of water.

To prepare starch agar, mix 10 g of powdered starch and 10 g of agar in 980 cm^3 of water and heat until agar is dissolved.

To prepare plain agar, mix 20 g of agar in 980 cm^3 of water and heat until agar is dissolved. Approximately 15 cm^3 of agar will make a thin layer in a 100 mm petri dish. Avoid making the layer of agar too thick.

If Clinistix are not available, cut small pieces of agar from the plates and heat in a test-tube with Benedict's or Fehling's solution.

The scalpels must be quite sharp. To conserve maize—and fingers—it would be well to demonstrate the proper technique for cutting the maize grains. Those cut at odd angles will still serve for the iodine test.

Moulds are sometimes troublesome. Careful attention to proper technique in handling petri dishes will help.

A. If all goes well, places where maize grains lay in Dish 1 give a positive test for sugar and Dish 3 gives a negative test, indicating the change of starch to sugar by something in the maize.

B. Where the maize grain lay in Dish 1, light spots occur in the dark background of starch–iodine complex, indicating some diminution in the amount of starch at these places. No such spots should occur in Dish 3. Evidently something in the maize grain reduced the amount of starch.

C. Putting A and B together leads to the conclusion that some starch has been changed to sugar.

D, E. Both these questions lead to the recall of enzymes.

F. Usually Dish 2 gives results similar to those given by Dish 3. Although enzyme production may have begun in the grains in Dish 2 before the grains were killed, usually little or none of the enzyme remains by the time the test is made.

G. Dish 4 is an overall control, indicating that the changes involve starch and not just something in the agar.

Some additional questions that may be used in discussion:

1. What advantage lies in the storage of food in the form of substances insoluble in water? [*The food 'stays put' and does not diffuse away.*]

2. Why must such substances be changed to soluble forms before they can be transported? [*Most plant transport involves passage through cell membranes; in general, solids do not diffuse through the membranes.*]

3. As it ripens, a banana becomes increasingly sweet. How might this be explained in the light of the investigation? Can you show that an unripe banana contains starch? [*During ripening, starch is changed to sugar. Use iodine.*]

4. Can the chemical change in the investigation proceed in the opposite direction? Why is young maize sweet? [*In both maize and garden peas sugars are rapidly changed to starch after*

picking; the sooner they are cooked and eaten the sweeter they are. Even in sweet varieties some conversion of sugar to starch occurs as the maize grain matures.]

AUDIOVISUAL MATERIALS

Models

Large-scale models are excellent devices for showing microscopic structure quickly. Perhaps the most useful for this course are those showing the microscopic structure of leaves, stems, and roots. Models give a three-dimensional impression that cannot be obtained from Fig. 3.10. On the other hand, models are misleading in some respects, especially as to relative size of microscopic and macroscopic structures, and they must emphasize the 'typical'. Therefore, pupils should have an opportunity to see a good variety of real plant materials both macroscopically and microscopically.

Films

Colour of Life. National Film Board of Canada, available through Guild Sound & Vision Ltd. 16 mm, colour, 24 min. (Illustrates the development of the maple seedling, by time-lapse photography; also shows seasonal changes and the process of photosynthesis in leaves.)

Pathways of Water in Woody Plants. Iowa State University, available from Ealing. (Movement of dye solution through a living woody plant provides a good basis for class discussion.)

Phototropic Response in Coleoptiles. Iowa State University, available from Ealing. (Time-lapse photography shows reactions of corn coleoptiles to various treatments.)

Plant Growth and Development. Ealing. Series of eleven film loops, using time-lapse photography to telescope plant processes. This set includes film loops produced by Educational Services, Inc.; Iowa State University; Heidenkamp Nature Pictures; and Walt Disney Nature Library.

Phototropism. BSCS Inquiry Film Loop, available from John Murray. (Why is it unsatisfactory to say, 'Plants bend towards light because they need it'?)

Water and Desert Plants. BSCS Inquiry Film Loop, available from John Murray.

TEACHER'S REFERENCE BOOKS

FOSTER, A. F., E. M. GIFFORD. *Comparative Morphology of Vascular Plants*. Freeman, 1959. (A somewhat advanced textbook on the structure of vascular plants. Good background for the first section of this Chapter.)

JAMES, W. O. *An Introduction to Plant Physiology*. 6th edition. Oxford University Press, 1963. (A good brief treatment of general plant physiology.)

LEE, A. E. *Plant Growth and Development*. A BSCS Laboratory Block. Heath, 1963. (Techniques for plant experiments.)

STREET, H. E. *Plant Metabolism*. Pergamon Press, 1963.

STREET, H. E., H. ÖPIK. *The Physiology of Flowering Plants: their Growth and Development*. Arnold, 1970.

4
THE FUNCTIONING ANIMAL

MAJOR IDEAS

1. Ecologically all animals are consumers. To obtain their energy, they capture or otherwise secure food and ingest it.
2. In animals digestion always occurs in a cavity of the body. From the cavity the digested foods—that is, foods changed to a diffusible form—pass through cell membranes into the cells of the body.
3. Since energy-release is basically aerobic in animals, oxygen is eventually required, and therefore all animals have means by which they obtain oxygen from the environment.
4. Most animals have a transport system by which substances are carried through the body. Auxiliary to transport through cell membranes and within cells there is, in larger animals, a vascular system, containing a fluid that is pumped by muscular movements of a heart.
5. Metabolic activities result in the accumulation of substances that are either useless or poisonous. These substances are excreted into the environment—in many animals by means of special organs or organ systems.
6. By means of homeostatic mechanisms involving chemical and nervous coordination, an animal maintains an internal steady state and copes with the vagaries of the external environment.
7. Nervous systems range from a very simple network of nerve cells to a complex system of highly specialized neurons coordinated by a brain.
8. Animals move by means of muscles. In most cases these are attached to the inside of an external skeleton or to the outside of an internal skeleton.

PLANNING AHEAD

Make a final check of materials for Chapter 5. If you have not been growing *Zebrina*, arrange for a supply (Investigation 5.1). If wood-lice cannot be collected locally, they should be ordered for Investigation 5.2. Have you decided what animals to make available to pupils for Investigation 5.4?

Looking farther ahead, check on *Coleus* for Investigation 1.1, *MHE* and fertilized chicken eggs for Investigation 1.3, *MHE*. For the latter you will need an egg incubator. If you have not already done so, calculate the number of eggs you will need, based on the number of groups. This will give you an indication of the incubator capacity you will require.

Has your technician had experience in rearing and handling fruit flies? If not, after reading over Investigation 2.2, *MHE* (pp. 58–62) and the commentary on it (pp. T175–81), order some cultures and arrange for him to gain some experience before your pupils approach this work.

GUIDELINES

Many teachers feel a frustrating sense of incompleteness throughout this chapter. Here one might expect the 'type animals' earthworm, rat, dogfish, and rabbit. Or one might expect human anatomy and physiology in a setting of hygienic and wholesome living. All these are important. But if the many aspects of biology that occupy so much space in this series of books are conceded to have value for secondary school pupils, then much that has been part of biology courses in the past must be omitted. For the omission of 'type animals' the authors submit no apology: depth studies of selected forms are important in the training of zoologists, but this is not our main aim.

Esssentially, this chapter is a survey of the ways in which animal structure is correlated with the requirements of 'the animal way of life': the intake of materials, the release of energy from foods, the disposal of excess and poisonous substances, the internal coordination of all metabolic activities, and the means of coping with the environment. Two major viewpoints pervade the chapter: the diversity of ways in which physiological requirements have been met among diverse animal groups, and the relationship between ecological conditions and physiological adaptations.

Such a view of animal structure and function is consistent with the ecological orientation of this series of books. The teacher should do whatever he can to reduce the abstractness by constantly providing examples of live animals, preserved specimens, and pictures. The textbooks aid in this effort by using the physiology of man as the terminal and most completely developed example of each major life function. Use your pupils as laboratory subjects wherever possible.

Chapter 4 does not include discussion of health and disease, but it will stimulate questions about diets, heart attacks, high blood pressure, cancer, nervous disorders, organ transplants, and so forth. Take advantage of the cases your pupils bring up in class and those that you find currently discussed in news media to demonstrate the relationship between the general problems of animal physiology and whatever subject has current interest.

Work on the chapter appropriately begins with an investigation of the structure and function of a living animal. This experience should be a base line to which a return should frequently be made throughout the study of the remaining parts of the chapter.

Once Investigation 4.1 is well under way, begin making study assignments. But before proceeding very far it is a good idea to review Chapter 1, *DLT* and appropriate portions of the Appendix, where animals used as examples in Chapter 4 of this book may be placed in their taxonomic relationships.

Because discussion of this chapter is not difficult to stimulate, some teachers prefer a large number of short assignments such as: pp. 100–4 (guide questions 1–4), pp. 104–7 (guide questions 5 and 6), pp. 109–12 (guide questions 7–9), pp. 112–16 (guide questions 10–12), pp. 116–19 (guide questions 13–17), pp. 122–7 (guide questions 18–21), pp. 128–32 (guide questions 22–24), pp. 132–5 (guide questions 25–28), pp. 136–41 (guide questions 29–32). However, by this time in the course, most pupils can handle longer assignments quite well. In such a case, assignments may be suggested as follows: pp. 100–7, 109–12, 112–19, 122–8, 128–35, 136–41. In either of these systems the investigations come at appropriate intervals.

TEACHING NOTES

INVESTIGATION 4.1

ANIMAL STRUCTURE AND FUNCTION (pp. 96–100)

The introduction (p. 96) sufficiently explains the intent of this investigation. Perhaps not a great deal of physiology is learned, but questions are raised that lead meaningfully into the following exploration of animal functioning. You will probably think of additional observations you would like your pupils to make. These, and questions relating to them, may be inserted at appropriate places.

Each part of the procedure is intended for a laboratory period, but the periods need not be on consecutive days.

Materials for Procedure A

If the American leopard frog (*Rana pipiens*) is used, an order for medium frogs will usually provide approximately equal numbers of both sexes. *Rana temporaria*, the indigenous grass frog, is also suitable for this experiment and is the species usually supplied by dealers—except in winter.

If the frogs arrive two or three weeks before use, they can be stored in a shallow plastic dish on the lower shelf of a refrigerator, a dozen or so per gallon-size container. A few bright pennies in each container may help to prevent 'red-leg', a fungus disease of frogs. The water should not be allowed to freeze and should be changed every two or three days. Frogs do not require feeding for several weeks, even months, under these conditions. The frogs should be removed from the refrigerator at least two hours (preferably two days) before the laboratory period. You or your technician should tie the bandage to the frogs' legs before the animals are distributed to the groups. The bandage, when used for one period at a time, does not seem to injure the animals. The frogs should be distributed in containers. Pupils should *never* be allowed to dangle the frogs on the cords. This is an excellent occasion for distinguishing between procedures necessary for legitimate learning and thoughtless or deliberate cruelty. The publications of UFAW are very useful guides in this respect (see Teacher's References at end of chapter).

Procedure A (pp. 96–98)

This part of the investigation enlarges upon some aspects of Investigation 1.3, *DLT*. That investigation should be reviewed before work is started.

Encourage a calm atmosphere—but do not expect too much. Chances are that about half the frogs will escape temporarily sometime during the day. Your own calm will be more effective in keeping such occurrences at a minimum than will any amount of exhortation. Frogs may be covered with battery jars to hamper mobility, but this also hampers visibility.

Two groups at a time should be assigned to the two aquaria to make the observations called for in instructions 13–16 (pp. 97–98), so that all groups do not try to do this simultaneously.

A. Frogs are essentially neckless.

B. We are tailless!

G. A frog's skin is loosely attached and relatively thin. A frog's skin quickly shrivels on drying out, but this should not be observed; related to it, however, is the common pupil observation that it is wet and slippery.

H. A frog has the same bony limb divisions that we have.

I. Several good points may be made, but the basic one is the lack of nails (or claws) on the digits.

K. The attachment of the legs to the pelvic girdle is such that the legs cannot be brought under the body and the knee joint cannot be rotated to give support from beneath.

L–Q. These questions all revolve around the coordination of movements of the nares and throat muscles. Air is drawn into the mouth through the nares by enlargement of the mouth cavity; then the nares are closed and the muscles of the mouth and throat force air into the lungs. Closure of the mouth is necessary for this series of actions. A frog lacks the ribs and diaphragm by which our thoracic breathing motions occur.

V. This question is related to the observations made in connection with L–Q.

W. The transparent nictitating membrane is frequently, but not always, observed during this procedure.

Materials for Procedure B

If frogs have been stored in formalin, they should be washed for at least one hour in running water before the laboratory period. Exposure to formalin changes the colours of some organs.

Procedure B (pp. 98–100)

A model man that can be 'dissected' should be used during this part of the investigation to bring out comparisons of frog and human structure. This is not only instructive in itself; it also carries along the comparative approach adopted in the remainder of the chapter.

Disposing of frogs in waste jars may be offensive to some pupils. At the conclusion of the investigation, you may have the frogs wrapped in paper towels and collected.

A. A frog's gall bladder seldom appears green. It is more often reported by pupils as bluish.

13. The rugosity in the inner stomach wall is sometimes overlooked as a pupil concentrates his attention on remains of his frog's last meal.

17. Very difficult. A frog's kidneys are almost shapeless.

C. Depending on the recent history of specimens, fat bodies vary from large to practically imperceptible. In the spring, when this investigation is usually conducted, fat bodies are usually more noticeable in males than in females.

D. The tubules leading to the bladder (called ureters but not strictly homologous to the ureters of mammals) are difficult to see; frequently no group in a class successfully makes this observation.

Discussion

Probably the most effective discussion will take place as you visit the groups during their work. But at least a major part of a period should be spent discussing the questions and the comparisons pupils have made between man and frog structure. This should uncover many individual differences among frogs (as well as among observers) and should do much to weaken a Platonic conception of *the* frog, which is, unfortunately and unintentionally, fostered by the careless language of almost all biological authors.

You may wish to review frog anatomy and physiology before leading this discussion. Most zoology texts have a brief treatment. But remember, it is not what authorities *say* but what pupils *see* that is important.

Acquiring Energy and Materials
(in part) (pp. 100–7)

p. 101, ¶ 2, lines 7–9. Some exceptions are noted in the Appendix to *DLT*. Tapeworms (p. 115) and *Oncicola* (p. 116) have no digestive systems. Note that these are both intestinal parasites. In a sense the specialized leaves of pitcher plants are digestive cavities, but they can hardly be said to ingest food. Leaves of Venus flytrap, however, really call to mind a problem of definition reminiscent of 'individual', p. 37, *WL*.

p. 101, Fig. 4.2. A sticky end to its tongue secures the prey.

p. 102, Fig. 4.3. A comparison of body plans reveals that only one figure shows an anus. Some pupils may anticipate a problem of waste disposal; a little research in zoology books can clear up this matter. Here, and in discussing subsequent systems, stress the physiological relationship between successive levels of organizational complexity. And point out that all existing levels of organization have been successful from the viewpoint of the survival of species.

p. 103, ¶ 2. Once the function of the gizzard is understood, pupils can understand the reason for placing sand and other grit in bird cages and poultry yards.

p. 104, Fig. 4.4. Most of the fibre digestion in the rumen is by numerous anaerobic bacteria—up to a thousand million per cubic centimetre of rumen contents. Presumably the bacteria benefit and the cow obtains much of its nutrition from the bacterial products, so the relationship is usually judged to be mutualistic. No sectional diagram can display the complexity of the digestive apparatus of a cow; only a three-dimensional model can do so.

p. 105, ¶ 3. Peptide bonds are broken by enzymatic hydrolysis—essentially the reverse of dehydration synthesis. (The same enzymes are used.) A look back to pp. 45–46 may be useful here. It is worth taking the time to picture hydrolysis, since all the digestive enzymes work in the same way.

p. 106, ¶ 2. Vomit sometimes tastes sour or bitter depending on its origin. In violent spasms some material from the duodenum may be vomited; this is bitter because of the highly alkaline bile. If, however, only stomach material is regurgitated, it is sour because of the hydrochloric acid content.

p. 107, Fig. 4.8. The enzyme nomenclature is that currently used. It is well to let 'ptyalin', 'trypsin', 'steapsin', etc., die.

INVESTIGATION 4.2

THE ACTION OF A DIGESTIVE ENZYME (pp. 108–9)

The teaching objective of the investigation is to have pupils understand some factors influencing enzyme action and how enzymatic activity can be studied. In Investigation 2.2, pupils saw a factor that destroys enzyme activity.

Though the use of saliva may be considered messy, it has the advantage of clearly relating enzyme action to a living organism. In the minds of pupils, diastase (which might be substituted) is divorced from life; if it were used, the investigation would become a purely chemical—not a *bio*chemical—one. Of course, some attention must be given to establishing a proper investigatory attitude, and good judgement must be used in selecting the pupils who contribute saliva.

Materials

Plain starch can be substituted for unsweetened biscuit, but it is difficult to chew.

For preparation of iodine–potassium iodide solution, see p. T22.

Use either Benedict's or Fehling's solution in testing for maltose. It is easiest to use commercially available solutions, but you can prepare your own if you wish.

Fehling's solution is made of two parts, *A* and *B*:

A. Dissolve 34.66 g of copper(II) sulphate, $CuSO_4.5H_2O$, in 500 cm^3 of distilled water.

B. Dissolve 173 g of potassium sodium tartrate (Rochelle salts), $KNaC_4H_4O_6.4H_2O$, and 50 g of sodium hydroxide, NaOH, in 500 cm^3 of distilled water.

For a combined Fehling's solution, mix equal volumes of the two solutions when you are ready to test.

Benedict's solution is made by heating 173 g of sodium citrate, $C_3H_4(OH)(COONa)_3$, and 100 g of sodium carbonate, Na_2CO_3, in 800 cm^3 of distilled water. Filter if necessary and dilute to 850 cm^3. Dissolve 17.3 g of copper sulphate, $CuSO_4.5H_2O$, in 100 cm^3 of distilled water. Pour this copper sulphate solution, with constant stirring, into the carbonate–citrate solution and add distilled water to make one litre.

A 3 per cent starch solution is made by boiling 3 g of soluble starch (laundry starch or corn-starch will do) in 97 cm^3 of distilled water.

Hydrochloric acid of pH 3 is made by first diluting 0.86 cm^3 of concentrated hydrochloric acid (36 per cent) to 100 cm^3 with distilled water. Add 10 cm^3 of this 0.1 M HCl solution to 990 cm^3 of H_2O.

To make HCl of pH 6, put 1 cm^3 of the pH 3 dilution in 999 cm^3 of H_2O.

You can make sodium hydroxide solution of pH 11 by first dissolving 0.42 g of solid sodium hydroxide in enough water to make 100 cm^3 of solution. Add 10 cm^3 of this 0.01 M solution to 90 cm^3 of water.

To make sodium hydroxide solution of pH 8, put 1 cm^3 of the pH 11 solution in 999 cm^3 of water. Because sodium hydroxide solution reacts somewhat with carbon dioxide in the air, stale sodium hydroxide solution may need a bit stronger concentration to equal the desired pH's. Adding the concentrated stock solution dropwise to the final dilutions should bring about the desired hydrogen-ion concentration. Test with universal indicator paper. Store in tightly sealed bottles until ready for use.

The exact pH's of these solutions are not critical. One acid and one base should be rather close to neutral, and the others should be rather far from neutral.

Procedure

The whole of Procedure A can be done as a demonstration, following which seven groups may each perform one of the operations in Procedure B. If pupils providing the saliva have been eating sweets their saliva will produce a positive sugar reaction.

Teacher demonstration: Make up a 5 per cent solution of maltose by dissolving 5 g of maltose in 95 cm^3 of distilled water. To about 10 cm^3 of this solution in a test-tube, add about 10 drops of either the combined Fehling's solution or the Benedict's solution. Heat gently until a red or yellow colour shows (copper(I) oxide). *Boiling* is not required.

Neither Fehling's solution nor Benedict's solution is specific for maltose (both give the same results with glucose and other reducing sugars), but that does not matter here. They *do* indicate that something happens to the starch after it is subjected to the action of saliva—if, as the procedure indicates, they give negative results with starch itself.

Procedure A (p. 108)

A. A negative test indicates a lack of action by saliva on starch. A positive test indicates the action on saliva on starch. On the basis of Fig. 4.8 this can be translated into the action of salivary amylase on starch to produce maltose. Pupils should be aware of the point at which interpretation from direct observational evidence ends and interpretation from other (more or less trustworthy) information begins.

Studying the Data (p. 109)

A. *In vivo* conditions for action of salivary amylase are a temperature of 37 °C and a pH of 7 or a little more.

Conclusions (p. 109)

A, B. In general, data show that the closer conditions come to the normal *in vivo* conditions, the greater is the activity of amylase.

C. In a human mouth both temperature and pH are probably much more variable than they are inside cells. Therefore, a wider tolerance for environmental vicissitudes might be expected of a salivary enzyme than of an intracellular one.

Acquiring Energy and Materials (contd.) (pp. 109–12)

p. 109, ¶ 2. Recall problem 3, p. 61.

p. 110, ¶ 5. In mammals, control of breathing movements is a function of the concentration of CO_2 in the blood (see p. 131, paragraph 4). A pupil can experience this regulation. He hyperventilates by taking several deep breaths of fresh air, then times the interval he can hold his breath. After breathing normally for five to ten minutes, he again times the interval he can hold his breath. Usually the time is longer after hyperventilation than after normal breathing. The increased ventilation does not increase the oxygen content of the blood, but it does remove more than the usual amount of CO_2 from the blood. It then takes longer to build up the concentration of CO_2 to the point where the respiratory centre is switched on. But the physical and chemical control of respiration is intricate, and the details still challenge physiologists. Some pupils might enjoy doing some research on the mechanics, as they are now understood, or on the unsolved problems. (*Caution—prolonged hyperventilation can cause dangerous black-outs.*)

Transporting Materials in the Body
(pp. 112–19)

p. 114, ¶ 2. An earthworm can easily be set up under a stereomicroscope. Place it in a moist petri dish and check from time to time to see that it remains moist. Pupils may then observe the pulsing of the dorsal blood vessel.

p. 114, Fig. 4.13. The diagram of the circulatory system of an earthworm is somewhat misleading, because the connections between vessels in the dorsal and ventral body walls cannot be shown in a longitudinal section.

p. 115, ¶ 3. Among reptiles there is a wide variation in the amount of ventricular septation; it is complete in the crocodilians.

p. 115, Fig. 4.14. Heart structure is simplified in these diagrams, the sinus venosus and conus arteriosus being ignored. The 'four chambers' of the comparative anatomist can only lead to confusion here.

p. 116, Fig. 4.15. Auriculoventricular valves (called 'tricuspid' on the right side, 'bicuspid' or 'mitral' on the left) point into the ventricle. When intraventricular pressure rises (ventricular systole) the auriculoventricular valves are driven closed. Heart action can be adequately visualized only in motion: see film loop *The Heart in Action* (p. T150).

p. 117, Fig. 4.17. 55 per cent.

p. 118, ¶ 1. The striking similarity between haeme and the porphyrin part of chlorophyll has been the subject of some interest and speculation among biologists. It sometimes interests pupils. At the least it highlights the biochemical unity of living things.

p. 118, ¶ 1. The reasoning for reduced activity in an anaemic person is: Activity requires an energy source; this energy derives from cellular respiration; cellular respiration requires oxygen; oxygen is delivered through oxyhaemoglobin. Thus deficiency of haemoglobin leads to deficiency of activity.

p. 119, ¶ 2. To most pupils blood clotting appears to be a simple process; upon investigation it turns out to be a complicated biological phenomenon. A blackboard diagram of the clotting mechanism will help to straighten out the complex of terms. One such diagram will be found in Schmidt-Nielsen, *Animal Physiology* (reference: p. 148). Some pupils may seek information on phlebitis and internal blood clotting.

INVESTIGATION 4.3

A HEART AT WORK
(pp. 120–1)

Materials

Daphnia can be collected from ponds and lakes during most months of the year. *D. pulex* may also be ordered from biological supply companies and quite often may be obtained at aquarium supply stores.

A *Daphnia* culture can be maintained for one or two weeks in pond water at 22 to 26 °C. The culture should be kept out of direct sunlight. If the culture is to be maintained for a longer period, a culture medium must be provided. Many types of media are suggested in literature.

A 50 cm^3 beaker is a good container for the *Daphnia*. It must be made of heat-resistant glass, since it will be transferred from ice water to hot water. A 250 cm^3 beaker provides a good water jacket.

If depression slides are not available, make a ring of wax dripped from a burning candle on to a clean slide. Place the drop of water containing the *Daphnia* within the ring; this will prevent the animal from being crushed when the cover slip is added.

If stereomicroscopes are not available, the low power of a monocular microscope may be used, but considerable difficulty in keeping the specimen within the field of view may be encountered.

Procedure

The counts must be made quickly. Therefore pupils should be familiarized with *Daphnia* in advance. At the same time assign temperatures to groups and have them decide which members are to perform each task during the laboratory period.

Two members of a group should make and record a count; then the other two members should verify the rate. Alternating pupils through the counting and data-recording routines provides maximum participation and instils greater responsibility for the results.

Give some practice in counting rapidly by

making pencil dots on paper in rhythm with the heartbeat. The dots should be made in a continuous line, back and forth in this manner:

. . . .
. . . .
. . . .

Assuming that room temperature is about 20 °C, suggested temperatures are: 15 °C, 10 °C, and 5 °C (ice water); and 25 °C, 30 °C, and 35 °C (hot water).

It is not necessary to have a separate hot-water supply for each group. One large container can be maintained at about 90 °C. If one assistant is made responsible for distribution of the hot water, movement about the room can be reduced and safety enhanced.

Studying the Data (p. 121)

The first point of this work is the variability of physiological data. Pupils are often unduly impressed by various 'normal' values for physiological measurements, such as body temperature and blood pressure. You can make a comparison between the heartbeat rate of *Daphnia* at room temperature and that of man by having pupils count their own pulses. These data, assembled on the blackboard, will not only provide a comparison with *Daphnia* but will also emphasize the variability of data and lead to a better understanding of the many 'normal' values found in textbooks of human physiology.

The second point to be stressed, if the data allow you to do so, is the effect of temperature on the physiology of a 'cold-blooded' animal. The normal heartbeat rate of *Daphnia* is 300–350 per minute. The rate usually declines with lowered temperatures and rises with raised temperatures, but at about 40 °C a decline occurs, and the heartbeat usually ceases at about 45 °C.

A. By the time Chapter 4 is reached, this question should be routine.

B. Inaccuracy of counting, temperature variation during the period of positioning the specimen, and variation in age and physical condition of specimens are all likely suspects.

C. The more data collected, the greater the range is likely to be, because the larger the number of counts, the greater is the chance that aberrant counts may have been made.

D. The class average is probably likely to be nearer the 'true' rate—the rate likely to be obtained from a second series of comparable data (see Investigation 1.1, 'Discussion' item D, p. 2, *PLW*)—because errors or differences noted in item B are compensated as data are averaged. In statistical terms, the more data, the greater the range (item C) but the less the standard deviation.

Conclusions (p. 121)

C. 'Warm-blooded' (homeothermic) vertebrates (e.g. dogs) are able to keep a steady rate of heartbeat despite wide changes in environmental temperatures, because the regulation of heartbeat occurs in the little-changing internal environment. 'Cold-blooded' (poikilothermic) vertebrates (e.g. frogs) are similar to *Daphnia* in that their internal processes, including heartbeat, are little insulated against fluctuations in environmental conditions such as temperature.

For Further Investigation

1. In addition to snails, you may use gnat larvae and *Tubifex* worms for comparisons.

2. Besides the suggested drugs other substances that might be tried are aspirin, tea, phenobarbitone, carbonated beverages, coffee, and tobacco.

Removing Materials from the Body (pp. 122–7)

p. 123, ¶ 1. Anything that an organism gets rid of *may* be a waste. It *is* a waste if the organism has too much of it. But our only criterion for 'too much' of a substance is that an organism habitually gets rid of it.

p. 125, Fig. 4.21. Fifty molecules of $CO(NH_2)_2$ contain as much nitrogen as 100 molecules of NH_3.

pp. 124–7. The removal of nitrogenous wastes is complex, and it is likely to be new to most pupils. Take some class time to go over this function. Stress the fact that the flow of blood through kidneys is controlled by nervous and endocrine systems, and in this way water content of the body is kept in balance. Pupils can generalize from this that the action of the kidney is correlated to ecological conditions by means of the coordinating systems.

p. 126, Fig. 4.22. This is a good example of the complementarity of structure and function.

Maintaining a Steady State
(pp. 128–41)

The idea of steady state (and its maintenance by homeostatic mechanisms) underlies the whole of the series of books. The idea is first referred to on p. 3, *WL*, and Chapter 5, *MHE*, is a great fugue on that theme. If the pupil has truly caught this idea, he will easily see the artificiality of the heading to this part of the chapter: it could just as well be the title of the chapter as a whole.

p. 129, ¶ 3. Pupils frequently confuse the functions of enzymes, animal hormones and plant hormones. This is a good place to discuss their similarities and differences.

p. 130, ¶ 4. Clotting of blood and coagulation of milk are dependent upon calcium ions. Some nervous disorders are associated with calcium imbalance. The permeability of cell membranes is calcium-dependent. So is the coalescence of cells; cells tend to fall apart from each other if calcium ions are lacking in the surrounding medium.

p. 131, Fig. 4.24. Refer pupils to Fig. 4.23, p. 127, for structural relation of adrenal glands to kidneys.

p. 131, ¶ 2. The adrenal cortex also contains hormones associated with sex. Pre-adolescents—even genetic females—having diseases of this gland may develop male secondary sexual characteristics.

p. 133, Fig. 4.25. Some biologists believe the nerve network of coelenterates to be a staining artifact.

p. 134, Fig. 4.26. Neuroanatomists do not seem to agree on axon–dendrite terminology. Therefore, these familiar terms have been omitted from this figure.

p. 136, ¶ 4. Among many time-honoured topics omitted from this series of books is consideration of structure and function of special vertebrate sensory organs. Even in the unlikely event that your pupils have not studied the structure and functioning of human eyes and ears, pausing to do so here might be unwise, because the point of 'Adjustment to the External Environment' might then very easily be overlooked.

p. 136, ¶ 5. We use instruments to enable us to become aware of stimuli for which we have no receptors, or which are too weak for our receptors to detect.

p. 139, Fig. 4.29. Muscles work in pairs, one in opposition to the other. Muscles only pull, never push.

p. 139. ¶ 4. Some disadvantages of being 'warm-blooded': In general, 'warm-blooded' animals require more food than 'cold-blooded' animals do. In extreme environmental conditions 'warm-blooded' animals may perish through inability to control their internal environments, whereas 'cold-blooded' animals under the same conditions may escape death through dormancy. Of course, some 'warm-blooded' animals in such circumstances abandon warm-bloodedness temporarily.

p. 141, Fig. 4.30. By exposing a maximum of body surface (especially nose, inner surface of ears, and the bottom of feet) to the surrounding air, the cat dissipates a maximum amount of heat. This indicates the air temperature is high. In cold air a cat usually depresses the ears and lies with feet tucked under the body and tail curled over the nose.

INVESTIGATION 4.4

CHEMORECEPTORS IN MAN
(pp. 142–5)

If time is pressing, one or more parts of this investigation may be omitted. If all materials are ready, Procedures A and B can be accomplished in one period. If you wish to combine Procedures B and C in a second laboratory session, then in the first session Procedure A can be repeated, with pupils of each pair alternating. This gives each pupil experience with all four tastes. Another alternative is to assign Procedure C as homework.

In all parts of this investigation, the power of suggestion is great. Pupils should be warned about it. Most will try to be objective if they are cautioned. This may be a good time to discuss the difficulty of experimentation that involves a report from the subject—the opposite to the difficulty mentioned in the background information (p. 142). Such a discussion provides a good transition to Chapter 5, especially to pp. 150–3.

Materials

All materials to be tasted have to be prepared with extreme care.

1. Glassware and other apparatus coming in contact with such materials must be clean and must not have been used for poisonous substances.
2. The materials used must not have become contaminated in any way.

3. Any materials remaining after use must not be returned to the original container.

For Procedure A weigh out a number of grams corresponding to the percentage and dissolve in enough water to make 100 cm^3 of total solution. The 1 per cent ethanoic (acetic) acid solution can be made by mixing 1 cm^3 of glacial ethanoic (acetic) acid and 99 cm^3 of water.

The toothpick swabs should be prepared before the laboratory period, because pupils tend to make them too large and loose. The balls of cotton should be wrapped very tightly and should not be more than 3 mm in diameter. The prepared swabs may be wrapped in paper and sterilized in an autoclave.

Small battery jars make convenient waste jars. If plenty of sinks are available, waste jars may not be needed.

Make the sugar solutions for Procedure B by mixing the most highly concentrated one first and then diluting appropriately. Prepare the 0.5 per cent solution by putting 0.5 g of sucrose in 99.5 cm^3 of water.

Vol. of Soln.	*Initial Concn.*	$+H_2O$	*Final Concn.*
20 cm^3	0.5%	+80 cm^3	= 0.1%
10 cm^3	0.5%	+90 cm^3	= 0.05%
20 cm^3	0.05%	+80 cm^3	= 0.01%
10 cm^3	0.05%	+90 cm^3	= 0.005%
20 cm^3	0.005%	+80 cm^3	= 0.001%

Follow the same procedure for making up the salt solutions.

Suggestions for additional solutions in Procedure C: grapefruit juice, garlic salt. All of the solutions should be dilute.

Procedure

Pupil assistants are very useful in this investigation. In Procedure A, for example, only one or two stock bottles of the solutions are required if pupil assistants deliver the solutions as needed.

A simple method for assigning solutions in Procedure B is to arrange for all pupils who were designated 1 (in Procedure A) to work with the sugar solutions, and all pupils who were designated 2 to work with salt solutions.

In Procedure C a system for labelling the 'unknowns' must be devised, so that the discussion following the experiment will be meaningful.

The pupils should have an opportunity to discuss the procedure with you before they begin. The instructions may seem complicated at first.

Studying the Data (p. 144)

Most of the answers to questions depend upon the nature of the data collected during the investigation. The distinction to be stressed is not between 'right' and 'wrong' answers, but between more or less carefully obtained data.

E. The assumption is that the nerve endings in the nose have a great deal to do with sensations reported as taste.

PROBLEMS (pp. 146–7)

Most of the problems can be answered, at least to some degree, by consulting standard books on physiology.

1. The general procedure for studying any hormone action is to remove the suspected source of the hormone; observe function in comparison with the functioning of intact animals and then inject the purified hormone, if available, to observe return to normal function.

2. The hormones of mammals are in many cases the same in all or a wide variety of species, and the antibodies are similar enough to provide immunity. Whole blood, however, is a vastly complicated biochemical mixture, and its proteins are highly specific. Blood even from one individual within a species may be incompatible with blood from another individual of the same species. This problem leads to the idea of blood types, though many more factors are involved.

3. Gaseous wastes diffuse away, commonly through leaves, and water is lost constantly by terrestrial plants. This much pupils already know. Some excess materials may be stored in parts that are discarded, such as leaves. But the problem of plant excretion still remains, and is occasionally discussed by botanists. As one wrote recently, 'Why do not plants urinate?'

4.

$$CO_2 + H_2O \xrightarrow[\text{anhydrase}]{\text{carbonic}} \text{carbonic acid } (H_2CO_3),$$

$$H_2CO_3 + KCl + NaCl \rightarrow KHCO_3 + NaHCO_3,$$

$$KHCO_3 \rightarrow K^+ + HCO_3^-,$$

$$\text{and } NaHCO_3 \rightarrow Na^+ + HCO_3^-.$$

Thus, most of the CO_2 is carried in the bloodstream as hydrogen carbonate ions. And about 20 per cent is apparently carried as carbaminohaemoglobin.

6. Iodine deficiency results in the production of fewer molecules of thyroxin. This results in symptoms of thyroxin deficiency, and in compensation for deficiency, the thyroid gland enlarges, forming a simple goitre.

7. Pupils should encounter haemocyanin, chlorocruorins, and haemerythrins in researching this problem. Most of the respiratory pigments contain one or more divalent metal atoms (iron, copper, vanadium), and it is thought that these atoms are the sites of oxygen attachment. In general there is a greater variety of respiratory pigments in primitive phyla than in more recently evolved phyla. The chemical relationship between haeme and chlorophyll was discussed on p. T144. The occurrence of haemoglobin in *Rhizobium* nodules is at present considered to be a biochemical accident or coincidence similar to the occurrence of cellulose in tunicates.

8. Both flushing and blushing result from vasodilation affecting the capillaries in the skin.

9. The sandwich contains carbohydrates, fats, proteins, mineral nutrients and cellulose. The pupil need merely consider the enzymes affecting (or failing to affect) these substances and the general story on pp. 105–7. But he might produce an account with greater detail by consulting the discussions of alimentary physiology in a text on human physiology.

10. Calcium is essential to the conversion of prothrombin into thrombin, but just how it acts does not seem to be known. Vitamin K is somehow essential to the production of prothrombin.

11. Carbon monoxide poisoning is essentially suffocation. Treatment is fresh air forced into the body by artificial breathing. In extreme cases, air containing 40 per cent oxygen and 7 per cent CO_2 stimulates breathing action.

12. Material for a superficial answer to this problem is contained in Chapter 4, but good pupils should be encouraged to delve further.

13. Again, there are various levels at which this problem might be answered. The basic point is that different methods of successfully maintaining life functions depend upon living in different environments. A sponge's methods are not successful where a mountain goat's are, and vice versa.

SUPPLEMENTARY MATERIALS

ADDITIONAL INVESTIGATION

EFFECTS OF CHEMICALS ON CAPILLARY CIRCULATION

There are numerous methods of demonstrating capillary circulation to pupils. This one may be used during work on Investigation 4.1.

Materials and Equipment

Frog, pithed
Dissecting tray
Scissors
Dissecting needle
Medicine dropper
Glass plate
Paper tissues
Stereomicroscope, ×40 to ×60
Solutions of drugs
Finger bowls

Procedure

Using scissors, make an incision through the skin and body wall parallel to the mid-line of the frog, along the ventral margin of pigmentation. The incision should extend about 2 cm from a point about 1 cm anterior to the hind leg. Apply pressure on the body wall on the side opposite the incision. By holding aside the intestines and ovary (if present) with a wooden handle of a dissecting needle, you can press the urinary bladder through the incision.

If the bladder is not well filled, gently force physiological saline solution into the cloaca, using a medicine dropper. Avoid introducing air into the bladder during this operation. It may be necessary to fill the bladder in this manner before it can be pressed from the body cavity.

Place the frog on a glass plate in such a position that light from below is transmitted through the bladder without obstruction. Cover the rest of the frog with a wet paper tissue. Every two or three minutes place a drop or two of physiological saline solution on the bladder.

A. What is the ratio of the diameter of the smallest visible blood vessel to the diameter of the largest?

B. How can you distinguish between *arterioles* (small arteries) and *venules* (small veins)?

C. Which vessels, if any, can be seen to pulsate?

D. Is the speed of blood flow the same in all vessels? If not, is it faster in the smaller or the larger vessels?

E. Red blood cells average 22 μm long, 15 μm wide, and 4 μm thick. What is the approximate diameter of a capillary?

You will be provided with solutions of one or more substances that are used as drugs. Using a medicine dropper, apply two drops of one drug to the surface of the bladder in place of the physiological saline with which you have been bathing it. One member of the group should apply the solution while another closely observes the capillaries through the microscope *before*, *during*, and *after* the application. Look for *dilation* (widening in diameter) or *constriction* (narrowing) of the capillaries. The dimensions of the red blood cells and the rate of flow are two standards for determining dilation or constriction. Observation should be continuous for at least three minutes to allow time for the drug to diffuse into tissues.

F. Record the name of the drug and the effect (if any) on the capillaries.

Before applying a second drug, wash the bladder thoroughly with physiological saline solution, using a medicine dropper to apply the solution and a crystallizing dish to catch it. It may be more convenient to use a goldfish for this experiment. If a fish is used, wrap the fish (except the tail) in wet cloth or paper and lay it in a petri dish. Wet the bottom of the petri dish and spread the tail out on the wet glass. Add a few drops of water every few minutes to keep the tail moist. Place the petri dish on the stage of the microscope. Adjust the fish on the microscope stage so that a strong light comes through the thin tissue of the tail.

Note: The UFAW Handbook, *The Care and Management of Laboratory Animals*, 3rd edition (E. & S. Livingstone, 1967), deals with handling (pp. 856–7) and pithing (pp. 996–7) of frogs.

Summary

G. Write a brief description of capillary circulation.

Assemble the data from the drug tests on the blackboard.

H. To what extent are the results obtained by groups using the same drug in agreement?

I. What factors may have caused any lack of agreement that exists?

J. According to the data, which drugs act as *vasodilators* (vessel dilators) in frogs?

K. Which act as *vasoconstrictors*?

L. For which are your data inconclusive?

M. Do you think these drugs would have the same effects on your own capillaries? Why or why not?

For the Teacher

If stereoscopes with high enough magnification are not available, ordinary monocular microscopes can be used, especially if they can be equipped with ×5 objectives.

Physiological saline solution for this investigation can be made by dissolving 7 g of NaCl in 1000 cm^3 of water. Or you can use the Ringer's solution prepared for Investigation 1.1 (p. T112). Suggested drugs are ethanol, sodium nitrate(III) (sodium nitrite), adrenaline chloride, acetylcholine bromide, and nicotine sulphate. These are readily available from the usual suppliers of chemicals. Good results have been obtained with nicotine extracted from cigarette filters. Acetylcholine, ethanol, and sodium nitrate(III) are vasodilators; nicotine and adrenalin are vasoconstrictors.

About half an hour before beginning the work, inject 3 to 6 cm^3 of physiological saline solution into the dorsal lymph sinus. This usually results in a well distended bladder. It is not uncommon in a pithed frog for circulation in the bladder to stop temporarily. The frog usually recovers in about fifteen minutes, but if it does not, another specimen must be used.

B. This question depends upon distinguishing capillaries. Blood flows into capillaries from arterioles and from capillaries into venules.

D. This is an application of Bernoulli's theorem and a good example of simple biophysics.

M. This is not a matter of yes or no, but of greater or lesser confidence. We might have more confidence in application to human beings if the work had been done on rats instead of frogs, still more if done on chimpanzees, and rather less if done on goldfish.

AUDIOVISUAL MATERIALS

Filmstrips

William Harvey. Encyclopaedia Britannica Filmstrips. Colour. (For the most part, historical content has been crowded out of Chapter 4. This filmstrip may help to restore the balance.)

Metamorphosis. Life Filmstrips. Colour. (Too often endocrine control is taught exclusively from a mammalian or (at most) a vertebrate viewpoint. This filmstrip deals with the hormones of insect metamorphosis; furthermore, it is a beautiful example of the way in which experimental procedure can be illustrated on a filmstrip.)

Circulation. McGraw-Hill Text Films. Colour. (Ties other systems to the circulatory and has good summaries of clotting.)

How the Nervous System Works. McGraw-Hill Text Films. Colour. (Good supplement to Chapter 4: all-or-none principle, acetylcholine at synapse, proprioceptors and enteroceptors.)

Films

Human Body: Skeleton. Coronet Films, available through Gateway Educational Films. 16 mm, colour, 10 min. (A good supplement to Chapter 4; shows, by fluorography, human skeleton in action.)

The Blood. Encyclopaedia Britannica Films. 16 mm, colour, 20 min. (Probably too much, but has good sequences that can be selected: blood-cell types and counting procedures, illustrations of capillary circulation and clotting.)

Digestion. Part 1. Muscular Movements. G.B. Instructional, distributed by Rank. 16 mm, colour, 17 min. (The mechanical processes, absorption, function of the liver.)

Digestion. Part 2. Chemical Changes. G.B. Instructional. 16 mm, colour, 19 min. (Structure and function of digestive glands, distribution of digested foods.)

The Kidney and Homeostasis. BSCS Inquiry Film Loop, available from John Murray. Super-8, colour. (Presents data from which pupils can derive the homeostatic function of kidneys. Best used before the textbook material is assigned.)

Water and Desert Animals. BSCS Inquiry Film Loop, available from John Murray.

Temperature and Activity in Reptiles. BSCS Inquiry Film Loop, available from John Murray. Colour. (This and the preceding film are excellent means for keeping ecological implications of physiology before pupils. Both raise problems for which pupils suggest hypotheses as data are presented.)

Fundamentals of the Nervous System. Encyclopaedia Britannica Films. 16 mm, colour. (Good as a review.)

The Heart in Action. Encyclopaedia Britannica Films. 8 mm, colour. (Illustrates the complex motions of the human heart.)

The Animal and the Environment. McGraw-Hill Text Films. 16 mm, colour, 28 min. (The self-regulating mechanisms, homeostasis, involved in breathing, heartbeat and kidney function are described and illustrated.)

The Senses. McGraw-Hill Text Films. 16 mm, colour, 28 min. (Sight, hearing and taste, and the nervous and chemical operations involved in these processes. Includes a good sequence on the experimental determination of taste in houseflies. Can serve well as a link with Chapter 1 of *MHE*.)

Pithing the Frog. Ealing. Super-8, colour, 3 min. (For the teacher.)

How Animals Breathe—Fresh Water. Ealing. Super-8, colour, 4 min. 10 s. (Good for extending pupils' ideas of breathing beyond themselves.)

TEACHER'S REFERENCE BOOKS

ADOLPH, E. F. 'The Heart's Pacemaker', *Scientific American*, March 1967, p. 32.

BAKER, P. F. 'The Nerve Axon'. *Scientific American*, March 1966, p. 74.

DE BRUL, E. L. *Biomechanics of the Body*. Heath, 1963. (A BSCS pamphlet that discusses the biophysics of the gross structure of the human body.)

MAYER, W. V. *Hibernation*. Heath, 1964. (A BSCS pamphlet that discusses in some detail the physiology of dormancy, particularly that of 'warm-blooded' animals under conditions of cold.)

OVERMIRE, T. G. *Homeostatic regulation*. Heath, 1963. (BSCS pamphlet No. 9.)

WOOD, I. E. 'The Venous System', *Scientific American*, January 1968, p. 86.

UFAW publications. Obtainable from Universities Federation for Animal Welfare, 230 High Street, Potters Bar, Herts.

Information leaflets on the care of animals:

The Rabbit
Guinea-Pigs
The Rat and Mouse
Hamsters and Gerbils
Cage Birds
Cold-Water Aquaria
The Goldfish
The Tortoise
Reptiles and Amphibians
A Few Bird Friends of Ours
Goats and Sheep
Cattle and Pigs
Horses and Donkeys
Poultry
Mammals in Britain

The UFAW Handbook on the Care and Management of Laboratory Animals: Part II (Rodents, Lagomorphs and Insectivores), and *Part IV* (Birds, Poikilotherms and Invertebrates). (These sections are particularly suitable for schools.)

The Humane Control of Animals Living in the Wild, and *The Use of Animals in Toxicological*

Studies. From Proceedings of a Symposium held in January 1969.
The Rational Use of Living Systems in Biomedical Research. (Proceedings of a Symposium held in October 1971.)
The Status of Animals in the Christian Religion.
An Introduction to the Anaesthesia of Laboratory Animals.
Humane Killing of Animals.
The Transport of Animals.

5
BEHAVIOUR

MAJOR IDEAS

1. The ability to detect and respond to stimuli from the environment (irritability) is a basic characteristic of living things.
2. Irritability undoubtedly involves complicated biochemical and biophysical actions at the molecular level; indeed, it must involve the whole physiology of the individual. But as a matter of convenience, behaviour is usually understood to consist of responses that are visible. In this view plants and protists exhibit behaviour, but it is debatable whether much clarity is gained by separating behaviour from the rest of the physiology of these organisms.
3. The scientist's observing and reporting are themselves examples of animal behaviour; hence, complete objectivity in the study of behaviour is extremely difficult to achieve. Furthermore, since different animals perceive different environmental stimuli, only the most painstaking analysis can determine to which of myriad possible environmental stimuli an organism is responding—and in some cases the stimuli may be completely unknown to the observer.
4. No completely satisfactory classification of behaviour exists. However, increasing levels of complexity can be distinguished, from simple tactic and tropic responses to reasoning ability.
5. Two approaches are common in behaviour research. In one, attention is principally directed to the mechanism of behaviour; in the other, attention mainly focuses on the way in which behaviour affects the life of individuals and populations. The former approach is physiological, the latter ecological.
6. From the viewpoint of mechanism, behaviours can be conveniently divided into two classes: innate and learned. While some behaviours are undoubtedly determined genetically, the way in which genes bring them about—especially the complex ones called instincts—is not clear. Learned behaviour has been intensively investigated in conditioning experiments, but much remains unexplained, from the problem of defining it to the phenomenon of insight (which is almost, but not quite, a monopoly of *Homo sapiens*).
7. From the viewpoint of ecological function, classifications are at present variable; investigations adopt the classification that is most convenient for the purposes of the investigation at hand. However, some fairly prevalent behavioural patterns can be recognized, such as territorial phenomena, social interactions, and so on.

PLANNING AHEAD

You need to make final plans now for the investigations in Chapter 1, *MHE*. By this time you should have obtained enough *Coleus* plants to supply the number of groups you plan to use in Investigation 1.1, *MHE*. Your plan for handling the schedule of events demanded by Investigation 1.3, *MHE*, should be complete; and while work on Chapter 5 in this book is proceeding, incubations of the eggs needed for the later stages should be started. You should check on the availability of Poppit beads for Investigation 1.2, *MHE*, and plan a substitute if you cannot obtain them.

If your technician required experience in

rearing, handling, and examining fruit flies, you should now be in the midst of this. If not, and if you have not been maintaining cultures, order them now. At the same time, order tobacco seeds for Investigation 2.3. These must be seeds that are expected to produce a 3:1 green–albino ratio in the seedlings.

Seeds for Investigation 2.5 (round and wrinkled peas) should be obtainable locally in the spring. Many more wrinkled varieties are available than round, but the variety Alaska can usually be obtained. Glucose-1-phosphate can be obtained from one of the usual biological supply companies, e.g. Harris Biological Supplies, and from specialized suppliers of biochemicals such as BDH.

GUIDELINES

Pupils come to biology with certain ideas about the word 'behaviour'. Whatever those ideas may be, they probably do not coincide with the concept that is encountered in Chapter 5. For the most part they are likely to be too narrow. Broadening the pupil's viewpoint must be one of the principal aims of the teacher. On the whole this is easier than narrowing a definition (as in the case of 'food'), but it requires constant attention.

No doubt most of the pupils' ideas of behaviour have human implications, and certainly major interest lies in human behaviour. But here the authors' views are similar to their views on sex education: the social science teacher can better handle the details of acceptable human behaviour in contemporary civilization; and the biology teacher can best contribute to the pupil's development by providing a broad biological perspective, which has value in itself and also provides a basis for a rational grasp of psyche and mores.

Undoubtedly every specialist feels that his field is slighted in school textbooks, but none with more justification than a behavioural scientist. It must be kept in mind that Chapter 5 is not *the* chapter on behaviour in this series of books, just as Chapter 3, *MHE*, is not *the* chapter on evolution. Behaviour is treated, at least implicitly, throughout *WL* and *PLW*, and in parts of Chapter 4, *LO*. Be sure your pupils realize also that Investigations 1.4 (Part C) in *WL*; 1.3 (in part) in *DLT*; 4.2 in *PLW*; 4.1 (in part) in *LO*; and 4.4, *LO*, have already involved them in the study of behaviour.

Today the study of behaviour represents one of the major frontiers of biology. Therefore you should not bypass this chapter. For good pupils who are learning to handle long assignments, it can readily be broken into two parts: one consisting of 'The Study of Behaviour' and 'Levels of Behaviour', with two associated investigations (pp. 149–70), and the other consisting of 'Some Patterns of Animal Behaviour', also with two investigations (pp. 170–84). For very slow pupils an opposite extreme of assignments might be pp. 149–53 (guide questions 1–4); pp. 153–60 (guide questions 5–8—include here also Investigation 5.2, pp. 167–70); pp. 161–6 (guide questions 9–14); pp. 170–7 (guide question 15); and pp. 177–84 (guide questions 16–20).

In the past, investigators of both human and non-human behaviour have often been hampered by inadequate knowledge of neurology and endocrinology, while investigators of nerve and gland function have suffered from a laboratory myopia that restricted the meaning of their researches. With increasing attention of interdisciplinary research, this old dichotomy is disappearing. Placing of the study of behaviour immediately after the chapters devoted to plant and animal physiology has been our way to emphasize the mutual interdependence of these fields of study.

TEACHING NOTES

The Study of Behaviour (pp. 149–53)

p. 150, ¶ 2. See last paragraph of 'Guidelines' above.

p. 151, ¶ 2. You may wish at this time to discuss fashions in scientific investigation. For thirty years *Drosophila* and maize were (not without good reason) *the* organisms of genetics. For an even longer period, experimental knowledge of learning in animals was basically a knowledge of learning in the white rat. Only recently have biologists begun to assess the probable bias by employing other animals (especially individuals from wild populations) in their researches.

p. 89, Plate IX. In controlling rodents, advantage has been taken of their colour blindness. Poison grain is coloured with hues that birds (which can distinguish colours) dislike. Thus, rodents indiscriminately eat the poisoned grain while birds in the area avoid it. What does the knowledge that most mammals are colour-blind do to the idea that bulls are incited by seeing 'red'?

p. 153, ¶ 3. Dogs are stimulated by harmonics that are too high for our ears to detect.

Studies indicate that while the upper limit of hearing for man is around 20 000 hertz, for dogs it is 40 000 Hz, and for dolphins (porpoises) it is at least 130 000 Hz. Young children can hear higher notes than their elders; and older ornithologists can in some cases no longer hear many of the spring birdsongs, a shortcoming which affects their ability to carry out breeding-bird population studies. An interesting sidelight on dolphin hearing: A researcher was rewarding a dolphin for whistling. The dolphin began at a higher pitch each time it whistled, being rewarded for each whistle until the sounds were too high for the researcher to hear. At this point, not receiving a reward, the dolphin came back down the scale until its pitch was within the range of the investigator, thus prompting rewards again.

Levels of Behaviour (in part) (pp. 153–5)

p. 154, ¶ 4. Plants, for example, exhibit what are called 'nastic movements', involving such things as opening of flowers, drooping of or changing the angle of leaves, and so on, responses which may be stimulated by light, temperature, etc., but which are not directly related to directional stimuli. Dandelions, daisies and other composites can be used to demonstrate nastic movements; their flower heads close when kept in the dark and open when illuminated. Night scented stock, some evening primroses and white campion do the reverse. They open at night and close during the day. In some cases the factor inducing the evening opening has been shown to be an increase in atmospheric humidity.

p. 155, ¶ 2. Not all biologists distinguish clearly between tropisms and taxes. 'Tropism' is properly restricted to non-motile organisms, principally plants. 'Taxis' applies to motile organisms and is not necessarily restricted to organisms that lack nervous systems; the term is also employed by entomologists.

p. 155, Fig. 5.4. Pupils may come up with numerous interesting explanations but probably few correct ones. At the base of each petiole, on the underside, are enlarged cells that ordinarily are turgid. Stimuli of various kinds—sudden contact, rapid temperature change, electricity, ethoxyethane (ether)—cause release of some diffusable chemical substance that stimulates these cells to lose liquid. The loss of liquid reduces their turgor to a point at which the cell walls collapse and this results in the folding of the leaf. Although folding is almost instantaneous, recovery of the plant may take some minutes. *Mimosa pudica*, incidentally, is fairly easy to grow; seeds may be obtained from biological supply companies.

p. 155, Fig. 5.5. Since *Paramecium* is a motile organism, its responses are obviously taxes. In B and C, the response is positive; in D, response is positive to the bubble of CO_2 but neutral to the bubble of air. Interpretation of A is somewhat tricky; although the organism responds positively to the drop of salt solution, it responds negatively to the high concentration in the centre of the drop. The textbook's terminology will not permit further description, but the teacher can provide additional descriptive terms. For example, since three of the situations involve chemical stimuli, the prefix 'chemo-' could be applied (as in chemotactic), while 'thigmo-' (touch) or 'stereo-' (solid) could be applied to the filter-paper stimulus.

INVESTIGATION 5.1

TROPIC RESPONSES IN PLANTS (pp. 156–8)

In its usage here, the word 'tropic' is usually pronounced with a long 'o'.

It is not necessary to have each group set up each part of this investigation. If a class is divided into six groups, for example, each part may be set up by two groups; every group then has another with which to check its results.

Materials

Maize grains should be soaked for three days before the investigation is set up. The larger the petri dishes, the better. Use non-absorbent cotton wool to pack the grains.

Paper tissues can be used instead of blotting paper, but the grade of paper used in desk blotters forms a better support for seeds and cotton wool. However, coloured blotters sometimes contain dyes that are harmful to the maize grains.

Various substitutes for flowerpots can be used. Milk cartons, with holes for drainage punched in the bottoms, are good. Any kind of small cardboard box, lined with sheet plastic or aluminium foil to keep the moisture for softening the cardboard, will serve.

Unless you plan to grow the seedlings further for some other purpose, the soil may be plain sand or fine vermiculite. Avoid a clay soil.

Zebrina (wandering Jew) is the preferred material for this procedure, but *Coleus* may be substituted.

Procedure

If taping the dish to a table seems unsatisfactory, try mounting the edge of the dish in a wad of modelling clay or plasticine.

Since radish seeds germinate quickly, not more than a week is needed to obtain results. The radish seedlings should be allowed to grow for only two or three days after they break through the soil.

Studying the Data (p. 158)

It is most important that the data obtained be the basis for discussion. This, of course, has been emphasized throughout, but Investigation 5.1 seems a good place to reiterate it. A corollary of this principle is that pupils should not be permitted to generalize on the basis of one experiment: if the radish seedlings in Procedure Part B turn away from the light, and if pupils can find reports of many other experiments in which the converse was reported, then pupils should be permitted to entertain some doubts concerning the validity of their own results.

Part A

E–F. Light may affect the shoots to some extent, and under the conditions of this experiment its effects cannot be separated from negative geotropism.

Part B

A. It is quite possible for individual differences among seedlings to occur. If so, this might be kept in mind for reference in Chapter 3, *MHE*. On the whole, however, cultivated plants show more uniformity than wild, as their breeders have artificially selected them for the characteristic of uniformity.

Levels of Behaviour (contd.) (pp. 159–66)

p. 161, Fig. 5.10. This is a famous demonstration of a 'releaser'. A parent herring gull has a yellow bill with a red spot on each side of the lower mandible, the contrasting spot representing a 'releaser' of tapping behaviour on the part of nestlings. On the models the less the contrast between spot and bill, the less the response on the part of the nestling gulls. Pupils who show particular interest in bird behaviour, including this research on releasers, might enjoy reading Tinbergen's book *The Herring Gull's World* (Collins, 1965).

p. 162, Fig. 5.11. Here merely a tuft of red feathers acts as a releaser, eliciting from a male a response that is associated with territorial behaviour (see p. 173). A whole, mounted immature robin, which has a *brown* breast, and a silvery model of a whole robin did not act as releasers. This study is included in David Lack, *The Life of the Robin* (Penguin, 1953).

p. 164, Fig. 5.12. The results seem to indicate that the rats used in this experiment learned faster when rewarded with food than they did without a food reward. The group that received no food reward until the eleventh day bears this out, but the record does not go far enough to indicate whether the apparent advantage of withholding the food reward is permanent.

p. 166, Fig. 5.14. The structuring of a situation to solve a problem, *if* the situation has not been encountered before, is certainly an example of insight. The 'if', however, is very important and very difficult to control. Lack of such control gives rise to many tales involving popular animals such as dogs and horses.

p. 167, Fig. 5.15. Although learning *per se* may not involve a high degree of intelligence, there seems little doubt today that porpoises are highly intelligent mammals. Considerable behavioural research involving porpoises and their relatives (the mammalian order of whales) has been carried out in recent years.

INVESTIGATION 5.2

BEHAVIOUR OF AN INVERTEBRATE ANIMAL (pp. 167–70)

Materials

Wood-lice are very easy to keep in the laboratory; you may want to set up a colony early in the school year. Large plastic containers about 20 cm diameter make excellent containers. Half fill each one with garden soil, scatter leaves on the surface of the soil, keep soil moist by pouring in some water every few days, cover dish with a piece of cardboard, and place in a fairly dark spot. Feed the crustaceans as suggested in this investigation. Wood-lice reproduce readily in their vivarium; and with a little ingenuity, you should be able to put them to use in other sections of your course—population ecology, life-history studies, reproduction, development, and so on, not to mention the soil ecosystem, since other organisms will be present as well.

Procedure

Part A (p. 168)

A. The animals usually scatter haphazardly in all directions.

B. Over a period of time, the majority move to the blotter end of the box. If any remain at the dry end, they usually cluster in one corner.

C. Most seek shelter beneath the blotter.

D. Usually the organisms tend to congregate in clusters.

The animals tend to move faster in the dry section. More detailed observations have indicated that wood-lice react to higher relative humidity with a slowing down of their movements. This tends to cause the organisms to congregate in moist areas, since they move more slowly there; in dry areas they tend to disperse, since they move more rapidly in such situations. Wood-lice prove very responsive to humidity (or lack of humidity). It is easy to demonstrate (and it often happens unintentionally) how short an exposure to a low relative humidity is required to incapacitate a wood-louse.

E. This may prove difficult to set up. If it is done properly (and the specimen performs 'properly'), the animal pursues an almost-straight-line path equidistant from the two lamps. This is because the organism is receiving equal stimulation from each lamp.

F. Again, if the animal behaves 'properly', it circles around to its left in a pattern called 'circus movement'.

G. Presumably the light; in their normal habitats wood-lice shun direct sunlight—an advantageous behaviour because of the drying effect. In this situation the animal turns left, away from the light, because the left eye is being stimulated by the light.

Part B (pp. 169–70)

A. The wood-louse moves up the T away from the light.

B. There should also be a lamp at the base of the T, as before. The wood-louse should turn down the arm away from the second light—that is, in a direction opposite to its normal turning.

Studying the Data

Part A (p. 170)

A. The answer has been intimated before. In a dry environment wood-lice become desiccated quite rapidly.

B. Again, this answer has been suggested before. Light = sunlight = drying effect.

C. This would seem to be an example of innate behaviour (see pp. 159–60). If one does not restrict the term 'taxis' to animals without a nervous system, then one could say that the wood-lice exhibit complex, negatively phototactic behaviour. If one is actually maintaining a colony of the crustaceans, it would be possible to try the investigation (or some modification) on very young individuals that have been separated from adults. Other ideas may also occur to you and your pupils.

Part B (p. 170)

A. Trial-and-error learning.

B. If, after the light at the end of the T arm was no longer used, the wood-lice still turned in a direction opposite their original turning direction, then their new behaviour was 'learned'.

D. This is a speculative question at this time, but it paves the way for Chapter 3, *MHE*.

For Further Investigation

In this experiment is it the light, or the heat of the light, or the low relative humidity near the light that the wood-louse is avoiding? Imaginative pupils should come up with numerous approaches to the problem: using a heat source without light, using light wavelengths not at the heat end of the spectrum, using dry heat vs. wet heat, and so on. This might make a good project for an eager pupil.

Some Patterns of Animal Behaviour (pp. 170–5)

p. 173, ¶ 6. 'Defending' and 'warning' certainly can have anthropomorphic connotations. But they can also be used to describe behaviour that has certain observable effects in other organisms. Your pupils may be quite ingenious in suggesting substitute terms, but on examination they are likely to have as many anthropomorphic connotations as the words they are intended to replace.

INVESTIGATION 5.3

A METHOD FOR STUDYING TERRITORIALITY (pp. 175–7)

To emphasize that behaviour tends to be diverse even under apparently identical conditions, two set-ups and two groups per class are desirable. Also, the materials for this investigation occupy considerable space and must remain in position for a considerable time; therefore, it may be necessary to use the whole class as a 'group'. In this case it may be possible to exchange observations between different classes.

Materials

Avoid round jars if possible; battery jars or tanks with straight sides can be obtained and these enable fish in adjacent jars to see each other clearly.

The best water to use is that in which the fish have been living. Tap water, however, can be 'conditioned' by allowing it to stand for about a week. Be sure the temperature of the water in the jars is the same as that of the water from which the fish are transferred.

Betta (Siamese fighting fish) and *Corydoras* (a catfish) may be obtained from pet shops.

Procedure

In Experiment F do not allow pupils to spend too much time on preparation of the models. The colouration need only be approximate. If the jars are set up on a Monday, Experiments A–F can be completed during the first week. Then the transfer of the three *Corydoras* to the large jar would occur on the following Monday. Removal of one *Corydoras* would occur on Monday a week later, and the last experiment would occur the next day. Taken together, Experiments A–E require approximately one and a half periods. Each of the other experiments requires only part of a period.

Conclusions

This investigation gives pupils a greater opportunity to organize ideas than did most of the preceding ones. After many months of experience in drawing conclusions through rather closely guided steps, most pupils should be able to see the plan involved in this series of experiments. Do not, however, allow a written report to be the end of the investigation. There should be differences in interpretation, and (if more than one set-up is involved) differences in observation. These should be fully aired in a class discussion.

Some Patterns of Animal Behaviour (contd.) (pp. 177–83)

p. 178, ¶ 1. The reaction of many men to birdsong by writing poetry is a matter on which there can be much discussion, mostly centring around the term 'communication'. One view might certainly be that the meaning the poet obtains from the singing is foreign to the motivation of the singing by the bird, and therefore no meaning has passed from bird to poet—hence no communication has occurred.

p. 178, ¶ 2. Other examples of communication by scent are the male moth seeking out the female by scent (p. 159). It is also suggested (p. 153) that dogs do some communicating by scent. Many mammals, including dogs, have urination posts which serve as a means of communication through scent.

p. 179, Fig. 5.21. You might raise the question of how such a colour pattern might have originated. This can lead on to discussing the matter of purposive action on the part of the moth and again pave the way towards Chapter 3, *MHE*.

p. 180, Fig. 5.22. See Shaw, 'The Schooling of Fishes' (reference: p. 186). Shaw postulates a number of advantages: confusing predators, locating food more readily, stimulating feeding, facilitating movement through the water as a group, and possibly giving some reproductive advantage.

p. 181, ¶ 1. For all their social 'virtues', ants have a very restricted nervous system and their behaviour is essentially innate. Man makes heavy use (ideally) of his cerebral hemispheres in his learning and reasoning; ants possess no such complex structures and certainly do no reasoning. However, a number of interesting behavioural investigations can be carried out with ants, an imaginative reference being S. H. Skaife, *The Study of Ants* (Spottiswoode, Ballantyne and Co., 1962).

INVESTIGATION 5.4

PERCEPTUAL WORLDS (pp. 183–4)

This is a relatively unstructured investigation. On a class-wide scale there are serious difficulties involved in such investigations, and the values to be derived from experience in developing hypotheses and devising appropriate experiments can probably best be achieved through individual and small-group work. However, something can be gained by occasionally setting up an open situation before an entire class.

After the pupils have grasped the nature of the investigation, you should present a list of suitable organisms. This might be supplemented with pupil suggestions. Animals already living in the laboratory and somewhat familiar to the pupils are perhaps best, for example *Daphnia*, brine shrimp, mealworms, planarians, frogs, and (after doing Investigation 5.2) wood-lice. Easily obtained organisms come next, for example earthworms, pond snails, locusts. But there is some gain in interest when less familiar animals are used, for example silkworms, cockroaches, salamanders, lizards.

A day should be allowed for consideration before choices are made. To avoid duplication, obtain first and second choices. Of course, all groups might use the same organism, but variety is desirable.

After the specimens have been obtained, allow each group at least ten minutes to become generally acquainted with the behaviour of the species it chose. If the organism is a quite unfamiliar one, time should be increased. The remainder of the period may be used for group planning.

You then need several days to review the plans and the equipment lists. Several more days are required for gathering the materials and equipment before a laboratory period can be scheduled. Whether one or two periods are needed will depend, of course, upon the plans. However, elaborate plans requiring more time should be avoided. Encourage simplicity and ingenuity.

It is essential that you guide pupils away from all inhumane procedures, even, if necessary, employing your veto power.

PROBLEMS (pp. 185–6)

1. Attempting to distinguish between 'reflex' and 'instinct' should introduce the pupil to the problem of definition in behaviour. If a reflex is a *direct* and immediate response to an environmental stimulus, as some suggest, then the human infant's grasping reaction is a 'reflex'. Yet the case of a young squirrel 'instinctively' burying a nut may be no different from a young child grasping, except perhaps in degree and situation. Certain behaviourists conclude that 'instincts' are 'mere bundles of reflexes', while others feel that instincts *per se* must be accompanied by consciousness. Pupils should be made to appreciate that the search for terminology must not become more important than the phenomenon being investigated. The past adaptive value, from an evolutionary survival standpoint, will be apparent to anyone who has watched films of primates in the wild or has observed them in zoos. Young primates grasp their mothers in order to be carried. Arboreal primates must early be able to grasp branches to keep from falling out of trees. It is interesting to observe female monkeys in zoos pushing their very young infants against the cage bars and leaving them hanging by their own hands. This behaviour strikes zoo visitors as being 'heartless' (or some such anthropomorphic term), but it is of adaptive advantage to the monkeys. The adaptive value of this behaviour for human infants today is presumably negligible.

2. One obvious approach would be to isolate a specimen of each of these organisms from others of its kind at 'birth' and see if the particular behavioural pattern becomes apparent in the isolated animal. Many investigations of this type *have* been carried out. In most cases the basic 'talent' would seem to have an instinctive foundation, with refinement being added through learning. For example, with respect to the singing of European chaffinches (illustration: p. 171), individuals raised by themselves can produce a very generalized 'chaffinch song', one which lacks, however, the detailed pattern of the typical song.

3. Pupils should be able to generalize that the greater the care of the young, the fewer the number of offspring. There are many sources of pertinent information including biology textbooks and encyclopaedias; but the information may be scattered.

4. The birth of a mammal introduces it to an entirely new set of environmental stimuli. Up until birth the foetus has been receiving oxygen from its mother, and its lungs are nonfunctional. At birth the oxygen supply from the mother is cut off. Then carbon dioxide builds up in the blood of the offspring and activates the respiratory centre in the brain, causing respiratory movements. Meanwhile, the respiratory muscles associated with the ribs and diaphragm may have become more responsive with the changed temperature and gas content of the new environment, facilitating the breathing movements. It would seem, thus, that the new respiratory behaviour is not learned. On the other hand, there has been an extended dialogue about the status of a baby chick's pecking behaviour. As soon as baby chicks hatch, they begin pecking at the ground. Is this behaviour learned, or is it innate? A baby chick uses a pecking movement of the head to crack the eggshell during hatching. This certainly couldn't have been learned, since the unhatched chick apparently has no way to learn it. Yet there are those who suggest that the embryo chick has picked up this pecking movement of the head from the beating of the embryonic heart, near which the head rests during embryonic development. Whatever the explanation, it *is* true that baby chicks learn *what* to pick up by trial and error, at first pecking at anything and not until later, through experience, pecking only at food.

5. There is limitless information on the migration of birds both in books and in periodical articles. Any ornithological text will have a good discussion, while excellent and comprehensive coverage is to be found in A. J. Marshall (Editor), *Biology and Comparative Physiology of Birds* (Academic Press, 1961), Vol. 2, pp. 307–39. Actually, the discussion in the textbook (pp. 171–5) represents a good starting point for this particular problem. A more comprehensive listing can be compiled from *Biological Abstracts*, if this publication is available.

6. If pupils, by this stage in their biological educations, can appreciate that the human being is related to other living things, then they may believe that studies of non-human behaviour might have some relevance to the study of human behaviour. The following two books relate the behaviour of other organisms very strongly to human behaviour: Lorenz's *On Aggression*, and Morris's *The Naked Ape* (references: p. T160). These two, by the way, have been best sellers, which emphasizes the amount of general interest in the subject.

7. Butting order in cows, pecking order in chickens and pigeons, status seeking in people—all are examples of social hierarchy. There is much literature, both popular and technical, on the subject. Two good periodical references are A. M. Guhl, 'The Social Order of Chickens', *Scientific American*, Feb. 1956, pp. 42–46; and S. L. Washburn and Irven DeVore, 'The Social Life of Baboons', *Scientific American*, June 1961, pp. 62–71. Among pertinent and readable books is H. Munro Fox, *The Personality of Animals* (Penguin Books, 1952), see pp. 104–10. Although we usually think of social hierarchy with respect to members of a single species, there are also examples in nature of social organization between members of different species that group together; for example, small winter birds and, in zoos, large wading birds, penguins, and monkeys. Indeed, if pupils have access to a zoo, they might ask keepers about existing social hierarchies among some of the captive mammals and birds. Any pupil who pursues the problem of social hierarchy will turn up all sorts of interesting information: injecting male hormones into a low-order chicken will raise her in the social hierarchy. A good milk cow at the top of a butting order in one dairy herd may cease to give milk if moved to another dairy's herd; lion tamers in a circus assume the top of the hierarchy in the lions' cage, if luck remains with them; and schoolchildren who move to other schools in mid-year have to work their way into the existing hierarchies, often (for the boys) at the expense of black eyes.

AUDIOVISUAL MATERIALS

Filmstrips

Symbiosis. Darwin's World of Nature Series. *Life* Filmstrips. (Concentrates on interspecific behaviour.)

Films

Social Behaviour in Chickens. BSCS Inquiry Film Loop, available from John Murray. (Observation of behaviour leads to the concept of social hierarchy.)

An Example of the Biological Significance of Colour. BSCS Inquiry Film Loop, available from John Murray. (Leads to design of an experiment to determine the role of colour in food selection by tortoises. Useful in leading up to Investigation 5.4.)

Mating Behaviour in the Cockroach. BSCS Inquiry Film Loop, available from John Murray. (The pupil develops hypotheses from observation of male and female behaviour. Illustrates the study of communication behaviour.)

Temperature and Activity in Reptiles. BSCS Inquiry Film Loop, available from John

Murray. (Relates the study of physiology to the study of behaviour.)

Plant Tropisms and Other Movements. Coronet Films, available through Gateway Educational Films. 16 mm, colour, 11 min. (Uses time-lapse photography to show a variety of kinds of movements in plants.)

Courtship in Birds. Ealing. Super-8 film loop, colour. (Behaviour of four large species is shown: black-footed and laysan albatrosses, whooping crane, and wild turkey.)

Behaviour. AIBS Film Series. McGraw-Hill Publishing Co. 16 mm, colour, 28 min. (Good for introducing the topic.)

TEACHER'S REFERENCE BOOKS

CARTHY, J. D. *Animal Navigation*. Allen & Unwin, 1956.

CARTHY, J. D. *An Introduction to the Behaviour of Invertebrates*. Allen & Unwin, 1957.

DALE, A. *An Introduction to Social Biology*. Heinemann, 1946.

FOLLANSBEE, H. *Animal Behaviour*. Heath, 1965. (BSCS Laboratory Block which contains much information on methods of studying behaviour in the school laboratory.)

FRAENKEL, G. S., D. L. GUNN. *The Orientation of Animals*. Dover, 1961.

GOODALL, J. *In the Shadow of Man*. Collins, 1971. (An enthralling account of ten years' work among chimpanzees.)

HILL, W. *Learning*. Methuen, 1964.

LORENZ, K. *Man meets Dog*. Methuen and Penguin, 1954.

LORENZ, K. *King Solomon's Ring*. University Paperbacks. Methuen, 1961.

LORENZ, K. *On Aggression*. University Paperbacks. Methuen, 1967.

MORRIS, D. *The Naked Ape*. Cape, 1967.

MUNN, N. L. *The Evolution and Growth of Human Behaviour*. Houghton Mifflin, 1955.

TINBERGEN, N. *Social Behaviour in Animals*. Butler & Tanner, 1965. (A well written coverage of the subject by an authority.)

WYNNE-EDWARDS, V. C. *Animal Dispersion in Relation to Social Behaviour*. Oliver & Boyd, 1962. (Elaborates the theory of population homeostasis through the development of social behaviour. The theory leaves much to be desired but the material used in explanation is most informative.)

MAN AND HIS ENVIRONMENT

Abundant evidence points to the conclusion that even on the time scale of Earth history, life is not a recent phenomenon. Even within their own short span of observation, pupils will have seen how mortality and natality are constantly changing the players as the play continues. And from this it is not difficult to extrapolate both backwards and forwards in time. The resulting concept of biological continuity is one of the most valuable acquisitions for a developing human mind. It is not only a key concept in the structure of modern biology; it is also a firm basis for a system of ethics. In a biology course that has a humanistic bias, this latter is a permissible consideration.

Perhaps the most important task in working with *Man and His Environment* is to maintain the sweep of ideas through the first three chapters without becoming entangled in too much detail. Without care this can be lost in the details of chick development or the manipulation of fruit flies.

Nevertheless such live materials are needed to maintain reality in the biology course, for the ideas are increasingly abstract. In Chapter 3 first-hand experience becomes very difficult to achieve. Everything possible should be done to use visual aids in defence of credibility.

The last two chapters are not, though they might appear at first sight to be, an afterthought. They are the goal towards which the course has been moving from the start. It therefore follows that these two chapters should not be pushed to one side at the end of the year's work.

1
REPRODUCTION

MAJOR IDEAS

1. Every individual organism has one or two precursors—parents. The discrediting of the spontaneous generation theory left only reproduction to account for the addition of new individuals to populations.
2. Reproduction (here considered as a process involving whole organisms as opposed to replication of molecules within individuals) is characteristic of living things but is not essential for the existence of any single individual. Yet, because individuals die, it is a process essential for the survival of the species and for the continued existence of the biosphere.
3. Asexual reproduction is found in all three kingdoms; it is most common among protists and least common among animals. Because the process does not allow for random recombination of chromosomes (an idea to be more fully developed in Chapter 2), organisms that reproduce only asexually are essentially in evolutionary blind alleys.
4. The basic point in sexual reproduction is the union of two cells (gametes), each normally from a different parent individual and usually morphologically distinguished as sperm cell and ovum. Upon these gamete distinctions rests the distinction between male and female.
5. Because chromosome number is constant in a species and fertilization doubles the number in the gametes, there must be some mechanism that reduces the number of chromosomes at some time between successive fertilizations. This mechanism is meiosis.
6. Characteristically a plant alternates a sexually reproducing gametophyte generation with an asexually reproducing sporophyte generation.
7. Such an alternation of haploid and diploid generations is unknown among animals. Further, most animal species have definite, separate sexes.
8. In both animals and multicellular plants, an adult organism develops from a zygote not merely by a proliferation of cells through mitosis but by a process of cellular differentiation. Thus embryonic development is basically a study of growth and differentiation of cells.
9. Adaptations such as timing mechanisms that maximize the chances of fertilization, internal fertilization, and protection and nourishment of the young are correlated with the production of fewer eggs as the chances of development to adulthood are increased.
10. Human reproductive patterns afford maximum opportunity for fertilization and provide for highly developed prenatal and postnatal care. The complex regulation of the menstrual cycle indicates a hormonal feedback system that is an excellent example of regulation.

PLANNING AHEAD

All should now be in readiness for the genetics investigations. The principal preparations for the remaining chapters require not so much the obtaining of materials as the thinking through of procedures. All the investigations in Chapter 3 are 'dry runs'; that is, the data have already been collected (or, as in Investigation 3.1, hypothesized), and the investigational work involves the manipulation of ideas rather than of equipment. If you are to be of maximum effectiveness as a teacher, you must understand the investigations so thoroughly that you can guide the thinking of your pupils, not merely provide answers.

GUIDELINES

Establishing the special place of reproduction among life processes is probably the best approach to the work of this chapter. If Flask 7 from Investigation 3.4 in *DLT*, has been saved (as was suggested), it can now be brought forth as a silent witness against abiogenesis.

Throughout the work on reproduction, it is important to have in the laboratory an abundance of living materials illustrating reproductive processes.

It is easy to assemble collections of organisms displaying the phenomena of asexual reproduction. Strawberry plants, onions, *Bryophyllum* and potatoes demonstrate various kinds of vegetative reproduction. Budding yeasts, fissioning *Paramecium*, and even budding *Hydra* are not difficult to obtain. Spores from bread moulds are easy to demonstrate. Among plants sexual processes are not very obvious, and recourse to prepared slides may be necessary, but flowering plants can certainly be provided in variety. For animals the problem of timing is difficult to overcome, but guppies and pregnant rats or mice can always be obtained. If this chapter is undertaken at a season when frogs in amplexus can be secured, any fertilized eggs obtained from them can be used as a supplemental study of embryological development.

The primary aim of the chapter is to establish a rational, objective and integrated view of reproduction as a basic (perhaps *the* basic) process in biology. Secondarily, the chapter provides a background for later development of concepts of heredity and evolution.

In pursuing these aims, pupils will undoubtedly gain perspective on their own personal interests in reproductive processes. During the last parts of the chapter, they may be encouraged by the scientific atmosphere to seek answers to personally perplexing questions. You should do what you can within your competence to deal with such questions. But this is a *biology* course. Just as the authors have stressed ecology rather than conservation, nutrition rather than dieting, and the dynamics of host–parasite relationships rather than 'health habits', so they insist that the biology of reproduction—not sex education—is the proper emphasis for the biology course.

A suggested division of the chapter for assignments is as follows: pp. 1–5 (guide questions 1–5), pp. 8–12 (guide questions 6–10), pp. 14–23 (guide questions 11–14), pp. 23–30 (guide questions 15–23), pp. 36–43 (guide questions 24–31). Fairly short assignments are desirable in this chapter, partly because Investigations 1.1 and 1.3 are long and must overlap several topics, if not each other. However, an alternative assignment scheme might be: pp. 1–14, 14–26, 26–43.

TEACHING NOTES

Why Reproduction? (p. 1)

p. 2, Fig. 1.1. Redi's experiment is not difficult to duplicate in warm weather. The conclusion derived from the results pictured seems very obvious, but it passes over the heads of some pupils unless you have other pupils explain it.

Types of Reproduction (in part) (pp. 2–5)

p. 2, ¶ 3. Asexual reproduction of potatoes is easy to demonstrate.

p. 4, Fig. 1.3. *Bryophyllum* is easy to grow in the laboratory.

p. 5, Fig. 1.4. Not only are slides of budding *Hydra* available for demonstration purposes from biological supply companies, but living *Hydra* kept in aquaria in the classroom can frequently be observed undergoing budding.

p. 5, ¶ 3. Note that the 'spores' of bacteria do not conform to this description. Bacterial spores are not really reproductive structures at all, but rather a means of circumventing unfavourable environmental conditions.

p. 5, ¶ 4. If 'fruiting' plants of species of *Lycopodium* are collected in the autumn and dried in a plant press, in spring the specimens will release visible showers of spores when tapped lightly. The spores can then be observed under the microscope. Or bread mould (*Rhizopus*) may be used to demonstrate sporangia and spores.

INVESTIGATION 1.1

VEGETATIVE REPRODUCTION (pp. 6–7)

This investigation is concerned primarily with the concepts of experiment, observation, control and conclusions. The pupils should be able to explain the rationale behind the experimental design and to discuss the concept of control. Some pupils may think that the original plant is the control in this experiment. Why isn't it? Does every experiment require a control? Recall Investigation 1.2 in *WL*, item E of 'Discussion', p. 10. (Cutting A is as near to being a control as any cutting in this investigation.) The ability of the pupil to observe will largely determine the quality of his conclusions. The conclusions should relate directly to *Coleus* as an organism capable of vegetative reproduction but one that does not normally use it as a mechanism for perpetuation of its species. Observations will lead to conclusions regarding regeneration.

Materials

Almost any container that allows good drainage may be used. The saucer permits watering from below, so that the plastic bag need not be disturbed.

Vermiculite may be substituted for sand. If possible the medium should be heat-sterilized in an oven before use.

Coleus is a common plant, easily obtained and likely to give good results. *Pelargonium* may be used, but results appear more slowly, and there is likely to be more difficulty with moulds. The plants from which cuttings are taken should be young and vigorous, with several side branches. The directions must be modified slightly if plants with alternate leaves are used.

Procedure

Good comparisons can be made if different species of plants are used by different groups, but replication should also be provided. One scheme is to have all groups in a class use the same species but have different classes use different species.

If groups must be large, each pair of pupils in a group may be made responsible for setting up one of the cuttings.

A quantitative aspect can be introduced into this experiment by having the pupils measure the growth, if any, on various cuttings and graph the data.

A. Wounds on *Coleus* and most other plants heal quite quickly unless the damage has been extensive. At the terminal portion of the branch, not just one but several new branches may grow.

B. In all probability, all four cuttings will still be alive; D is more likely to die than is either A or B. The evidence of pupils for whether the part is alive or not will be varied and will form the basis for an interesting classroom discussion.

C. If roots develop, they come either from the callus (a mass of poorly differentiated cells covering a wound) at the bottom of the branch or from the stem immediately above it. They first appear as somewhat shiny swellings on the stem and later break through the epidermis and resemble ordinary roots.

D. The cut surface at the bottom of the branch is most likely to be covered with a layer of callus.

E. Depending on the situation, the tip of the cutting may have wilted or be in the process of growth.

F, G. See C and D above.

H. The cut surface will probably be dried out, and the cut tip may even shrink back to the node where the leaves arise. Any new growth of stem or branches occurs from axial buds, but this growth does not usually occur until well developed roots have formed.

I–N. See comments above. Root development is unlikely in Cuttings C and D.

Studying the Data (p. 7)

A. The healing of a wound (callus formation) on the surface of a living plant involves regeneration. This may not be very obvious to pupils; but if new branches have grown (as usually occurs in Cutting B), this should be obvious evidence for regeneration.

B. This question must be answered with respect to the experimental cuttings. Which types of broken branches have the best chance of survival if they become embedded in the earth?

Present the following statement to pupils for discussion: Vegetative reproduction as a result of accidental fragmentation is probably not important to the *Coleus* population, but it is of importance to willows growing on the banks of streams and to many plants that grow in shallow water.

Types of Reproduction (contd.) (pp. 8–12)

p. 8, Fig. 1.6. A demonstration of the sexual reproduction of *Rhizopus nigricans* by plus and minus strains can be performed by the pupils. The necessary living materials can be obtained from biological supply companies.

p. 9, ¶ 1. The mention of mitosis as a means of increasing cell number is not accidental here. This is a good time to review the material on mitosis, beginning on p. 16 in *LO*, before introducing the process of meiosis.

p. 10, ¶ 3. The branch of human knowledge in which truth is established by reasoning alone is mathematics. Such branches of human knowledge as theology and philosophy must be excluded because of the difficulties of establishing what is 'truth' in these areas.

p. 11, Fig. 1.8. The diploid number of chromosomes is four.

p. 12, Fig. 1.9. Crossing-over has little significance at this point. But since it occurs regularly in meiosis, it deserves to be mentioned as part of the process. Throughout the course pupils have encountered facts that were not immediately explainable. And if they are getting the spirit of the course, they may quickly ask, 'Why does crossing-over occur?' Put in that very common form (and better pupils may no longer be using such a form), the question is, of course, basically unanswerable. Historically, crossing-over was at first merely a cytological curiosity, as it must be here to the pupils. However, it did turn out to have significance; and in the next chapter its significance will be presented to your pupils.

INVESTIGATION 1.2

MEIOSIS (pp. 12–14)

If you have used the model technique in connection with mitosis in Chapter 1 of *LO* (see pp. T118–19), the general scheme of the procedure will already be familiar to the pupil. Otherwise some time will be required to establish the identity of the model materials.

Materials

Pipe cleaners may be used as substitutes for Poppit beads, but they are not as suitable as they were for simulating mitosis, because they have to be cut and the pieces twisted together to show crossing-over. Further, when the 'chromosomes' are grasped at the 'centromeres' and pulled towards the poles, pipe cleaners do not assume characteristic shapes. If pipe cleaners must be used, they may be dyed to obtain two colours.

If enough materials are available, three pairs of homologous 'chromosomes' may be constructed. A very short pair will clearly show the relationship between chromosome length and the probability that crossing-over will occur.

Procedure and Discussion

In this investigation the manipulation of the materials achieves the purpose, but a number of points can be made in discussion, either during the procedure or afterwards. One worth considering (because of its bearing on matters in Chapter 2) is the random distribution of paternal and maternal chromosomes during synapsis. Since no directions are given for arranging the colours, there is likely to be considerable variation among groups when the 'chromosomes' are placed in homologous pairs. The consequent separation of paternal and maternal chromosomes may be pointed out, without necessarily developing any genetic implications at this time.

When work with this model of meiosis has been completed, pupils should be able to contrast and compare the processes of mitosis and meiosis as well as to review the usefulness of biological models.

Note: This investigation may be repeated when recombinations are dealt with (Chapter 2). The beads then represent genes. If the beads are marked with washable ink, the genetic effects of crossing-over can readily be seen. For example, the events shown in Fig. 2.17 can be duplicated; or crossing-over can be demonstrated at different distances from the centromere; or, with strands of 15 or more beads, double crossing-over can be shown.

B. The size and manipulability of the model permit easy viewing of the events under consideration. The action can be stopped at any stage, reversed, or repeated for complete comprehension. This would not be the case with living material.

C. The model has the disadvantage of all models: it is not the real thing. A model is a form of analogy and has, in some degree, all the advantages and disadvantages attached to the verbal form. (Refer to Investigation 2.1 and pp. 59–60, *WL*.) A simulated series of events cannot anticipate the variations in a living system, and by its very nature the model takes some liberties and shortcuts to illustrate aspects of the process under consideration. At every point where a model differs from the biological reality—for example, size, colour, materials, metabolic activity—danger of misunderstanding lurks.

Patterns of Reproduction (in part) (pp. 14–30)

p. 14, ¶ 2. Characteristics associated with evolutionary history are likely to be of importance in classification primarily because they are basically stable within the genetic complements of the organism and can be used to categorize organisms with similar features. But, further, if a classification is to reflect ideas of genetic relationship, characteristics associated with evolutionary history are the materials from which a phylogeny can be constructed.

p. 15, Fig. 1.11. In *Ulothrix*, gametes are morphologically all of one kind. Therefore, they cannot be considered either ova or sperm cells; they are simply gametes, often referred to as isogametes.

p. 15, Figs. 1.12, 1.13. These provide a basis for a discussion of relative advantages of motile and non-motile gametes.

pp. 14–23. The three examples of plant reproduction used in this section are designed to communicate two ideas: (*a*) the alternation of haploid and diploid, sexually and asexually reproducing generations in plants, and (*b*) the reduction, in the course of evolution, of the haploid gametophyte generation and the concomitant dominance of the diploid sporophyte generation. The second of these purposes dictates the choice of *Selaginella* as an example in place of the more familiar forms. Nevertheless, some suggestions for work with ferns are given on pp. T170–3.

p. 16, Fig. 1.14. You should not expect the details of this figure or of the similar Figs. 1.16 and 1.21 to be memorized. They should be compared and used to establish the evolutionary decline of the gametophyte generation.

p. 17, ¶ 3. *Selaginella* is not likely to be familiar to pupils. But it is fairly easily cultured in a terrarium or greenhouse, so living material should be available in the classroom. Biological supply houses and botanical gardens provide stocks of some species.

p. 19, ¶ 2. In angiosperms the distinctions between ovum and sperm are not clear because both gametes are reduced to mere nuclei. On the basis of motility, the nuclei produced by the pollen grain can be considered male. However, the chief basis for distinction lies in the homology of pistil structures with megaspores of more primitive tracheophytes and of pollen grains with microspores, which (in *Selaginella*, for example) produce undoubted sperm cells.

p. 20, ¶ 2. Refer to Fig. 2.5, *DLT*. The diversity in pollen grains may provide an interested pupil with a fine topic for microscopic investigation. This can be started now with house or greenhouse plants; it can be extended to wild flowers later in the spring.

p. 21, ¶ 1. The relationship between plants and insects is mutualistic since plants benefit from insect visitation in being pollinated; the insects benefit by obtaining food (nectar). This form of mutualism has been investigated extensively with respect to mechanisms, but little quantitative assessment of 'benefit' has been made in the sense discussed on pp. 87–88, *WL*.

p. 24, ¶ 1. An oyster's sex is perhaps best described as 'indeterminate'. When the gonads produce ova, the animal is female; when they produce sperms, male.

p. 24, ¶s 1–2. Interested pupils may investigate the processes of spermatogenesis and oogenesis. which help to explain why sperm cells are smaller and more numerous than eggs. But they should be cautioned that the usual description—in close association with meiosis—does not apply to these processes in most plants.

p. 25, ¶ 2. To most students hermaphroditism and parthenogenesis (pp. 25–26) are highly abnormal. But what is abnormal in one group of organisms may be normal in another. A definition of 'normal' may be necessary.

p. 25, ¶ 5. The contrast between internal and external fertilization provides examples of the correlation of structural adaptations and function, such as the reduced production of eggs and behavioural changes that increase the efficiency of fertilization.

p. 25, ¶ 7. Parthenogenesis occurs in the context of sexual reproduction and can be described meaningfully only as a variation of it. However, in parthenogenesis the basic point of sexual reproduction, the union of two cells to form a new individual, is lacking. Thus, there is room for your pupils to debate, but be sure that they realize that it is a verbal, not a scientific, issue.

p. 27, ¶ 2. *Hydra* seldom produces sexual structures in the school laboratory; the stimuli for sexual reproduction seem to be the relative concentrations of oxygen and carbon dioxide in the water. Prepared slides of *Hydra* bearing gonads, and of budding *Hydra*, can be obtained from biological supply companies.

p. 27, Fig. 1.26. Note *Hydra oligactis*, as shown in the figure, always has two parents because the sexes are separate. But (p. 27, line 4) most other *Hydra* species are hermaphroditic; thus a pupil can reasonably conclude that a sperm cell might fertilize an ovum on the same individual from which the sperm cell came, producing a zygote that had a single parent.

p. 28, ¶ 3. Earthworms kept in a terrarium can occasionally be found in copulation. Earthworm cocoons, resulting from the secretions of the clitellum, can also be found.

p. 29. You may find in your classes one or two pupils who are tropical fish fanciers. If so, they may be able to set up aquaria with breeding fish.

INVESTIGATION 1.3

CHICK EMBRYOLOGY (pp. 31–36)

This investigation is a good test of the extent to which (*a*) a class has developed group coordination, and (*b*) individuals have developed manual dexterity and the habit of careful attention to directions. The investigation requires a considerable degree of skill, and it is to be expected that some groups will be unsuccessful.

You must become familiar with all techniques before attempting to lead a class through them.

Preparations

Fertile eggs may be obtained from hatcheries or from poultry farmers who keep cockerels in their flocks. In ordering, make some allowance for infertile eggs.

To secure embryos that correspond in development to those seen in the standard illustration of chick embryos at 24 hours, 33 hours, 48 hours, etc., keep the eggs cool (*but not below 10 °C*) until they are incubated. When eggs are stored for more than a week after they are laid, viability of the embryos is greatly reduced. During incubation of the eggs, temperature should be kept between 37 and 39 °C; and humidity, 50 per cent or more. Because sufficient oxygen must pass through the shells to the embryos, some air space must surround each egg. Using a soft pencil, mark each egg with the date and time at which incubation is started. To prevent the embryos from sticking to the shells, rotate the eggs daily. The marks will enable you to keep track of the rotation.

Set up a schedule for incubating the eggs. In the 48-hour eggs, variations of even a few hours can produce considerable differences in the appearance of the embryos. Therefore, when putting eggs for this stage into the incubator, consider the hour when each class meets. The exact hour at which incubation starts is less important for later stages, however. Note that incubation of some eggs must begin *twenty-one* days before use.

Suggestions for improvising an incubator may be found in R. E. Barthelemy, J. R. Dawson, Jr. and A. E. Lee, *Equipment and Techniques for the Biology Teaching Laboratory* (D. C. Heath & Co., 1964). Before use, thoroughly test the incubator (whether improvised or purchased) for its ability to maintain a steady temperature over the required time.

Materials

Plastic refrigerator dishes may be substituted for crystallizing dishes.

For the opening of early-stage eggs, cotton wool forms a better nest than paper tissues; but it is more expensive.

Scissors must have fine points. Ordinary dissecting scissors are not satisfactory.

Petri dishes may be substituted for solid watch glasses.

The chick physiological saline is different from that used for frog material. To prepare it, dissolve 9 g of sodium chloride in 991 cm^3 of distilled water. Assign a pupil laboratory assistant to maintain the solution at a temperature of 37 °C and to deliver it to groups as needed.

Procedure

Part A can be done as a demonstration. Two or three eggs, set up around the room, can be observed by small groups of pupils while other work is in progress. Parts B and C each requires a full laboratory period. Or both can be done in one period if Part B is assigned to some groups and Part C to others; the materials are then exchanged between groups for observation by all.

Part A. The unincubated egg (p. 31)

A. Albumen is nearly homogeneous in appearance except for the chalazae, two stringy masses projecting from the yolk along the long axis of the egg. An especially observant pupil may also note that near the yolk the albumen is somewhat more dense than it is peripherally.

C. This space between the membrane and the shell is occupied by air.

D. The middle of a laboratory investigation is not the time for book research. The answer to this question—albumen, shell membrane, and shell—should be deduced from observation.

E. The pupil is not in a position to provide detailed answers, so accept any reasonable function. Protection, for example, is an acceptable answer for all three layers. The albumen is not a food source, although many pupils may suggest this. Some may guess it to be a water source. For the most part problems of respiration, dehydration, etc., will probably not occur to pupils at this time.

Part B. The two-day embryo (p. 32)

A. If development is normal, approximately the top half.

B. Pupils may have a number of ideas; but because the yolk sac is highly vascular, many should suggest that it transfers good material from the yolk to the embryo.

D. Normally, the heart begins regular contractions after about 44 hours of incubation. The actual rate must be determined by the pupil.

E. The answer varies with observational ability, but most pupils should be able to see most of the structures in Fig. 1.31.

12. In the 33-hour egg, the embryo is oriented more or less in a straight line. By 48 hours the anterior end has greatly enlarged and there is a flexing of the body to one side, the heart being prominent in the concavity of this flexure.

F. It is difficult for pupils to trace the path of the blood. It passes from the heart into the ventral aortas, along the dorsal aortas, and out through the omphalomesenteric arteries to the plexus of vessels on the yolk. It then returns to the heart from the extra-embryonic vitelline circulation.

Part C. The five-day embryo (p. 34)

A. It should not be difficult for a pupil to ascertain that the amnion is liquid-filled.

C. Normally at five days the entire yolk is covered by the sac.

D. The wording of this question may encourage more imagination than observation. At this stage, the limb buds have just become visible and appear as somewhat flipper-like appendages, like neither wings nor legs. The only real basis for distinction is position.

8. Pupils can make a wide variety of observations. One should concern the increased torsion and flexure of the five-day embryo. The somites have become much more numerous, better defined, and larger in the anterior half of the body; the arterial arches have increased from two to four; the arteries and veins are better defined; and, in addition, pupils should notice both size and position changes.

Part D. Later stages of development (p. 34)

A. Primarily, oxygen and carbon dioxide.

B, C. The yolk sac is connected to the embryo in the abdominal region by means of a yolk stalk. The yolk sac becomes progressively smaller, finally being drawn into the body and incorporated into the small intestine, where it is still present several days after hatching.

D. The food in the yolk is absorbed by the blood vessels on the surface of the yolk sac and carried by blood vessels through the yolk stalk, the liver, the heart, and ultimately the rest of the developing embryo. There is an opportunity here for pupils to go into some detail by recalling ideas from Chapter 4, *LO*.

Pupils frequently wish to preserve their specimens. This can be done in 70 per cent ethanol. The small embryos on the filter paper rings can be preserved in vials; larger embryos, in wide-mouth jars. Because water in specimens dilutes the preservative, drain out the ethanol after two days and replace.

Studying the Data (p. 35)

This investigation constitutes the principal attention to embryology in this course. Consequently discussion should be thorough, and the points scattered through pp. 23–43 should be woven into it. Opportunities for discussion should occur after each of the parts of the investigation, and a concluding comparative discussion should be based on the written summaries. Following are some comments on the questions which precede that item:

(*a*) The extra-embryonic membranes allow the chick egg to develop on land.

(*b*) The shell and albumen would be of less importance if the egg developed within the hen.

(*c*) The circulatory system must develop early in order to carry food through the entire embryo and to remove waste products.

(*d*) Segmentation is seen primarily in the somites and in the nervous system.

(*e*) Pupils should have seen a dorsal hollow nerve cord but probably did not recognize the notochord. The aortic arches indicate the presence of pharyngeal pouches. However, any reasonable combination of ideas that indicates the pupil has synthesized his observations should be acceptable.

Patterns of Reproduction (contd.) (pp. 36–43)

p. 36, ¶ 1. *Note:* the eggs of *ovoviviparous* animals develop internally, utilizing only food in their yolks, and the young are essentially hatched within the body of the female before encountering the environment. This is common among fishes and reptiles. Encourage your pupils to get examples and details, not merely dictionary definitions.

p. 36, ¶ 2. Just as the development of social hierarchy (see BSCS film loop *Social Behaviour in Chickens*, reference: p. T159) conserves energy within a closely associated group of animals, so also does seasonal sexual behaviour. The behavioural patterns associated with cyclic reproduction are well developed in birds that fly to high-latitude breeding grounds in spring and, after raising their young, fly equator-wards again in autumn.

p. 36, Fig. 1.35. This and Fig. 1.38 are better than nothing, but much more detailed anatomical illustrations should be available. Better still is a 'skeletorso' with interchangeable inserts of reproductive parts.

p. 37, ¶ 4. Emphasize the extirpation experiment as a common way to elicit data about the physiological processes of specific organs.

p. 38, ¶ 3. The interactions discussed in this and subsequent paragraphs are excellent examples of homeostasis. The hormonal relationships among the pituitary gland, the ovary, and the uterus should be tied to what pupils already know about regulation from their study of Chapter 4, *LO*.

p. 38, Fig. 1.37. These diagrams illustrate the text description, but you will probably have to provide most pupils with some help in interpreting them.

p. 40, ¶ 3. The events of fertilization and implantation institute a new sequence of controls, which cause the cessation of the mentrual cycle and the maintenance of pregnancy.

p. 41, ¶ 3. Abundant materials on descriptive human embryology are available. Pupil interest is high, so you should at least take time for some presentation such as filmstrips. Comparisons should be made with features that the pupils have themselves observed in chick embryos.

p. 41, Fig. 1.41. Occasionally a physical break occurs within the placenta, between the two circulatory systems, and then there may be direct contact between the blood of mother and infant. This is of considerable importance if the blood types differ, particularly in the case of the Rh types. Some pupil might want to investigate this and report to the class.

p. 41, ¶ 3. *Note:* the lack of a fully formed placenta prevents the development of marsupial embryos to the extent attained in placentals. When nourishment becomes critical, the young are born in a foetal condition. However, muscles of limbs and mouth are developed enough for the young to crawl into the mother's pouch and attach themselves to the teats. From this time on they are nourished with milk.

p. 42, ¶ 3. If pupils ascertain the gestation periods of various mammals, an interesting classroom discussion can ensue. Correlations between birth size and length of gestation period can be made. Generally, the smaller the animal, the shorter the period of gestation, but there are exceptions that can challenge pupils' ingenuity in providing possible explanations. The following book gives the gestation periods of a great many mammals: *Systematic Dictionary of Mammals of the World*, 2nd edition, by M. Burton (London Museum Press, 1965).

PROBLEMS (pp. 44–45)

1. One way would be to remove the placenta and analyse it for the presence of progesterone. The pupil might ask, however, 'Could not the progesterone found in the placenta have originated somewhere else and been stored there?' Pupils should be encouraged to provide an answer for that question. They might be led to suggest that if all other known sources of progesterone were extirpated and progesterone were still found in the placenta, this would be evidence of its production there. Pupils should then be asked what tissues would have to be extirpated. Obviously, the ovaries would be the first to consider.

2. A pupil might come up with the suggestion that sperm cells of birds have more resistance to high temperatures than those of mammals. Most pupils probably do not know that birds have extensions from the lungs, air sacs. A pair of these air sacs extend abdominally and lie close behind the testes. Undoubtedly, they cause some reduction in testicular temperatures.

3. Principally, propagation by cuttings ensures offspring that have the same traits as their parent. Because of genetic recombination, this is not necessarily true with seeds, which are produced sexually, even if both parents have the desired traits. In addition, most plants produced by cuttings mature faster than those produced from seeds. In grafting, man takes advantage of established rootstocks to further quicken early growth of cuttings from plants with desirable characteristics. Frequently varieties that have excellent root growth have poor fruit or flower characteristics, and vice versa. Grafting combines the desirable traits. For example, most citrus varieties are budded (a kind of grafting) on sour orange stock. This is a large subject, and an interested pupil can carry it far, both in the library and in the greenhouse or field.

4. Whether *Paramecium* presents a special case of sexuality or not depends upon one's definition of sexuality. If sexuality is an exchange of nuclear material that provides genetic variability, then conjugation in *Paramecium* is certainly included. If, however, the pupil defines sexuality in terms of sperm and ova, then in *Paramecium*, as in *Rhizopus* and *Ulothrix*, there is no way to distinguish mating types as male and female.

5. This can be a very extensive investigation. Some examples: the stigmas becoming receptive before the pollen matures; the anthers maturing and shedding their pollen before the stigmas are receptive; the flower being constructed so that there is little chance of its pollen being deposited on the stigma; the pollen failing to germinate or the growth of the pollen tube being inhibited.

By this time the pupils should be aware that the inclusion of the word 'always' in a question makes it somewhat difficult to answer. However, most cereals and garden peas are normally self-pollinated.

If a plant is self-pollinated normally, it is almost certain to produce seeds. An isolated plant of this type can propagate its species, but it could not if it were an obligatory cross-pollinator. On the other hand, cross-pollination introduces a larger degree of variability into a population, allowing genetic changes that may permit survival under quite adverse conditions; self-pollinated species in the same situation might become extinct.

6. (*a*) Yuccas are insect-pollinated plants depending upon pronuba moths for fertilization. (*b*) In the hollies usually planted for ornament, male and female plants are separate. Fruits are developed only if a male and female plant are relatively close together. In a pair, no berries develop on the male plant; if both plants are of the same sex, neither bears fruit. (*c*) Apomixis is the answer to this question. Apomixis is the formation of seed in the pistil of a flowering plant without fertilization. It is thus comparable to parthenogenesis and has the same genetic implications. The families Compositae, Rosaceae, and Gramineae have species that reproduce apomictically. In some species it seems to be the only method of reproduction. (*d*) Night-blooming plants are usually pollinated by night-flying insects such as moths. (*e*) Bees kept near flowering plants ensure maximal fertilization for insect-pollinated species. (*f*) Peas are normally self-pollinated. Obviously, any plant dependent upon wind pollination or some other mechanism independent of insects might produce seeds when grown in an insect-free greenhouse.

7. Pupils will find that vertebrates can be divided into amniotes (reptiles, birds, and mammals) and anamniotes (amphibians and fish). Embryonic membranes are basically an adaptation to a land environment. Amnion and chorion are usually the best developed of all membranes; yolk sac and allantois vary in extent of development, but in some groups are quite rudimentary.

SUPPLEMENTARY MATERIALS

ADDITIONAL INVESTIGATION

ALTERNATION OF GENERATIONS IN FERNS

This is a predominantly observational investigation using a combination of fresh plant materials and prepared microscope slides. It provides comparisons with the descriptions of alternation of generations in *Selaginella* and in flowering plants.

Introduction

Ferns are of no great economic importance. But the importance of an organism is not established on the basis of its economic value alone. Some forms of life may be more worthy of study than others, but all are products of a long, slow process of change and evolution; each form presents unanswered questions about ancestry, biochemical processes, and behaviour. When you look at a fern, it may be a good idea to remind yourself that before you is a modern descendant of plants that dominated the Earth for millions of years before man arrived to appreciate—or ignore—them.

Materials and Equipment
(for each group)

Intact specimen of fern sporophyte, fresh or dried
Living fern sporophytes in pots, 2 or 3 per class
Hand lens
Scalpel
Microscope slide
Medicine dropper
Cover slip
Monocular microscope
Fern gametophyte
Stereomicroscope

Procedure

Examine your specimen of a fern sporophyte, comparing it with a living plant growing in a pot.

A. Where is the stem located in the growing plant?

B. With a hand lens observe the leaves (fronds). The veins are neither netted nor parallel but are said to be *dichotomous* (refer to p. 37, *DLT*). What does this term mean?

C. Notice that some of the leaves have small, brown dot-like or elongated structures, called 'sori' (singular, 'sorus'). On which surface of the leaves are the sori to be found—on the upper or the lower?

D. Do the leaves that bear sori look like those that do not?

With a scalpel, scrape off a few of the sori. Deposit them in the middle of a clean slide; add a drop of water and a cover slip. Examine with a microscope, using low power. The small stalked structures you may see are spore cases. If you find any, examine one closely and notice the row of thick-walled cells across the top and around one side. Draw the whole spore case. If some of the cases have been broken in handling, spores may be scattered on the side; other spores may be observed inside the broken cases. Draw two or three spores as you see them under high power.

As a spore case reaches maturity, its cells begin to dry out. The row of thick-walled cells straightens, tearing the case open and exposing the spores. As the case continues to dry out, the row of thick-walled cells acts as a spring and snaps back, throwing the spores into the air. This is one of the fastest movements found anywhere in the plant kingdom. The spores are so

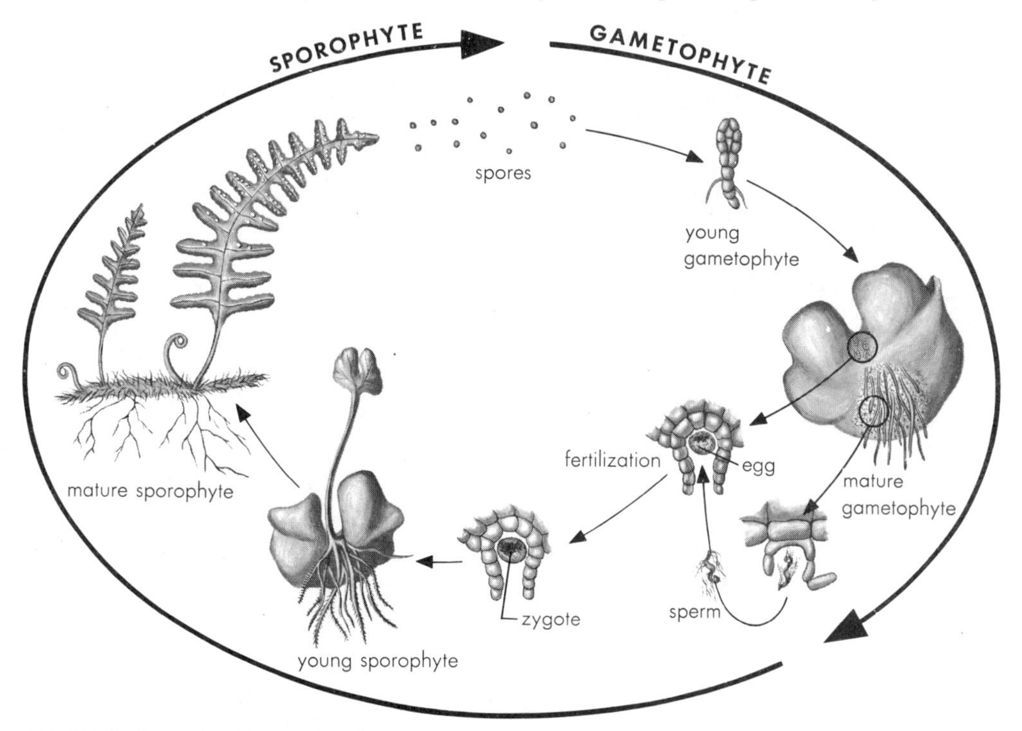

Fig. T-17. Reproductive cycle of a fern.

small that they may float in the air for a long time.

E. Volcanic islands and lava flows are frequently populated by ferns and mosses well in advance of the appearance of seed plants. Explain.

Spores that settle from the air in favourable locations germinate and develop gametophytes. Examine a fern gametophyte under the stereomicroscope.

F. Describe the size and shape of the gametophyte.

Observe the thread-like structures on the lower surface; these are not vascular, so they are called rhizoids rather than roots.

Although it is small and relatively inconspicuous, the gametophyte is important in the fern reproductive cycle, for it produces both male and female gametes. The male organs (*antheridia*) appear as small, dome-shaped, multicellular structures among the rhizoids on the under-surface of the gametophyte. Antheridia generally develop before the female organs (*archegonia*) appear. The archegonia are also produced on the under-surface of the gametophyte, but near the notch. They are flask-shaped, with enlarged basal portions buried in the tissues of the gametophytes and necks projecting about the surface. Each basal portion of an archegonium contains an egg.

Remove an antheridium and crush it in a drop of water on a clean slide. Add a cover slip and examine under low power of the monocular microscope.

G. Can you see any released sperms? What kind of movement, if any, do they show?

H. Sperms liberated from the antheridia swim to the archegonia; fertilization takes place in the archegonia. In what kind of habitat must gametophytes grow for fertilization to occur?

I. In most ferns both antheridia and archegonia are produced on the same gametophyte, though sperms and eggs do not ordinarily mature at the same time. How is this likely to affect the parentage of the offspring?

J. The zygote divides repeatedly to form a

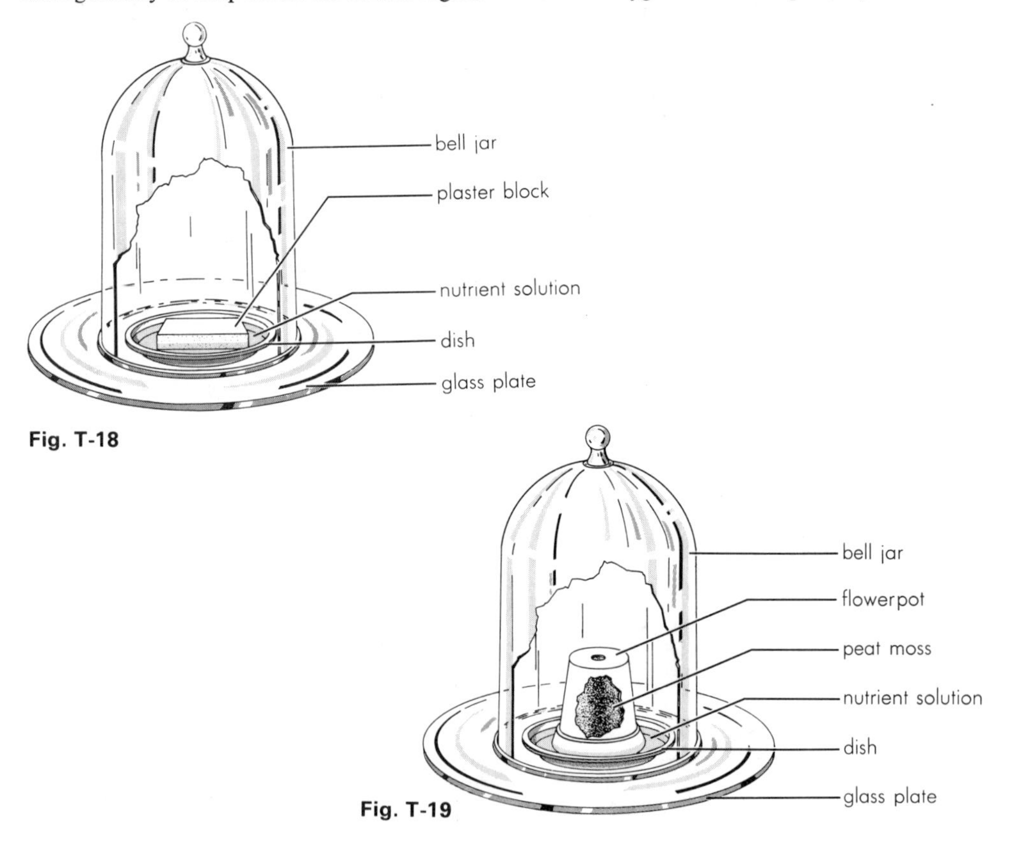

Fig. T-18

Fig. T-19

young sporophyte, which generally grows for several years before it begins to produce spores. Soon after the roots and leaves of the young sporophyte are well developed, the parent gametophyte dies and decays; in other words, the sporophyte remains dependent upon the gametophyte for a very short time. How does this compare with the relationship between gametophyte and sporophyte in mosses?

For the Teacher

Polypodium vulgare is suggested for this investigation, but almost any fern of a convenient size can be used. The specimens for each group should be of the same species as the potted plants, which are included so that the position of the rhizome (underground stem) can be shown.

Ejection of spores from sporangia may be demonstrated if mature sporangia are available. Test sporangia from different individuals until some are found at the proper stage; keep these plants in a plastic bag until class time. Place a sorus under a stereomicroscope and focus a hot, bright light on it, which will cause the sporangia to dry out. They first dehisce; then, as drying continues, the annuli snap back, ejecting the spores.

Fern gametophytes may be grown by several methods. One involves planting spores on blocks of plaster of Paris. Mix plaster of Paris with water until it attains a viscous, but easily stirred, consistency. Pour it into a small aluminium pie dish (smaller in diameter than the dish you will use). When it hardens, wash it thoroughly with a nutrient solution. The solution may be made from a commercial fertilizer (1 g dissolved in 1 litre of water). Set up the apparatus shown in Fig. T-18. Lay fern fronds bearing mature sporangia on a piece of white paper, and place under a lamp to dry. The spores that collect on the paper may then be blown gently on to the moist surface of the plaster block. Place the apparatus in indirect but not dim light. Young gametophytes should appear in about ten days; mature antheridia and archegonia should develop in about two months.

A similar method of growing fern gametophytes involves use of a clean, preferably new plant pot stuffed tightly with fibrous (not granulated) peat moss (Fig. T-19). Collect and sow spores on the surface of the pot in the manner described for the plaster block.

AUDIOVISUAL MATERIALS

Films

Mitosis and Meiosis. Indiana University Films, available through Boulton-Hawker Films. 16 mm, colour, 16 min 30 s. (Very good for recalling mitosis and contrasting meiosis with it.)

Asexual Reproduction. Indiana University Films, available through Boulton-Hawker Films. 16 mm, colour, 10 min. (Includes time-lapse photography of fission, budding, and spore formation.)

Theories of Development. AIBS Film Series. McGraw-Hill Text-Film Division. 16 mm, colour, 28 min. (Outlines history of theories and depicts development of frog.)

The Chick Embryo from Primitive Streak to Hatching. Encyclopaedia Britannica Films. 16 mm, colour, 13 min. (Includes time-lapse photography of the formation of the heart, beating of the heart, and circulating blood.)

Plants That Grow from Leaves, Stems and Roots. Coronet Films, available through Gateway Educational Films. 16 mm, colour, 11 min. (Presents many examples of vegetative reproduction.)

Flowers: Structure and Function. Coronet Films, available through Gateway Educational Films. 16 mm, colour. (Stamens, anthers, pistils, stigmas, ovaries, pollen, and fruit in both macroscopic and microscopic views. Pollination by insects and wind. Time-lapse sequence of pollen tubes growing from pollen grains.)

The Human Body: Reproductive System. Coronet Films, available through Gateway Educational Films. 16 mm, colour, 13 min 30 s. (Similarities and differences in male and female reproductive organs. Animation and photomicrography, including fertilization of the human ovum.)

The Liverwort: Alternation of Generations. Coronet Films, available through Gateway Educational Films. 16 mm, colour, 16 min. (Complete life cycle from gametophyte to sporophyte and back. Vegetative reproduction through gemmae. Transfer of sperm from antheridium to archegonium.)

Reproduction among Mammals. Encyclopaedia Britannica Films. 16 mm, black-and-white. (Uses the domestic pig for illustrative purposes, including development of sperms and eggs, fertilization, and embryonic stages.)

The Fish Embryo: From Fertilization to Hatching. Encyclopaedia Britannica Films. 16 mm, colour, 12 min. (External fertilization and the tracing of the zygote from the first cell cleavage to the formation of the young fish.)

Meiosis: Sex Cell Formation. Encyclopaedia Britannica Films. 16 mm, colour, 16 min. (Explains the role of meiosis in gamete formation.)

Amoeba—Fission. Encyclopaedia Britannica Films. 8 mm film loop, colour, 3 min. (The process of binary fission in *Amoeba.*)

Earthworm, Part 3, Reproductive System. Ealing. Super-8 film loop, colour, 3 min 10 s. (Chiefly anatomical, showing relation of male and female systems in a hermaphroditic animal.)

Algal Syngamy, Isogamy in Chlamydomonas. Ealing. Super-8 film loop, 3 min 30 s. (Fusion of gametes in an alga with no morphological distinction between eggs and sperm cells.)

Pollen Tube Growth. Ealing. Super-8 film loop, colour, 2 min. (Excellent time-lapse photography of pollen-tube growth, and formation of nuclei.)

Mating Behaviour in the Cockroach. BSCS Inquiry Film Loop, available from John Murray. Super-8, colour. (Relates behaviour to the reproductive process.)

TEACHER'S REFERENCE BOOKS

BRITISH MEDICAL ASSOCIATION. *How Your Life Began.* A *Family Doctor* booklet. (Discusses sex determination, some simple human genetics and embryo development.)

COHEN, J. *Living Embryos.* 2nd edition. Pergamon, 1967. (An introduction to the study of vertebrate animal development.)

DALLAS, D. M. *Sex Education in School and Society.* National Foundation for Educational Research, 1972. (A critical review of current methods and objectives. The work of schools and other social agencies is evaluated, and details are given of available resources, teaching aids and information.)

FLANNAGAN, G. L. *The First Nine Months of Life.* Heinemann, 1963. (A profusely illustrated account of the development, from egg to birth, of the human being.)

JAMES, W. O. *Elements of Plant Biology.* Allen & Unwin, 1966. (Covers reproduction in plants.)

SIMPSON, G. G., C. S. PITTENDRIGH, L. H. TIFFANY. *Life.* Routledge & Kegan Paul, 1965. (Plant and animal reproduction.)

TANNER, J. M., G. R. TAYLOR. *Growth.* Life Science Library. New York. Time, Inc., Book Division, 1965.

VILLEE, C. A. *Biology.* Saunders, 1967. (Plant and animal reproduction.)

WOOD, C. *Sex and Fertility.* Thames & Hudson, 1969. (Discusses fertilization, pregnancy, embryology, birth and the control of fertility.)

2
HEREDITY

MAJOR IDEAS

1. That Gregor Mendel was able to develop a fruitful inheritance theory, which was fundamentally different from all previous theories, stemmed in the main from the fact that he had both a good mathematical and botanical background.
2. The principles of probability theory are used by biologists to test various theories of heredity and to make predictions concerning future generations.
3. Mendel's experimental results led him to formulate a particulate theory of inheritance. By assuming that heredity is determined by particles transmitted from generation to generation through the gametes, he was able to explain the phenomena of dominance, segregation, and independent assortment. His choice of gene pairs clearly illustrates the scientific rule of extreme economy.
4. Soon after their rediscovery, Mendel's principles were linked with nineteenth-century advances in cytology to form a chromosome theory of inheritance. Abundant experimental work during the first quarter of the twentieth century substantiated his theory.
5. The 'proof' of a scientific theory consists in its continued ability to account for new evidence as it arises; a theory can never be finally proved.
6. Twentieth-century experimental work has revealed new principles of heredity: linkage, nondisjunction, non-dominance, multiple alleles, continuously varying traits, etc. All these phenomena were unknown to Mendel, but they proved to be explainable in terms of his theory.
7. The characteristics of an individual organism are the product of an inextricable interaction of heredity and environment.
8. Improvements in cytological and statistical technique have greatly increased knowledge of human heredity without resort to breeding experiments: human chromosomes have been accurately numbered and described; chromosome anomalies have been linked to clinical syndromes; and metabolic defects—traceable to gene defects—have been elucidated. Yet many human 'traits' of great interest still have not yielded to genetic analysis.
9. The origin of new traits in organisms has been traced to changes in nuclear materials—mutations. In recent years gene mutation has been linked to alterations in the structure of DNA molecules. These alterations occur 'naturally', but they can be caused to occur at increased rates by treatment of cell nuclei with high temperatures, certain chemicals, or (especially) ionizing radiations.
10. During the last quarter century, genetic experiments with moulds, bacteria and other micro-organisms have resulted in an understanding of the way in which genes function in heredity. The 'coding' of nucleotide bases in DNA is transferred via messenger-RNA to the assembly of specific sequences of amino acids to form proteins (including enzymes, which mediate biochemical reactions).

PLANNING AHEAD

Because laboratory work and problems in Chapter 2 require a large amount of time, it is fortunate that the need to plan ahead is now considerably reduced. You should have reviewed the investigations for Chapter 3.

Check your materials for the investigations in Chapter 3. It is possible to do Investigation 3.1 with only one cat skeleton and one human skeleton. For Investigation 3.3 you may find purchasing blood-typing kits more convenient than assembling the materials yourself.

GUIDELINES

The teacher who dives boldly into Chapter 2, assigning large blocks of material, setting up brief discussion periods, and passing quickly from one investigation to another, is doomed to disappointment. Genetics is a subject which requires careful preparation. To lead pupils through Mendel's results is easy; to get pupils to memorize the definitions of 'allele', 'non-dominance', 'mutation', 'lethal', etc., requires more persistence. But to really understand genetics is difficult; the more fully you are aware of the difficulty, the greater the likelihood of a reasonable degree of success in teaching it.

The two main difficulties encountered are terminology and the necessity for mathematics. The teacher can alleviate the first difficulty by moving slowly from one set of terms to the next. He can cope with the second difficulty by repetition and an abundance of examples. Pupils, however, are, in the main, interested in genetics, especially human genetics, and this interest can be used to overcome other difficulties.

Several specific topics are troublesome and require extra teaching effort. Especially if you are somewhat inexperienced, be alert for signs of confusion in the following: the meaning of 'proof' in science; the fact that many allelic genes may exist in a *population*, even though an individual normally carries only two alleles; the idea that most mutations are harmful. In addition, reasoning from data is difficult for many pupils; the following topics in Chapter 2 deserve considerable teacher explication: Mendel's reasoning, Sutton's reasoning, Morgan's reasoning about the Y chromosome, the reasoning involved in mapping genes, the reasoning in the *Neurospora* experiments, and the nature of the genetic code.

In view of these difficulties, we suggest the following assignments: pp. 47–51 (guide question 1), pp. 52–58 (guide questions 2–5), pp. 58–64 (guide questions 6 and 7), pp. 64–69 (guide questions 8–11), pp. 70–75 (guide questions 12–15), pp. 75–80 (guide questions 16 and 17), pp. 81–85 (guide questions 18–22), and pp. 85–90 (guide questions 23 and 24). With very good classes some advantage of logical organization might be gained by combining as follows: pp. 47–58, 58–64, 64–73, 74–80, 81–85, and 85–90. In this scheme the *Drosophila* work is still assigned by itself, but it is expected that you will expand considerably on 'Mechanism of Gene Action'. This scheme incorporates each investigation in an assignment, which has not previously been the practice in recommending assignments. In this chapter, however, the investigations contain a high degree of content and integrate directly into the text discussions—an ideal striven for, but less perfectly attained, in previous chapters. Investigation 2.2. deserves special assignment consideration, not only because of its length, but also because of the statistical content.

In chemistry and physics, solving problems is recognized as an important method by which pupils gain understanding. In biology, this principle has not been frequently adopted except in genetics. Select and adapt to secure series of problems that are commensurate with your pupils' abilities. It is more desirable to supply many simple problems that illustrate a limited number of principles through the use of different traits in different organisms, rather than to cover a wide range of principles with a limited number of difficult or sophisticated problems.

A final point: many pupils have a natural interest in human genetics, and those from country areas may also have an interest in animal breeding. It is well to capitalize on pupil interests, but they should not be allowed to capsize the course. It is easy to wallow without headway in an endless recital of human disorders and in vague speculation concerning the role of heredity in producing them.

TEACHING NOTES

Inheritance (pp. 47–48)

p. 47, ¶ 1. Suggestion: Introduce the topic by projecting a slide showing a human family—the more diverse the children, the better. Ask what characteristics have been inherited. Ask what inheritance is. Then, if you have slow readers, follow with a co-operative oral reading of the introductory paragraphs. Pupils may then be somewhat motivated to begin the arduous journey (28 pages!) back to their main interest—human genetics.

The Work of Mendel (in part) (pp. 49–51)

p. 49, Fig. 2.3. Being fresh from Chapter 1, pupils should grasp the caption without difficulty even though only the term 'self-pollination' was used there.

p. 50, ¶ 2. In this and succeeding paragraphs it is necessary to keep in mind that a pea seed *is* the next generation; the seed characteristics of the parental plants were those of the seeds from which they grew. This is most likely to cause difficulty when two traits are considered together. Observe a plant in the field. The height of the plant is a characteristic of one generation; the colour of the seed on that plant is a characteristic of the next generation.

p. 50, Fig. 2.4. Some books list flower colour as a trait studied by Mendel. But this is not an additional trait. Mendel noted that white flowers regularly accompanied white seed coat and that coloured flowers (plus a reddish tint on the stem in the axils of the leaves) regularly accompanied coloured seed coat. This did not bother Mendel in his development of the idea of independent assortment. It might be considered an extreme example of linkage, and you might mention it when pupils come to that topic (p. 70), but it is regarded as an example of pleiotropism, which is not discussed in this series of books.

p. 51, Fig. 2.5. This table should be the focal point for the discussion of the first reading assignment. It is important that pupils know what a ratio is and be given an opportunity to struggle with the significance of the data. On the basis of later experimental work, some irreverent biologists have declared that Mendel must have 'cooked' his figures. More charitable writers have suggested that Mendel intended his figures to be illustrative only. It really doesn't matter. Error of detail does not detract from the discovery of an inclusive principle; on the other hand, no degree of accuracy in figures would have compensated for faulty reasoning that might have led to unsupportable conclusions. Those who wish to discuss this are recommended to read: *Experiments in Plant Hybridisation* by Gregor Mendel, Introduction by R. A. Fisher (Oliver & Boyd, 1965); *A History of Genetics* by A. A. Sturtevant (Harper and Row, 1965).

INVESTIGATION 2.1

PROBABILITY: TOSSING A PENNY (pp. 52–53)

The ideas developed in this investigation are not only necessary for understanding Mendel's theory; they are basic to an understanding of modern science. Science deals largely (some scientists would say entirely) with probabilities, not with certainties. For example, the principles of probability are at work in the disintegration of radioactive atomic nuclei and in the collisions of molecules in gases, as well as in the distribution of genes from one generation to the next.

Both parts of the 'Procedure' can be carried out in one period. 'Studying the Data' requires the pooling of results—and some discussion—which can best be done co-operatively in class.

Studying the Data (p. 53)

C, D. Increasing the number of tosses decreases the average size of the deviation. Point out the relationship between this conclusion about size of sample and the practice, in several past investigations, of combining group data.

The Work of Mendel (contd.) (pp. 53–58)

p. 53, ¶ 3. 'In science the simplest explanation that *fits all the facts* is always preferred.' You should make sure that this paramount principle of reasoning is not overlooked. The 'Principle of Extreme Economy' far antedates modern science. Perhaps its most famous form is the statement by William of Occam: 'Entia non sunt multiplicanda praeter necessitatem' ('Entities should not be multiplied beyond necessity'—Occam's razor), but as the doctrine of sufficient cause it can be traced to Anaximander at the dawn of Western philosophy.

p. 54. 'Testing the reasoning': This section should be accompanied by an abundance of simple problems.

p. 55, ¶ 2. With respect to an array of traits, self-pollination is *not* equivalent to crossing siblings, because siblings usually differ greatly not only in an array of traits due to recombination in the gametes from which they were formed but also in recombinations in the gametes they themselves form. But with respect to a single trait, 50 per cent of all gametes in an F_1 generation have one allele, and 50 per cent have the other, whether produced by one individual or by more than one.

p. 56. 'The theory applied to two traits': This whole section requires much teacher explication. It is a rare pupil who can grasp this entirely on his own. Use problems freely. Corn cobs that graphically illustrate dihybrid ratios can be obtained from biological supply companies.

p. 57, Fig. 2.7. This graphic representation contains the danger (as does Fig. 2.6, p. 55) that the pupil may *count* the gametes and peas instead of interpreting them as ratios. Note that the two characteristics are both of the embryo. Difficulty arises if one attempts to combine embryo and mature-plant traits, such as seed colour and plant height. (See note to p. 50, paragraph 2.)

INVESTIGATION 2.2

MENDELIAN MONOHYBRID CROSS IN *DROSOPHILA* (pp. 58–64)

Work with *Drosophila* is highly desirable, but difficulties may occur. However, mastering the difficulties is eminently worthwhile because these little beasts, in addition to their starring role in genetics, have many non-genetic uses—for example, to show insect metamorphosis and to illustrate population growth under various conditions.

Materials

Etherizer. Many types of etherizers are in use. For simplicity, economy and ease of handling, the authors recommend the type shown in Fig. T-20. For each etherizer obtain a 5 cm diameter glass funnel, a 25 mm diameter glass specimen tube with a cork and a small piece of cotton wool. Make a hole in the cork large enough to insert the funnel and tie the cotton wool to the funnel with thread. The cotton wool acts as absorbent material for the ether.

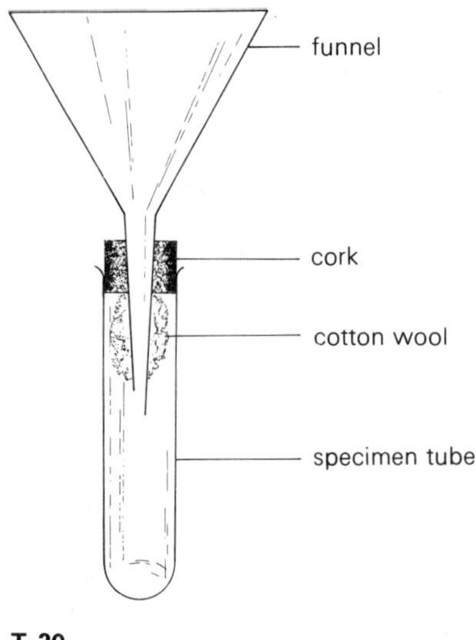

Fig. T-20

The cheapest technical grade of ether (ethoxyethane) is quite adequate. *Caution:* When pupils are working with ether, the room should be well ventilated. Warn them about dangerous explosive fumes: *No flames of any kind* should be allowed in the room.

If you do not wish to risk the use of ether, you can reduce temperature to immobilize fruit flies for examination. Place the fly containers in ice water for a few minutes, and then pour the flies on small metal plates resting on ice cubes in crystallizing dishes. Some teachers have been quite successful with this substitute for etherization.

Maintaining cultures. Purchasing cultures to be used directly by your pupils is possible but expensive. It is much better to maintain stocks of the flies in your laboratory. After a little instruction, a small group of pupils can handle the work as a long-term project.

For routine culturing of *Drosophila*, the following items of equipment are needed:

- Culture bottles ($\frac{1}{2}$-pint milk bottles, 250 cm³ wide-mouth collecting bottles, or any wide-mouth jars of similar size), about 12 for five or six classes
- Glass vials (about 80 mm × 23 mm), for individual crosses, 1 gross for each cross
- Absorbent cotton (cheapest grade), 400 g
- Cheesecloth (cheapest grade), for enclosing cotton plugs
- Pyrex flask (1000 cm³ or larger) with wicker covering. The cover allows you to pour directly from the flask while the medium is hot. (A kitchen saucepan may be substituted)
- Iron support stand
- Meker burner (preferred) or bunsen burner
- Wire screen (with asbestos), to prevent the medium from scorching while cooking
- Large Pyrex funnel with a 15–20 cm length of rubber tubing attached and closed with a pinch clamp
- Paper towelling (2 cm × 5 cm strips for each culture bottle and narrower strips for each tube)
- Graduated cylinder (100 cm³ or larger)
- Balance
- Glass-marking crayon, adhesive labels
- Filter paper (large), for weighing materials on the balance
- Pressure cooker or autoclave for sterilization
- Measuring cup
- Ring stand and clamp

Many kinds of media have been used to culture *Drosophila*, and it is generally agreed that most are satisfactory if yeast is added. If you have only limited laboratory assistance, it is best to buy premixed medium. But if you wish to prepare your own, the two formulas given below are recommended, one for its simplicity of measurement and storage and the other for its economical use of agar and its inclusion of a mould inhibitor.

The first medium is prepared from a dry mixture as follows:

Sugar (sucrose), 3 parts by volume
Cornmeal, 2 parts by volume
Brewer's yeast, 1 part by volume
Granulated agar, 1 part by volume
5–6 cm^3 of 0.5 per cent propionic acid, as a mould inhibitor if required

Mix these ingredients thoroughly. To make up the medium, add about 400 cm^3 of the dry mix to 1 litre of cold water.

The second medium is prepared as follows:

Water, 540 cm^3
Maizemeal, 50 g
Rolled oats, 25 g
Dextrose, 50 g
Brewer's dry yeast, 5 g
Agar, 3 g
0.5 per cent propionic acid, 3 cm^3

A $\frac{1}{2}$-pint milk bottle filled to a depth of 2.5 cm requires about 45 cm^3 of medium; therefore the quantities given above suffice for about 12 bottles.

Place the liquid mixture resulting from either the first or second formula over a burner, and bring to the boil. Boil gently for about five minutes or until the foaming stops. After boiling, transfer the mixture from the flask to the glass funnel mounted on the ring stand. Extend the rubber tubing into a culture bottle or tube and regulate the flow of medium with the clamp. Dispense medium into each bottle to a depth of 2.5 cm and into each tube to a depth of about 2 cm. With a glass rod or a pencil, push one end of a doubled strip of paper towelling to the bottom of the container while the medium is still soft. It is sometimes necessary to gouge a fermentation vent through the medium to the bottom of the container; otherwise, when the live yeasts are introduced, CO_2 may build up rapidly under the medium and push it towards the top of the bottle or tube, trapping the flies. The vents may not be necessary if the paper strip is inserted deeply. The strips also provide additional surface for egg laying and for pupation. Plug containers with cotton stoppers wrapped in muslin. (These stoppers may be used again and again if sterilized each time.)

After the culture containers have been filled and plugged, they must be sterilized in a pressure cooker or autoclave. The standard time for sterilization is fifteen minutes, at about 15 lb of pressure. This procedure will usually kill bacteria and mould spores in the medium. Then the culture containers may be stored for several weeks in a refrigerator or freezer. If a refrigerator or freezer is available, all the containers needed for Investigation 2.2 can be made up at one time.

About twenty-four hours before flies are to be introduced, inoculate the medium (thawed, if it has been frozen) in each container with 6 drops of a milky suspension of living yeast. The dry, packaged yeast obtainable in supermarkets is quite satisfactory for preparing this suspension.

Stock and experimental cultures may be kept on shelves or on the desks in the laboratory. They should not be kept in a refrigerator or near a radiator or on a window ledge in the sun. If the room is quite warm (25 °C), *Drosophila* develop rapidly; if it is cool (15 °C), they require a longer period of time to complete the life cycle. If the temperature of *Drosophila* cultures exceeds 28 °C for a long time, the flies become sterile.

Change cultures every three or four weeks to avoid contamination by mites (minute parasites that live on flies) and moulds. Medium should never be allowed to dry out, and water or yeast suspension should be added when moisture is required. At least two sets of each culture should be maintained in case one fails. Old culture bottles should be sterilized and cleaned when they are no longer needed.

Obtaining stock. Pure strains of *Drosophila* may be obtained from biological supply companies, or, in some instances, from geneticists in nearby colleges or universities.

The flies are supplied in small tubes. Use these to start your own stock cultures. Before transferring the flies from the tubes to culture bottles, be sure there is no excess moisture on the medium or the sides of the bottles: flies can get stuck rather easily. Place a funnel in a bottle. Before removing the cotton wool stopper from a tube, force the flies to the bottom by tapping the tube on a table. When the flies have collected on the bottom of the tube, quickly remove the cotton wool stopper and invert the tube into the funnel. Holding the tube funnel and bottle firmly together, force the flies into the bottle by gently tapping the bottom of the bottle on a rubber pad. At least five pairs of the flies should be used as parents in making stock cultures.

Add a little yeast suspension to the stock tube and attach the tube to the side of the culture bottle with a rubber band. Eggs and larvae still in the tube will thus be saved to continue development and to provide more stock materials.

Mutants. The wild-type fly is the standard phenotype from which all mutant types are inherited departures. Each mutant type is given a descriptive adjective. Such names as 'white' (for white eyes, a recessive) or 'Bar' (for 'barred' eyes, a dominant) are used.

For convenience a symbol is assigned to each mutant gene (see Fig. 2.18). The symbol is sometimes the initial letter of the mutant name—for example, ebony body is represented by the symbol *e*. But the number of mutant genes is so large that they cannot all be represented by single letters; so additional letters are used, such as the one immediately following the initial letter (*ey* for eyeless) or suggestive letters—especially consonants—from the rest of the name (*dp* for dumpy). By convention the names and symbols of recessive mutant types begin with small letters, while names and symbols of dominants begin with capitals. *Drosophila* geneticists no longer use the alternative-sized letter for the alternative member of a pair of alleles. Instead they substitute a plus sign (+) as a superscript after the gene symbol to represent the wild-type allele. Example: the dominant wild-type allele of b (black) is not B but b^+; and the recessive wild-type allele of B (Bar) is not b, but B^+.

The kinds of mutant you use with your classes should be ones that are easily distinguished by your pupils from the wild type. Ebony and dumpy mutants are good.

Procedure

The experimental work should be preceded by at least one practice period in which pupils learn to etherize and handle flies properly and to distinguish between male and female flies. Begin by showing the Ealing Film Loop *Handling Drosophila* (no. 81.0754) two or three times.

Etherizing. After they are etherized, flies can be put on to the examination plates for observation. Care should be taken to avoid overheating the flies with a strong light. If the light source must be brought close to the flies, fluorescent lamps should be used. Or to filter out the heat, you can pass light from an incandescent lamp through a round flask filled with weak copper sulphate solution.

Do not leave flies in the etherizer more than one minute. Etherized flies recover in about five minutes. If the flies begin to recover before examination is completed, they may be returned to the etherizer for a second dose. Warn pupils that the flies are more easily killed by a second dose of ether than by the first; therefore, re-etherizing should not exceed thirty seconds. Take care that the pupils do not over-etherize their flies and kill them, thus limiting the supply of flies for the use of subsequent classes. If flies are over-etherized, their wings turn up vertically over their backs. If this happens, you might as well put them in the killing-bottle (Fig. 2.9).

Making the cross. After mating, female *Drosophila* store the sperms they receive and use them to fertilize a large number of eggs over a period of time. Therefore the females used in any experimental cross must be virgin. Flies used in maintaining stock cultures need not be virgin, since the females could only have been fertilized by males from the same genetic stock.

Females usually do not mate until twelve hours after emerging. If *all* the adult flies are shaken from a culture bottle, the females emerging in that bottle during the next twelve hours are likely to be virgin when collected. Some geneticists recommend collecting the females after a ten-hour period.

If a temperature of 20 to 22 °C is maintained, mating, egg laying and the development of larvae to a good size should take no more than a week. After seven or eight days the parent flies should be shaken out of the container. This will prevent confusion of the parent flies with the offspring when counts are made. Once flies begin to emerge, they may be classified and counted for ten days. Counting longer than this again runs into the danger of overlapping generations. The first F_1 flies may have mated before being counted and, hence, may have left some eggs in the same container.

Records. Adequate records are essential to the success of an experiment and an understanding of the results. Impress upon each pupil the importance of properly labelling culture containers, and tubes, and of keeping a daily record of all observations relating to the experiments.

Studying the Data (p. 62)

B. If no slip-ups have occurred, there should be only one phenotype.

C. The results of the P_1 mating illustrate the principle of dominance.

D. As long as a sex-linked trait is not used, the F_1 generation should not be affected by either parent carrying the mutant trait.

E. All the F_1 generation are genetically the same with respect to the trait being considered; therefore there is no need for precautions as to parentage. The flies of the P_1 generation have already been removed (p. 61, item 8).

F. *With respect to the trait being studied*, two phenotypes should occur among the F_2 flies. Pupils may notice other variations. If so, two points may be made: (*a*) the success of Mendel stemmed in part from his refusal to be distracted by variations other than the ones he was studying, and (*b*) variability among offspring is to be viewed later as an essential condition of evolution.

H. The combined data are more likely to approximate a 3:1 ratio of dominant to recessive than the data of a single group. This illustrates the first principle of Investigation 2.1.

Conclusions (p. 64)

B. If the data are *not* significantly different from the expected ratios, then the experiment indicates that the fruit fly trait involved in the experiment was inherited according to the prediction of Mendelian theory.

C. If the difference is significant, then pupils should seek (as they have done several times before) to suggest factors in their procedure and technique that might account for this. Otherwise they must conclude that the fruit fly trait involved in the experiment was not inherited according to Mendelian theory. Of course, such a conclusion should never be based on a single experiment.

Further Developments in Mendel's Principles (pp. 64–73)

p. 64, ¶ 2. Like most textbook statements, this is an oversimplification. Sir Francis Galton, Darwin's cousin, fully appreciated the value of mathematics in biological studies. He was also a pupil of heredity but apparently did not know of Mendel's work.

p. 66, ¶s 3, 4. This is another contribution to the pupil's concept of the scientific enterprise. It is, of course, applicable in many places, but this seems to be a good opportunity to discuss the nature of theory. It should be linked to the discussion of the theory of spontaneous generation in Investigation 3.4, *DLT*, and should be returned to when you are discussing the theories of evolution and natural selection in Chapter 3 of this book.

p. 69, ¶ 3. A review of meiosis may be needed at this point.

p. 70, ¶ 2. Refer pupils to Plate I, p. 88.

p. 70, ¶ 5. The further apart two genes are on a chromosome, the more potential breakage sites exist between them. This is essentially an intuitive matter. But suppose that breakage sites occur at 1 μm intervals; then two genes located 8 μm apart have more chances of separation by crossing over than two genes that are only 3 μm apart.

Even if the mathematics of recombination is not grasped, the idea that crossing-over produces new combinations of characteristics should be stressed. This becomes an important point in the discussion of evolution (pp. 107–8).

p. 71, Fig. 2.17. (45% AB + 5% Ab + 5% aB + 45% ab) × (45% AB + 5% Ab + 5% aB + 45% ab) = 20.25% AABB + 4.50% AaBB + 41.00% AaBb + 4.50% AABb + 0.25% AAbb + 4.50% Aabb + 0.25% aaBB + 4.50% aaBb + 20.25% aabb.

p. 72, ¶ 2. *Note:* It is possible under abnormal circumstances for an individual to have three alleles for a specific gene locus. This would occur if, in the formation of one parental gamete, nondisjunction had occurred in the chromosome bearing that locus. Thus a zygote containing three such chromosomes (trisomic) would be produced.

p. 73, Fig. 2.19. During the discussion of Investigation 2.3 you may want to point out a further complication in this situation. If hair is plucked from the back of a Himalayan rabbit and the animal is then placed in a low-temperature cage, the regenerated hair is dark; if the plucked rabbit is placed in a high-temperature cage, the regenerated hair is usually white.

INVESTIGATION 2.3

SEEDLING PHENOTYPES (pp. 74–75)

The nature–nurture argument in its various forms has had a long history and as a study of interaction between genetic constitution and environmental conditions, discussion of the matter has importance, particularly with respect to the survival of populations and to evolution. This investigation presents the problem in its modern context. Some pupils will see its relevance during future discussions, such as those on human skin colour and human intelligence on pp. 77 and 78.

Materials

Tobacco has the advantage in this experiment of requiring very little space for germinating large numbers of seeds. This is particularly important in providing for darkness; a single box on a window ledge will hold the petri dishes of several classes. Moreover, it has been found that the small size of the tobacco seeds arouses a great deal of interest. And at this stage of the course, many pupils have developed considerable pride in their laboratory dexterity—the small seeds are a challenge to them.

Procedure

The seeds should be counted out before the laboratory period and delivered to the groups in small tubes.

Eight to ten days are normally required before counting begins. If the experiment is set up on a Monday, the first count usually can be made eight days later (on a Tuesday), and the last count then occurs on a Friday. If the experiment is set up on a Friday, the first count can be made on the tenth day (a Monday). Other schedules involve an inconvenient break for a weekend.

Groups of four are suggested. One pair of pupils can count the seedlings in Dish A and the other pair those in Dish B, each pupil checking the count of his partner. Exchanging dishes between pairs on alternate days provides a further check and emphasizes group responsibility.

While counting is in progress, you should check the counts of Dish B (on Day 1 especially). The distinction between the yellow and green cotyledons is sometimes rather difficult to make. And take care that the white radicles, which emerge from the seed coat first, are not counted.

Studying the Data (p. 75)

A. If all goes well, there should be some green seedlings in Dish A and none in Dish B.

B. The only difference should be the presence of light in Dish A and its lack in Dish B.

C. Not unless all the seedlings in Dish A are green.

D. The percentage of green seedlings on Day 4 in Dish B should be greater than 0.

E. The change from dark to light environment.

F. Not unless 100 per cent of the yellow seedlings became green.

G. Since all the seedlings were exposed to the same environmental conditions, it is reasonable to conclude that the differences are hereditary.

Conclusions (p. 75)

A. The data from Dish B on Day 2, taken in isolation, might be considered as supporting this hypothesis.

B. The data from Dish A, taken in isolation, might be considered as supporting this hypothesis.

C. Lack of chlorophyll in tobacco seedlings can result either from a genetic or from an environmental factor. Or, in language less likely to come from pupils: the environment may greatly alter or even entirely suppress a genetic potential; individual phenotypes may be different even though their genotypes are the same.

After conclusions have been reached by each group separately, data from all groups may be pooled on the blackboard. A ratio of green: yellow can then be calculated, related to Mendel's results, and used to arrive at a decision on the kind of inheritance involved in cotyledon colour.

Suggested questions for class discussion: What must have been the genotypes of the parent plants? What must have been their phenotypes? When investigating this trait, could you follow the Mendelian P_1–F_1–F_2 sequence of generations? (The last question brings up, somewhat prematurely, the matter of *lethal* genes. Actually, chlorophyll formation is inhibited in homozygous recessives only for a time. Some albino seedlings ultimately form chlorophyll and grow to maturity. Thus a P_1 cross is possible. And this, of course, brings up the relativity of the term 'lethal'.)

Human Heredity (pp. 75–78)

p. 76, ¶ 4, lines 4–7. These statements can be explained thus. (*a*) If either parent shows the trait, then at least half the offspring will have the dominant gene. (Of course, the statement does not apply to a single family with one or two children only.) (*b*) A gene for a sex-linked trait is on the X chromosome—and any offspring receiving that chromosome from the father must be a daughter. (*c*) Half of a carrier's X chromosomes have the recessive gene. If one of her gametes unites with a gamete containing a Y chromosome, the offspring will be a male and will show the trait; but a carrier's daughter has one X chromosome from the father, and this chromosome cannot have the gene for the trait unless the father shows the trait. A sex-linked trait carried by a woman was said to go to 'all of her sons' instead of to 'one-half her sons'.

p. 77, ¶ 2. Human blood types are encountered again in Fig. 3.15 and Investigation 4.3.

p. 77. *Note:* Both $\mathbf{I}^a$ and $\mathbf{I}^b$ are dominant to **i** but are non-dominant with respect to each other.

p. 77, ¶ 3. Some very recent research has cast doubt on this explanation of phenylketonuria as being the result of a simple recessive gene.

p. 78, Fig. 2.21. This photomicrograph provides material for a good discussion. Note the X and Y chromosomes in the lower right-hand corner.

INVESTIGATION 2.4

HUMAN INHERITANCE (pp. 78–81)

This investigation can be done as a homework assignment and checked in a class discussion. In some cases it may be best to discuss the first four questions in class before the pupils begin their independent study.

If charts for testing colour blindness are available, you may have pupils make a study of this trait by testing their parents and siblings. But stress the tentative nature of results. Butterworths (Edinburgh) Ltd., London Road, Meadowbank, Edinburgh 7, can supply Ishihara colour blindness test charts.

Procedure (pp. 79–80)

Percentages of male and female infants

A. Females: 1, males: 2.

C. $pX = 0.5, qY = 0.5$.

D. $pX = 1, qY = 0$.

E. 50 per cent male, 50 per cent female.

F. Assumption (*b*).

G. Fewer male births than female must be expected.

H. Since the data are exactly opposite to the prediction, Assumption (*a*) is probably incorrect.

Inheritance of red–green colour blindness

A. This gene must be recessive, because it does not appear in the F_1 of Pedigree A.

B. The trait appears to be sex-linked because of its transmission only to males in the F_1 of Pedigree B.

Inheritance of haemophilia

A. Queen Victoria, Alice of Hesse, Irene, Alexandra, Alice, Beatrice, Victoria Eugenia.

B. 0 per cent.

C. (*a*) 0 per cent, (*b*) 100 per cent.

D. 50 per cent. Note that in the case of Queen Victoria only one out of four were haemophiliac, but, of course, we do not know how many zygotes failed to survive.

E. (*a*) 0 per cent, (*b*) 50 per cent.

The Source of New Traits (pp. 81–85)

p. 81, ¶ 2. The term 'mutation' is used somewhat ambivalently. Basically it refers to a phenotypic change. But it sometimes seems to refer to the postulated genetic change that is associated with the phenotypic change, the observable chromosomal change, or the inferred gene change. Normally, this does not seem to interfere with pupil comprehension, but you should keep it in mind when wording test or exam questions.

pp. 82–84. 'Theory of Gene Mutations': Before beginning this section, pupils should review p. 47, *LO*.

p. 82, ¶ 5. It is difficult to over-emphasize the importance for biology of discoveries in molecular biology during the past fifteen years. But DNA has become a kind of status symbol, and the authors believe that the memorization of the 'anatomy' of the DNA molecule is as inexcusable in secondary school biology today as was the memorizing of crayfish appendages and cockroach mouthparts yesterday. The important thing to emphasize is the way in which the structure of the molecule fits the function that hypothetical genes must perform.

p. 84, ¶ 4. Refer to Investigation 2.3 if the idea of lethal genes (not necessarily the term) came up during your discussion of chlorophyll-less tobacco seedlings.

p. 84, ¶ 5. This is a very difficult idea for pupils to grasp; it seems to contradict the idea of 'random'. But the idea is very important for understanding the mechanism of evolution in Chapter 3.

Mechanism of Gene Action (pp. 85–87)

p. 87, ¶s 4, 5. For the symbols, refer pupils to Fig. 2.24.

p. 87, ¶ 6. Stress the idea of *theory* here and revert to the earlier discussion concerning the nature of theory (p. 66). Pupils may find the concluding sentences more difficult to swallow than all the myths of Greece. But plausibility is not a necessary ingredient of a theory (many people still find heliocentricism implausible). A theory need only explain and integrate verifiable observations, and this the DNA–RNA–protein theory does—at least for the present.

INVESTIGATION 2.5

GENETIC DIFFERENCES IN PEAS (pp. 88–90)

By associating a clear, macroscopically visible trait with a microscopically visible one and this, in turn, with a biochemical one, this investigation may bring home to the pupil the chemical basis of heredity.

Materials

In the spring, when this investigation is normally done, pea seeds are easy to obtain. However, most varieties are wrinkled, and a little searching may be necessary to obtain a round variety; ask for Pilot or Alaska. Round and wrinkled varieties can also be obtained from biological supply companies, but at greater expense.

Directions for preparing glucose agar: To 400 cm^3 of distilled water add 8 g of agar (be sure it is plain agar) and 2 g of glucose-1-phosphate (the Cori ester). Boil vigorously until the foam becomes quite coarse. Pour into petri dishes. Cover and keep in refrigerator until one hour before use. The glucose-1-phosphate is expensive, so it is desirable to use small petri dishes and keep the agar layer thin. If these directions are followed, the amounts given here are sufficient to prepare forty dishes.

Be sure pupils use a separate medicine dropper—three altogether—for each of the two extracts and for the iodine–potassium iodide solution.

For preparation of iodine–potassium iodide solution see p. T22.

Procedure

As given in the text, this exercise takes three days to complete. However, a two-day schedule can be used, as follows: *Day 1.* The initial weighing of peas and the observation of the starch grains. Have the preparation of the extract done out of class by a special group. *Day 2.* The observations of the glucose agar and (between observations) the weighing of the soaked peas.

On Day 1 the bottles are weighed wet because they will be wet when the final weighing is done.

Dried peas are very hard, and grinding them with mortar and pestle is difficult. It may be necessary to break the peas with a hammer and squeeze them in a pair of pliers before mortar and pestle are effective. Ideally a small liquidizer should be used (40 cm^3 of water for each 10 g of peas). Be sure to clean the machine thoroughly between grindings.

The extract must be stored in the refrigerator, since the enzyme deteriorates rapidly at room temperature. But be sure to remove the extract from the refrigerator an hour before use, since action of the enzyme is very slow at low temperature.

When the test is made on the agar, the warmer the room, the better. In standing overnight, most of the solid materials will settle out of the extract, and the almost clear supernatant can then be decanted. However, clear extract entirely free of starch grains can be obtained quickly with a centrifuge.

Studying the Data (p. 90)

A. If all goes well, the data should show that wrinkled peas absorb more water than do round peas.

B. If we assume that the peas that absorbed the most were able to do so because they had lost the most, the wrinkled peas lose more water than the round ones as they mature.

C. The grains of starch from round peas are simple ovals. The starch grains from wrinkled seeds are round in shape and compound, being subdivided into quadrants or possessing grooves. Whether or not pupils think they can make predictions depends on how consistent the various groups were in observing.

D. Again, if all has gone well a consistently greater enzyme action should be obtained from the extract from wrinkled peas. The starch forming enzyme involved in this investigation is starch phosphorylase. It combines glucose-1-phosphate molecules to form the amylose fraction of natural starch.

Summary (p. 90)

C, D. Further breeding results would indeed be necessary to confirm that these three biochemical and structural characteristics are the result of single gene action. However, such an assumption is not unreasonable even with the pupils' limited data, since the enzyme is concerned with starch formation, there was an observed difference in the starch grains, and the ability of starch to hold water (as in puddings) is well known. Of course, pupils may be willing to make the assumption on the basis of the reading they have done in previous pages; this is not unreasonable, but give it second place to independent arguments.

PROBLEMS (pp. 92–93)

1. Bull = heterozygous for polled
Cow A = homozygous for horned
Cow B = homozygous for horned
Cow C = heterozygous for polled

2. A herd of roan cattle only could not be independently maintained. If any breeding were allowed, both red and white offspring would soon occur. Of course, the farmer could maintain a herd by adding only the offspring of red and white cows (not considered a part of the herds) and preventing the roan animals from interbreeding.

3. The farmer should eliminate from his flock both parents and all offspring whenever a black sheep appears.

4. F_1 = all white; $F_2 = \frac{3}{4}$ white and $\frac{1}{4}$ yellow.

A cross between an F_1 individual and a homozygous white individual would produce only white offspring.

5. The results of these breeding experiments can be explained on the basis of non-dominance.

6. (*a*) A or O children
(*b*) A or O children
(*c*) A, B, or AB children
(*d*) A, B, or AB children
(*e*) A, B, AB, or O children

7. (*a*) The F_1 generation consists entirely of individuals that are tall and have red, round fruit.

(*b*) **P** = Round, **T** = Tall, **R** = Red.

$\frac{1}{64}$ **PPTTRR** $\frac{1}{32}$ **PPTtRR** $\frac{1}{32}$ **PpTTRR**
$\frac{1}{16}$ **PpTtRR** $\frac{1}{32}$ **PPTTRr** $\frac{1}{16}$ **PPTtRr**
$\frac{1}{16}$ **PpTTRr** $\frac{1}{8}$ **PpTtRr** $\frac{1}{64}$ **PPttRR**
$\frac{1}{32}$ **PpttRR** $\frac{1}{32}$ **PPttRr** $\frac{1}{16}$ **PpttRr**
$\frac{1}{64}$ **ppTTRR** $\frac{1}{32}$ **ppTtRR** $\frac{1}{32}$ **ppTTRr**
$\frac{1}{64}$ **ppttRR** $\frac{1}{16}$ **ppTtRr** $\frac{1}{32}$ **ppttRr**
$\frac{1}{64}$ **PPTTrr** $\frac{1}{32}$ **PPTtrr** $\frac{1}{32}$ **PpTTrr**
$\frac{1}{16}$ **PpTtrr** $\frac{1}{64}$ **PPttrr** $\frac{1}{32}$ **Ppttrr**
$\frac{1}{64}$ **ppTTrr** $\frac{1}{32}$ **ppTtrr** $\frac{1}{64}$ **ppttrr**

(*c*) Round, tall, red: 27/64.
Round, dwarf, red: 9/64.
Pear, tall, red: 9/64.
Pear, dwarf, red: 3/64.
Round, tall, yellow: 9/64.
Round, dwarf, yellow: 3/64.
Pear, tall, yellow: 3/64.
Pear, dwarf, yellow: 1/64.

8. The mutant could be crossed with flies bearing traits produced by genes having known loci. Crossing-over with a known gene would produce breeding results from which the location of the mutant genes could be determined.

9. Since sexual reproduction is lacking, a new trait would depend upon a somatic mutation in a meristem cell, from which bud and shoot carrying the mutation might develop. The likelihood of this occurring is small. But this problem may lead a pupil to the topic of somatic mutations.

10. (*a*) XXX ('metafemale'): anatomical female with subnormal sexuality and subnormal mentality. (*b*) XXY (Klinefelter's syndrome): male with small testes that do not produce sperm and with some development of mammary glands. (*c*) XYY ('supermale'): many appear to be normal males of large stature, but some show excessively aggressive behaviour.

Also known are individuals with a single X chromosome (Turner's syndrome): these are sterile females. A case of an individual having a single Y chromosome was reported in 1968, but this requires confirmation.

11. Non-dominance could be interpreted by a bloodline theory of inheritance. Also, phenotypic characteristics that are the result of the interaction of several genes could be so interpreted. However, Mendel's results, all of which fortuitously involved simple traits exhibiting the dominance phenomenon, were impossible to interpret by any kind of blending theory.

12. The problem involves the production of mutations, one of which, by chance, might be the wanted characteristics. Anything, then, that might increase the mutation rate would be desirable (see p. 85).

13. This problem concerns somatic mutations and mosaicism. See any good genetics text. But pupils should be able to make a stab at the last question without a reference.

14. See the notes on Fig. 2.19 (p. T181).

15. In mathematics, proof depends upon establishing the internal consistency of a system, beginning with postulates that are given or accepted as a basis. In science, proof is dependent on consistency with verifiable observations.

SUPPLEMENTARY MATERIALS

ADDITIONAL PROBLEMS

Most of these are very simple, intended only to give practice in simple genetic thinking, not as extension for highly capable pupils.

1. In garden peas, tall vine is dominant and short vine is recessive. If a homozygous tall plant is crossed with a homozygous short plant, what genotypes are possible in the F_1 generation?

2. In garden peas, inflated pod is dominant and constricted pod is recessive. If a plant homozygous for inflated pod is crossed with a plant homozygous for constricted pod, what ratio of phenotypes would you expect to find in the F_2 generation? What ratio of genotypes would you expect?

3. In guinea pigs, short hair is dominant and long hair is recessive. A short-haired male and a short-haired female produced mostly short-haired offspring, but a few were long-haired. Show how you can determine the genotypes of the parents. [*By considering all the possible genotypes and eliminating those that could not produce the observed results.*]

4. In laboratory mice, the normal grey colour is dominant over the albino (all-white) colour. Starting with pure-bred albino and pure-bred grey as parents, what is the ratio of phenotypes in the F_2 generation?

5. A pea plant that was homozygous for axial flowers was crossed with a plant that was homozygous for terminal flowers (see Fig. 2.4). What ratio of genotypes would you expect in the F_2 generation? What ratio of phenotypes would you expect?

6. In the following cases, **Z** stands for a certain dominant gene and **z** stands for a certain recessive gene. What ratios of genotypes would you expect from the following crosses: (*a*) **ZZ** × **zz**, (*b*) **Zz** × **Zz**, (*c*) **Zz** × **zz**, (*d*) **Zz** × **ZZ**?

7. In a certain species of plant, one pure-bred variety has hairy leaves and another pure-bred variety has smooth leaves. A cross of the two varieties produces offspring that all have smooth leaves. Predict the ratio of phenotypes in the next (F_2) generation.

8. In maize, yellow seed colour is dominant and white is recessive. A certain cob had a mixture of yellow seeds and white seeds. What colour seeds could the parents have grown from? [*Parents may have both been yellow (Yy × Yy) or one yellow and one white (Yy × yy).*]

9. In a certain animal, a breed is known that always has a hairy tail; another breed is known that always has a naked tail. How would you determine which trait was dominant?

10. In garden peas, axial flower position is dominant and terminal flower position is recessive; tall stem is dominant and short stem is recessive. A plant that is known to be pure-bred for tall stem and axial flowers is crossed with a plant having short stem bearing terminal flowers. What is the phenotype of the offspring? What is the genotype of the offspring? Predict the kinds of offspring (phenotypes) that would appear in the F_2 generation and their ratios. [*Offspring are tall, with axial flowers, genotype TtAa. In the F_2 generation we expect 9 tall with axial flowers to 3 tall with terminal flowers to 3 short with axial flowers to 1 short with terminal flowers.*]

11. In antirrhinums red flower colour (**R**) is non-dominant to white (**r**), the heterozygotes being pink; and normal (broad) leaves (**B**) are non-dominant to narrow (grass-like) leaves (**b**), the heterozygotes having leaves of medium breadth. If a red-flowered, broad-leaved plant is crossed with a white-flowered, narrow-leaved one, what will be the phenotypes and their expected ratios in the F_2 generation?

12. Several different genetics situations are described below. In each case, you are to decide which of the following kinds of inheritance is involved:

(*a*) non-dominance
(*b*) linkage with little or no crossing-over
(*c*) two genes on different chromosomes
(*d*) several to many genes on different chromosomes
(*e*) multiple alleles
(*f*) linkage with a high percentage of crossing-over

Situation 1: You have a number of plants belonging to an F_2 generation. About one-quarter have no thorns, about one-quarter have long thorns, and about one-half have short thorns, all about the same length. What kind of inheritance is involved? [(*a*)]

Situation 2: In another set of plants, also of an F_2 generation, you find the following phenotype ratios: 9 red flowers and broad leaves; 3 red flowers and narrow leaves; 3 white flowers and broad leaves; 1 white flower and narrow leaves. What kind of inheritance is involved? [(*c*)]

Situation 3: You discover a species of bird. In the population some have long crests on their heads, some have scarcely any crests, and many have crests showing all possible intermediate lengths. What kind of inheritance is involved? [(*d*)]

Situation 4: In a certain species of insect, three eye colours are known: red, orange, and yellow. If pure-bred red-eyed individuals are crossed with pure-bred orange-eyed, all the offspring are red-eyed. If pure-bred orange-eyed are crossed with pure-bred yellow-eyed, all the offspring are orange-eyed. If pure-bred red-eyed are crossed with pure-bred yellow-eyed, all the offspring are red-eyed. What is the most likely method of inheritance? [(*e*)]

Situation 5: In another case in which you study two traits of a plant at the same time, you find in the F_2 generation that three-quarters have hairy stems and green seeds while the other one-quarter have smooth stems and brown seeds. What kind of inheritance is involved? [(*b*)]

Situation 6: Among the many varieties of popcorn, one has an average ear length of 6 cm (ranges from 4 to 9) and another has an average ear length of 16 cm (ranges from 12 to 19). When the two varieties are crossed, the offspring have ears with an average length of 11 cm (ranges from 5 to 19). What is the most likely method of inheritance? [(*d*)]

Situation 7: The F_2 generation of a dihybrid cross in a plant showed the following ratio of phenotypes: 66 per cent had hairy stems and round fruit; 9 per cent had hairy stems and pear-shaped fruit; 9 per cent had smooth stems and round fruit; 16 per cent had smooth stems and pear-shaped fruit. What kind of inheritance is involved? [(*f*)]

AUDIOVISUAL MATERIALS

Charts

In teaching the basic ideas of progeny ratios to be expected from various crosses, repetition is valuable. A good way to secure this—in addition to the use of numerous problems—is through the prominent display of classroom charts that illustrate the ratios.

Filmstrips

Introducing Genetics, a set of six colour filmstrips: *Dominance*; *Incomplete Dominance, Segregation, Punnett Square*; *Independent Assortment and Linkage*; *Genetics and the Cell*; *New Trait Combinations*; *Population Genetics*. A teaching manual is included. Ward's Natural Science Establishment, available from Harris Biological Supplies. (This above-average filmstrip series presents the fundamental principles of genetics in a programmed sequence.)

DNA: A Key to All Life. *Life* Filmstrips. Colour. (A beautiful set of DNA illustrations based on an article that appeared in *Life* magazine.)

Films

DNA: Molecule of Heredity. Encyclopaedia Britannica Films. 16 mm, colour, 16 min. (Biochemical mechanisms of inheritance are illustrated by animation. *Neurospora* experiments and human genetic disorders. A difficult but worthwhile film.)

Genetics: Improving Plants and Animals. Coronet Films, available through Gateway Educational Films. 16 mm, colour, 13 min. (A well organized film that clearly shows some of the practical applications of genetics, thus adding a dimension to the pupil's study of this chapter. If you have time for one film only, this probably should be it.)

Biochemical Genetics. AIBS Film Series. McGraw-Hill Publishing Co. 16 mm, colour, 28 min. (Linkage, crossing-over, and gene-mapping are difficult ideas, and this film can help to clarify them.)

TEACHER'S REFERENCE BOOKS

AUERBACH, C. *Heredity, an introduction for O-level students*. Oliver & Boyd, 1965.

CLARKE, C. A. *Human Genetics and Medicine*. Arnold, 1970.

DARLINGTON, C. D. *Genetics and Man*. Allen & Unwin, 1964. (Contains stimulating discussions of human sexual selection.)

DARLINGTON, C. D., K. MATHER. *The Elements of Genetics*. Allen & Unwin, 1949. (Continuous variation.)

HASKELL, G. *Practical heredity with Drosophila*. Oliver & Boyd, 1961. (An introductory account.)

LAWRENCE, W. J. C. *Plant Breeding*. Arnold, 1968. (Introduces plant breeding and indicates the genetic principles on which the subject is based.)

MENDEL, G., Editor J. H. BENNETT, with commentary and assessment by R. A. FISHER. *Experiments in Plant Hybridization*. Oliver & Boyd, 1965. (Contains an analysis of Mendel's pioneer work.)

WALLACE, B., T. DOBZHANSKY. *Radiation, genes, and man*. Holt Library, 1964. (Useful for extra reading and discussion.)

WINCHESTER, A. M. *Heredity. An introduction to genetics*. Harrap, 1964.

3
EVOLUTION

MAJOR IDEAS

1. The idea of evolution, both of the Earth itself (inorganic) and of its inhabitants (organic), is very old.
2. The life of Charles Darwin affords vivid evidence of the manifold forces that may have some influence on the work of a great biologist: reading, travelling, thinking, observing, collecting, conferring, experimenting, writing, conjecturing, documenting . . .
3. In his theory of natural selection, Darwin provided a plausible explanation for the mechanism of organic evolution. After a partial eclipse early in this century, Darwin's theory now receives firm support through studies in many sophisticated areas of modern biology, including genetics, physiology, biochemistry, behaviour, and population dynamics.
4. Evolution is a directed change in one or more characteristics of a population. Essentially it is a statistical concept involving a shift in the means of a series of frequency distributions such that a plot of the means has consistent direction.
5. The mechanisms of heredity maintain a basic stability in the characteristics of populations. If this were not so, it would be impossible to recognize change when it occurred in a species.
6. Mutations are the basic source of change, but where recombinations of genes are possible, as they are in sexual reproduction, the chance of genetic change is greatly multiplied.
7. The theory of natural selection provides an explanation for the directional nature of evolution. Basically it rests upon differential reproduction among populations: Those that produce more offspring capable of living to maturity and of reproducing themselves tend to survive; those that produce fewer offspring of lower viability and fertility tend to die out.
8. The theory of natural selection is an adaptational explanation of evolution. It requires both a genetic constitution and an environment with which the genetic constitution interacts. Among species that live in highly stable environments, evolution proceeds slowly.
9. In a widespread species, distance—geographic isolation—prevents completely random mating throughout the population; speaking figuratively, we could say that eddies develop in the species' gene pool. As a result, recognizable discontinuities may occur, either through response to local environmental differences or, in small isolated populations, through random genetic drift. In such cases, the concept of subspecies is useful.
10. A new species originates when a population (whether originally recognizable as a subspecies or not) becomes reproductively isolated from other related populations. In most cases, reproductive isolation develops when genetic changes accumulate in populations that have become separated by a geographical barrier.
11. In some cases (mostly in angiosperms), new species are known to have originated suddenly through the mechanism of polyploidy. The continued existence of the necessarily small original population of such

a new species is dependent upon reproductive success and upon interaction with the environment—natural selection.

PLANNING AHEAD

Throughout the course major emphasis has been placed on planning ahead for laboratory work. In a course that places such stress on laboratory work and that makes such heavy demands on the teacher's time to implement a heavy laboratory programme, this planning emphasis is justified. The need to plan ahead for laboratory work is now drawing to a close. Check on your needs for the investigations in Chapter 4.

GUIDELINES

If the idea of evolution seems to be new at this point, then the previous six hundred and forty-four pages have not been well understood. This is not *the* chapter on evolution; rather it is the chapter that deals with ideas concerning the mechanisms by which the biosphere has evolved. This is an outcome of reproduction and genetic mechanisms, and it is the logical culmination of the previous two chapters. It might also be considered the climax of a biology course—if man did not exist.

The three major headings of the chapter provide a convenient partitioning for assignments: pp. 94–102 (guide questions 1–6), pp. 103–14 (guide questions 7–14), and pp. 114–24 (guide questions 15–21). The two investigations are firmly embedded in these assignments.

The first assignment is a narrative with no investigations, but it is of significance not only for an understanding of modern evolutionary thought, but also for an appreciation of the making of a scientist. It is a variation on the historical motif that has been used to introduce a number of previous chapters. In the previous instances the intention has been to illustrate the dependence of scientific progress on technological advances, the piecemeal development of scientific concepts, and the international character of the scientific enterprise. Here, however, the intention is to illustrate the growth of an idea in the mind of a scientist and to depict him as a human being. Few biologists fit the purpose so well as Darwin. From his youth a man of many faults, leading (after his one great adventure) an outwardly dull and prosaic life, neither an amateur nor a professional by today's standards, ignorant of mathematics and remote from the universities, a turgid writer—what a wonderful antidote to the popular vision of the scientist as a superhuman figure in white laboratory coat! No matter that there can be no Darwin in the milieu of twentieth-century science; the uniqueness of every genius–environment complex provides a corollary.

The second assignment deals with the genetic aspects of evolution—the factors of stability and of change; and with natural selection—the factor of guidance. It includes two fairly difficult investigations, which can be done in class or can be assigned as homework. Since these investigations involve mathematics, the procedures as well as the results should certainly be discussed in class and misunderstandings cleared up. Unless this follow-up is made, the assignments may prove worthless as well as discouraging.

The third assignment delineates the process by which species presumably come into being—a combination of genetic action, natural selection, and isolation.

TEACHING NOTES

Charles Darwin and Evolution
(pp. 94–102)

p. 94, ¶ 2. Darwin was nearing his twenty-seventh birthday when he encountered the Galapagos Islands. He was born on 12 February 1809.

p. 95, Fig. 3.2. The tree cacti pictured in the lower photograph are in the same genus (*Opuntia*) as the prickly-pear cactus common in the United States, but they assume a very different life form.

pp. 96–97, Fig. 3.3. Thirteen of the fourteen Galapagos finches are shown here. They differ mainly in bill shape and size and in associated feeding habits. The six dark species in the foreground (p. 97) feed on seeds of different sizes or on cacti. Those on p. 96 and the one on a branch on p. 97 are tree species, feeding on insects of different sizes or on fruit and buds. The finch on p. 97, holding a cactus spine in its bill, belongs to the species that extracts insects from under the bark of the tree cacti pictured in Fig. 3.2. About these various finches Darwin commented: '... one might really fancy that from an original paucity of birds in this Archipelago, one species had been taken and modified for different ends.' There are several publications on these birds, including one by Lack (reference: p. T196).

p. 96, ¶ 1. For the term 'niche', refer to pp.

92–93, *WL*. This chapter constantly harks back to the first three books of the series.

p. 96, ¶ 2. It is fitting that today there is a Darwin Research Station in the Galapagos.

p. 96, ¶ 3. The phrase 'study of nature' here does not mean merely bird watching or butterfly collecting. The study of nature is what is usually meant by the word 'science'.

p. 97, ¶ 1. Here and in the next several pages an attempt is made to portray 'Gas' Darwin as a real person, with faults as well as talents. Too seldom do we, as science teachers, attempt to bring distinguished scientists alive for our pupils. Perhaps this is why, in some respects, scientists have been separated in the minds of many from the rest of society. Chapter 5, *LO*, also begins with biographical sidelights on Darwin, and in Chapter 3 of the same book (p. 86) he is shown working with his son. Pupils who wish a further—and very provocative—acquaintance with Darwin might read his autobiography: *The Autobiography of Charles Darwin*, edited by Nora Barlow (Collins, 1958).

p. 99, Fig. 3.4. There is no certain answer to this question, of course. From the standpoint of general appearance, the leghorn breed would appear most similar to the red jungle fowl, though in other matters, such as colouration and behaviour, other breeds may retain more genes from the wild ancestry. Among the most unusual of the breeds pictured here is the Yokohama, a long-tailed fowl whose tail-covert feathers, never moulted, may exceed 20 feet in length. Such a character might appeal to chicken fanciers but would probably be an extreme handicap back in ancestral jungle country.

During the 1920s the *National Geographic Magazine* published excellently illustrated accounts of the various breeds of domestic fowl as well as of horses, cattle, and pigeons. Old issues are now difficult to obtain, but many libraries have bound volumes of the magazine.

p. 100, Fig. 3.5. Wallace became more famous for his later work on biogeographical distribution. It is coincidental that he, like Darwin, was an avid beetle collector.

p. 102, ¶ 1. At the memorable 1 July Linnean Society meeting, Darwin's paper was presented by Charles Lyell, and Wallace's paper by Joseph Hooker. For some time Darwin had been corresponding with the American botanist Asa Gray about evolution. A short sketch embodying Darwin's evolutionary ideas had been sent to Gray in September of 1857, and this was presented at the Linnean meeting to establish the priority of Darwin over Wallace. The 1 July meeting caused little stir. Amusingly enough, the president of the Linnean Society at this time regretted in retrospect that no significant papers had been presented during his term of office. Yet a century later another Linnean Society president looked back on the Darwin–Wallace presentation as 'the most important event in the history of our Society . . .' It was the publication of Darwin's book in 1859 that precipitated intellectual tumult.

p. 102, ¶ 5. In attempting to explain how hereditary variations occur, Darwin modified the ancient Greek idea of 'pangenesis'. According to his hypothesis, various parts of an organism's body produce tiny particles called 'pangenes', which are carried to the gametes by the blood and thus affect the heredity of the next generation. This was really a kind of 'acquired characteristics' theory, and it has long since been discarded.

The Process of Evolving (in part) (pp. 103–5)

p. 103, ¶ 2. You may want to use at this point the BSCS Inquiry Film Loop *The Peppered Moth: A Population Study*. It is most effective if you have pupils who can be counted upon not to have read ahead in their textbook, but it is good in any case.

pp. 105–8. Review meiosis, mutation, the action of multiple alleles, and the role of genes in continuously varying traits before discussing 'The Stability Factor' and 'The Change Factor'.

p. 105, ¶ 3. The rediscovery of Mendel's work on genetics at the turn of the present century seemed, indeed, to sound the death knell for Darwinian evolution, because the emphasis was on stability rather than on change. And De Vries' work on macromutations in evening primroses demonstrated great genetic changes, rather than the minor changes necessary for the process of Darwinian evolution. Only more recently (from the 1930s) has it been appreciated that most genetic changes *are* minor and do support Darwin's conception of evolution ('micro-evolution', as we call it today).

INVESTIGATION 3.1

THE HARDY–WEINBERG PRINCIPLE (pp. 105–7)

This investigation demonstrates the application of mathematics to problems of genetics and evolution. By relating it to Investigation 3.2, you can emphasize the relationship between genes in populations and the action of natural selection. If a class has a good background in mathematical reasoning, the investigation may be assigned for home study and then discussed in class. In most cases, however, it is advisable to work with the class co-operatively.

The case of Hardy and Weinberg is like that of Darwin and Wallace and many others in the history of science—an example of nearly simultaneous discovery by two persons unknown to each other.

Procedure (p. 106)

To make PTC papers, dissolve 1.3 g of PTC in boiling water and make up to one litre, or make a 1 per cent solution of PTC in propanone (acetone). Soak filter paper in the solution. Remove, dry, and cut into strips 1 cm by 2 cm in size. The strips have an infinite storage life. The ratio of tasters to non-tasters of PTC is approximately 70 per cent to 30 per cent of the population.

A. 50 per cent of egg cells will have allele **T**, and 50 per cent will have allele **t**.

B. 50 per cent of sperm cells will have allele **T**, and 50 per cent will have allele **t**.

C. (*a*) 25 per cent; (*b*) 50 per cent; (*c*) 25 per cent.

D. **T** = 0.5, **t** = 0.5. Note that the frequencies of **T** and **t** are the same in F_1 and F_2 generations.

E. The percentage of egg and sperm cells that have allele **T** will be 70 and allele **t** will be 30.

F. The frequency of allele **T** = 0.7 and allele **t** = 0.3.

Conclusion (p. 107)

In a large, randomly mating population without selection pressure and without mutation, gene frequencies remain the same generation after generation. Your pupils are not likely to put in the qualifications, but these are very important: evolution involves a change in gene frequencies, and it is only the existence of mutations and of selection pressures that prevents the Hardy–Weinberg principle from stabilizing gene frequencies. At this point you can easily show the pupils that the calculations are based on random mating; further, pupils can easily see, when it is pointed out to them, that mutations at a given locus will change gene frequencies. However, the pupils may not yet be able to appreciate the effect of selection. When they have completed Investigation 3.2, come back to this point.

The Process of Evolving (contd.) (pp. 107–10)

p. 107. 'The Change Factor': see note to p. 105, paragraph 3 (p. T190).

p. 107, ¶ 6. A species of organism that reproduces only asexually is sometimes said to be 'close to the end of the evolutionary road'. While this is somewhat hyperbolic, the change factor for such species is indeed limited, because in such a species a mutation can pass only to the direct line of descendants from the individual in which it occurs. More importantly, however, such a species lacks the opportunity for increased variation by recombination that is available to species in which genes from two parents are necessary for the production of any offspring.

p. 108, ¶ 3. *Note:* In the light of twentieth-century genetics Lamark's theory has additional weaknesses. All the evidence of modern genetics shows that traits of sexually reproducing organisms are passed on to offspring through gametes only and that the somatic (body) cells, which are involved in the 'use or disuse' of structures, have no effect upon the genetic character of gametes. This is not to say, of course, that the environment in which DNA molecules in gametes find themselves may not have an influence on the expression of the particular DNA. Furthermore, the retention of non-functional, vestigial structures can be more readily explained through genetics and natural selection than through use and disuse.

p. 109, Fig. 3.8. There has been much argument over how valuable protective colouration really is. It would seem obvious that white hares are better camouflaged in snowy country than are brown rabbits. But no one survival factor can be considered in isolation from others.

p. 110, Fig. 3.9. The Irish elk, whose antlers resembled those of a moose, was actually more closely related to the modern European fallow deer. The modern New World 'elk' is also a type of deer rather than a true elk, its more appropiate name being 'wapiti', whereas the American moose is equivalent to the present European elk.

INVESTIGATION 3.2

SICKLE CELLS AND EVOLUTION

(pp. 111–14)

Preface this investigation with a review of capillary circulation and the shape of normal human blood cells. Sickling is associated with a simple biochemical defect: beta-haemoglobin of sickle cells contains, at the six-position, valine instead of the normal glutamic acid.

To obtain the most value from the investigation, it is necessary to work through it point by point. The aim is not to get correct, neatly written answers to all the questions; the aim is to see how the reasoning proceeds.

Procedure

If most of the pupils in the class have some facility in reasoning and can put their thoughts into words easily, they can be asked to work through the investigation independently. Class discussion can then centre on points of controversy or confusion. In most cases, however, it is recommended that you work step by step with the class. If you do this, stop when you arrive at item L and resume discussion the following day. This provides pupils with an opportunity to formulate hypotheses for item L. Even if they read on and find out which way the hypotheses should point, they will have some experience in wording hypotheses.

B. Note that the pedigree eliminates the possibility of sex linkage. Under conditions of the question, only non-dominance is possible.

E. 25 per cent.

F. If marriage is random with respect to this trait, 16 per cent (0.16).

G. This is a repetition of item E.

H. Since offspring with sickle-cell anaemia die in childhood, they can leave no descendants. Therefore parents who produce children with sickle-cell anaemia have a reduced chance of descendants unless, on the average, they have more children than normal parents.

I. Two.

J. It should reduce the frequency of the sickling gene.

K. Darwin, of course, would have said nothing about gene frequencies but would have stated the matter in terms of the natural selection of individuals.

M. Any hypothesis for item L must depend upon a mutation rate toward sickling that equals the rate of elimination by natural selection. Therefore the hypothesis is weakened by this information.

N. The information neither supports nor weakens a hypothesis based on differential fertility: It is negative evidence. But the tendency is to assume a 'normal' situation in the absence of contrary evidence.

O. Since the malaria parasite lives within red blood cells at one point in its cycle, it has a close association with them and with haemoglobin. Hookworms are not discussed specifically in the text (though illustrated in Fig. 1.33, *DLT*). But from knowledge of the roundworm group, pupils should suspect that the connection of hookworms with haemoglobin is less intimate than that of *Plasmodium*. Nevertheless, hookworms are blood feeders, so this is merely a matter of differential probability.

P. The wording might be: The frequency of the sickling gene does not decline, because persons who carry it have a greater resistance to malaria (or hookworm) and therefore have greater survival than homozygous normal individuals.

Q. Sickle-cell individuals with malaria:

$$\frac{12}{43} = 27 \text{ per cent.}$$

Normal individuals with malaria:

$$\frac{113}{247} = 45 \text{ per cent.}$$

R. When the chi-square test is applied, the expected frequencies of sickle-cell individuals with malaria and without malaria are the frequencies in the sickle-cell individual's population that would correspond to the frequencies in the normal population.

Calculations of expected numbers per class:

Sickle-cell individuals with malaria:

$$\frac{125}{290} \times 43 = 18.5.$$

Sickle-cell individuals without malaria:

$$\frac{165}{290} \times 43 = 24.5.$$

Normal individuals with malaria:

$$\frac{125}{290} \times 247 = 106.5.$$

Normal individuals without malaria:

$$\frac{165}{290} \times 247 = 140.5.$$

Calculation of chi-square:
Sickle-cell individuals with malaria:

$$\frac{(12-18.5)^2}{18.5} = \frac{41.25}{18.5} = 2.23.$$

Sickle-cell individuals without malaria:

$$\frac{(31-24.5)^2}{24.5} = \frac{41.25}{24.5} = 1.68.$$

Normal individuals with malaria:

$$\frac{(113-106.5)^2}{106.5} = \frac{41.25}{106.5} = 0.39.$$

Normal individuals without malaria:

$$\frac{(134-140.5)^2}{140.5} = \frac{41.25}{140.5} = 0.29.$$

$$\chi^2 = 4.59.$$

The degrees of freedom equal (number of columns minus one) times (number of rows minus one). For the table on p. 113, this is $(2-1)\times(2-1) = 1\times 1 = 1$. In the chi-square table on p. 63, the line 'χ^2 for two classes' gives the values for one degree of freedom.

The chi-square value indicated that the difference in the percentages of sickle-cell and normal individuals with malaria (27 per cent compared with 45 per cent) could be expected to occur by chance between 1 and 5 times in 100 ($4.59 > 3.841$). This expectation is usually considered to indicate a 'significant' difference. Therefore, we can reject the null hypothesis that there is no difference between sickle-cell and normal individuals, and conclude that having the sickle-cell allele reduces the likelihood of having malaria.

S, T. The difference is so great that a statistical test is not really needed, but the data give pupils another chance to go through the chi-square calculations with much simpler arithmetic than before.

Calculations of expected numbers in each class:

Sickle-cell individuals with malaria:

$$\frac{16}{30}\times 15 = \frac{240}{30} = 8.$$

Sickle-cell individuals without malaria:

$$\frac{14}{30}\times 15 = \frac{210}{30} = 7.$$

Because the two groups were of the same size, the calculations for the normal individuals are the same.

Calculations of chi-square:
Sickle-cell individuals with malaria:

$$\frac{(2-8)^2}{8} = \frac{36}{8} = 4.50.$$

Sickle-cell individuals without malaria:

$$\frac{(13-7)^2}{7} = \frac{36}{7} = 5.14.$$

Normal individuals with malaria:

$$\frac{(14-8)^2}{8} = \frac{36}{8} = 4.50.$$

Normal individuals without malaria:

$$\frac{(1-7)^2}{7} = \frac{36}{7} = 5.14.$$

$$\chi^2 = 19.28.$$

The chi-square value indicates that the difference between sickle-cell and normal individuals with respect to malaria could be expected to occur by chance less than one time in a hundred ($19.28 > 6.635$). This expectation is termed 'very significant'. The null hypothesis is confidently rejected, and this constitutes support for the hypothesis in item P.

U. This is a somewhat ambiguous question. Pupils may have been impressed with the Hardy–Weinberg principle and answer No. But this situation is different from that in Investigation 3.1. There the original frequency was calculated on the basis of *both* original populations. Here the 22 per cent applies to only one of the original populations. The frequency of the sickling gene in the American Indian and European populations was presumably close to zero, so combining the three populations reduces the *frequency* of the gene. It does not, however, decrease the *number* of such genes, which depends upon the number of individuals bearing them. This can be reduced only if there is some factor which reduces the number of individuals that bear the genes relative to the number that do not.

V. No.

W. Because of the selection pressure against the allele, we would expect its frequency in the total population to decrease.

Conclusions (p. 114)

A. The main point here is that natural selection operates to eliminate homozygotes (through defective physiology) and to conserve heterozygotes (through protection against the debilitating effects of a parasite).

B. Lack of malarial parasites in the environment tends to change the frequency of the sickle-cell allele from the frequency that can exist in an environment where the parasites are abundant.

C. With respect to oxygen supply, the sickling gene certainly is harmful. But when malaria is a factor in the environment, the sickle-cell allele in the heterozygous state is an advantage to survival. When malaria is not a factor in the environment, the allele is a detriment to survival. Thus 'harmful' and the other terms are relative (to the environment) rather than absolute terms. The terms also vary between homozygous and heterozygous states. The hybrid has an advantage that neither homozygote has.

Speciation (pp. 114–24)

p. 114, ¶ 1. The reference to the species concept is so important that you might well make it a class review assignment.

p. 115, ¶ 1. Modern industrial processes tend to produce a smaller amount of soot, so presumably a decrease in soot—hence in dark trees—would favour increased survival of the light-coloured moths and a decline in number of dark-coloured moths.

p. 116, Fig. 3.13. The fossil history of the Equidae is interestingly summarized in *Horses*, by G. G. Simpson (Oxford University Press, 1951). Recall that pupils have themselves worked on another trait, involved in horse evolution (Investigation 4.1, *PLW*).

p. 117, Fig. 3.14. Obviously, the horses in Fig. 3.13 represent genera from different levels of geological time, whereas the honeycreepers are seven contemporary species. Thus, no pictured honeycreeper could be ancestral to any of the others, but they might all have a common ancestor. In the case of the horses, of course, we are looking at variations through time, while with the honeycreepers we are viewing variations through space. As a reflective sidelight, it would seem likely that there is greater variability inherent in honeycreepers than in horses. There could be a number of reasons for this, including island distribution, more frequent breeding, and greater number of offspring per pair.

p. 118, ¶ 2. Silver foxes—refer pupils to Plate II, p. 88.

p. 118, ¶ 4. The text is somewhat vague here concerning the nomenclature of subspecies, and the vagueness continues in Figs. 3.18 and 3.19. Subspecies are represented by trinomials. Thus the black snake subspecies in the north-eastern United States is called *Coluber constrictor constrictor*, that in the northern part of the Middle West *Coluber constrictor flaviventris*.

p. 119, Fig. 3.17. This figure and also Fig. 3.18 are derived from the work of Dr Walter Auffenberg. When Fig. 3.17 was constructed, the samples from Florida were analysed according to the separate localities from which they were taken; all other samples were lumped by states.

p. 120, ¶ 2. Yellow wagtails—refer pupils to Plate III.

p. 121, Fig. 3.19. If a criterion for species separation involves nonbreeding between two somewhat similar groups in the same area, then one could argue that *fuscus* and *argentatus* are actually two separate species of gulls, since they occur together but do not interbreed—they are reproductively isolated. This figure is based on the work of Dr Ernst Mayr. See *Animal Species and Evolution* by E. Mayr (Harvard University Press, 1963).

p. 122, ¶ 5. Though the salmon in the ocean are not really a freely interbreeding population, the population from each stream is not considered to be a separate species for the following reasons. From the practical standpoint of identification of specimens, there are few, if any, morphological differences that a taxonomist could use to distinguish separate species. This could be so because isolation has not existed long enough to produce discernible differences, or because all of the streams have exerted basically the same selective forces (i.e. the environments are all essentially the same). In similar cases, such fish populations are known to be capable of interbreeding under artificial conditions, suggesting that all the fish populations may best be regarded as a single species despite apparent reproductive isolation at present.

p. 123, ¶ 3. Although not discussed here, there are instances in which man has succeeded in creating new species through artificial means. The best-known example is that of the Russian plant geneticist G. D. Karpechenko, who in 1928 crossed a radish with a cabbage (both species in the cabbage family) and eventually produced a self-reproducing new population—a new species. But unfortunately for agriculturists, it had the root of a cabbage and the head of a radish. Even earlier, in 1925, Clausen and Goodspeed in the United States had crossed two species of *Nicotiana* (tobacco) and eventually produced a fertile hybrid—a new species.

PROBLEMS (pp. 125–6)

1. Sexual reproduction, involving recombinations through (*a*) crossing-over in meiosis and (*b*) genes from two parents, increases the possible variability of offspring, thus affording more 'choices' for changing selective forces to work on. Self-fertilization and parthenogenesis, on the other hand, tend to promote the genetic *status quo* (like mother, like daughter).

2. Man selects for a particular variety of traits in his domestic animals (and plants, too), traits that may or may not have survival value in the wild. Natural selection, on the contrary, selects only the set of traits best suited to survival in a particular environment; where environments differ within the geographical range of a species, some variation (subspecies) may, indeed, occur.

3. In the case of polydactyly, even dominant genes have to be 'selected for'. The possession of extra digits is of no present survival value. In the case of populations where 97 per cent of the individuals have Type O blood we are dealing with a relatively small, isolated population in which the recessive gene for Type O blood has attained a high frequency, perhaps by random genetic drift (see p. 144). However the **i** gene is frequent in the general population of the United Kingdom also. At present there is no evidence that any of the ABO blood types have survival advantage. Perhaps the main message of the problem is: Whether a gene is dominant or recessive has nothing to do with its survival value.

4. It should be remembered that although sexual reproduction increases variety, there are other sources of change available to asexually reproducing organisms, such as gene and chromosomal mutations. If dominant, a gene mutation may show up much more rapidly in an asexually reproducing population than in a sexually reproducing one, which throws some doubt on the statement about the 'end of the evolutionary road'. But many such mutations are detrimental in an existing environment.

5. There is generally a relationship between clutch size and chances for survival. The important question is how many eggs survive to become reproductive adults. This number is undoubtedly lower as latitude becomes higher. Even if populations of adults at high latitudes were larger, this does not mean that these populations would replace the populations of lower latitudes. After all, the latter populations are probably already in balance with the carrying capacity of the land there.

6. If this problem is discussed with pupils in a class, it might be interesting to find out how many of the pupils are nearsighted or have diabetes. A hundred and fifty years ago, many people with such genetic traits undoubtedly did not survive to produce children, due to the action of natural selection. Now medical science countermands natural selection in many cases, allowing such genetic traits to become more widespread. But this is only the beginning of the discussion . . .

7. With this one the pupil is pretty much on his own.

8. A good rule of thumb to keep in mind relative to the evolutionary process is that, in general, the more stable the environment, the less the evolutionary change. Oceans tend to afford a more stable environment than land and tend to have fewer isolating barriers. Thus, many marine organisms have shown little change through time. One would want, on the one hand, to review the sources and degree of variability within a particular species population and, on the other hand, to consider the stability of the species' environment. A changing environment associated with a variable species might result in rapid changes; static organisms in static environments ordinarily show little change.

9. The valley apparently represents a geographical barrier for the grasshopper populations. One might guess that at either end of the valley, where the two colour phases can fairly easily come together, there could be intermediate grades. Since the populations seem fairly separate in the main part of the valley, it appears unlikely that they are flying across.

10. There are various possible adaptations for a particular niche. In the case of a dark environment, one possibility is blindness, while another is bioluminescence combined with non-blindness. From the standpoint of natural selection, it makes little sense to have a population both blind and bioluminescent. Cave animals have evolved directly from organisms that lived outside of caves, in streams or on land. Although such organisms may have mutations resulting in blindness, few exhibit mutations resulting in bioluminescence. Once bioluminescence got started, mutations toward blindness might well have been subject to negative selection.

11. The more one considers this, the more consequences come to mind. Obviously, there would be an increase in mutation rate, thus affording greater variability for natural selection to work with. Also, changes in the environment induced by such a war might favour a new set of characteristics in organisms. Population size of certain organisms would be reduced, and populations might be isolated, leading to increased genetic drift. Many individuals might become sterile, which would have a serious effect on survival of certain species. If certain species are exterminated by such a war, it should be apparent that the 'forward-direction-only' nature of evolution would ensure that such species would never again show up on Earth. On the other hand, elimination of some species might permit others, formerly unsuccessful, to flourish.

AUDIOVISUAL MATERIALS

Filmstrips

Darwin Discovers Nature's Plan. 'Darwin's World of Nature', Part 1. *Life* Filmstrips, 1959. (Helps to vivify the story of Darwin's life.)

The Enchanted Isles: The Galapagos. 'Darwin's World of Nature', Part 2. *Life* Filmstrips, 1959. Colour. (Excellent series of striking pictures that supplement the textbook description of the Galapagos; gives special attention to the finches and their influence on Darwin's thinking.)

Evolution Today. 'Darwin's World of Nature', Part 9. *Life* Filmstrips, 1960. Colour. (Summarizes present views of the history of life, emphasizing the continuing discovery of fossil evidence.)

The Galapagos: Darwin's Finches. Ealing. Film loop.

Environment: Part 1. Rocks and Fossils. Part 2. The Age of Dinosaurs. Part 3. Mammals and Man. Diana Wyllie Ltd.

Films

The Peppered Moth: A Population Study. A BSCS Inquiry Film Loop, available from John Murray. Super-8, colour. (Best used before the account of these moths is encountered in the textbook.)

Mimicry. A BSCS Inquiry Film Loop, available from John Murray. Super-8, colour. (The pupil is led to hypothesize concerning the factors that might have given rise to selected examples of mimicry.)

TEACHER'S REFERENCE BOOKS

BARNETT, L. *The Wonders of Life on Earth.* Time, Inc. 1960. (A popularly written and beautifully illustrated consideration of evolution and life.)

BENNETT, A. J. 'Mendel's Law', *S.S.R.*, 1964, 158, **46**, pp. 35–42.

DARWIN, C., A. R. WALLACE. *Evolution by Natural Selection.* Cambridge, 1958.

FORD, E. P. *Mendelism and Evolution.* 8th edition. Methuen, 1965. (An introduction to some aspects of evolutionary genetics.)

HIMMEL FARB, G. *Darwin and the Darwinian revolution.* Chatto and Windus, 1959. (Provides an interesting account of Darwin's theory.)

LACK, D. *Darwin's Finches. Scientific American* offprint. W. H. Freeman, 1953. (An account of the finches of the Galapagos.)

SMITH, J. M. *The Theory of Evolution.* 2nd edition. Pelican, 1966.

4
THE HUMAN ANIMAL

MAJOR IDEAS

1. Anatomically, man is a vertebrate animal, a mammal of the order Primates; he is distinguishable from other primates by a number of structural features. These peculiarities lie not in unique structures, but rather in the *degree* of development of basic primate structures.
2. Man's anatomical equipment allows him no single outstanding physical accomplishment, but it provides him with unparalleled versatility.
3. The physiological peculiarities of man are mostly temporal, centring upon his slow development to maturity. They are important primarily because they provide a foundation for the evolution of behavioural characteristics, on the basis of which man's unique culture has arisen.
4. During the past few decades paleontologists have uncovered an unlikely (but still meagre) amount of fossil evidence for the evolution of hominids. Efforts to elucidate the pongid–hominid divergence have not been as successful; the great apes and man have apparently evolved along separate lines since the Miocene.
5. On the basis of the biological definition of species, all living hominids undoubtedly belong to one species. As in other wide ranging species, partial geographic isolation has resulted in the development of varieties.

PLANNING AHEAD

Chapter 4 contains the last laboratory investigations of the course; there are no investigations in Chapter 5. Consequently, the kind of planning ahead that has been principally discussed in the preceding chapters is now at an end.

GUIDELINES

'Man is an animal' is a statement of biological fact. Obviously this statement does not preclude other viewpoints. Science is not the whole of man's experience, and its limitations should be understood. A difference between man and other animals that is even sharper than the one discussed in the third paragraph of p. 128 of this book is afforded by the idea of human spirit, or soul. But it is appropriate for biology teachers to recognize the limits of biological science, and also the need for considerable knowledge of philosophy and theology in discussing this question. Therefore, the discussion of man in Chapter 4 is confined to matters that are biological *sensu stricto* and is not allowed to transgress into the more remote parts of anthropology.

If there is such a thing as 'pure' science, the search for human fossils certainly qualifies as an aspect of it. Yet this is one form of science that captures the interest of almost everyone; the best indication is the space that newspapers are willing to devote to the subject. Not everything that is given space in newspapers is worthy of classroom use, but discussions of evidence for man's origin are almost always lively.

Considering that man is the most widely distributed macroscopic terrestrial organism, he shows remarkably little subspecific differentiation. Yet what little there is has been a matter of inordinate concern. It is, therefore, appropriate to conclude discussion of the human animal with this biologically trivial topic.

Work on the chapter can begin as appropriately with Investigation 4.1 as with study and discussion of pp. 128–33 (guide questions 1–8). Pp. 136–42 (guide questions 9–16) then form a second study assignment. Investigations 4.2 and 4.3, both of which are primarily concerned with variations in human populations, can be taken up at any convenient time. Investigation 4.3, with its fairly simple but self-motivating procedure, is an excellent concluding laboratory experience for the year.

TEACHING NOTES

The Uniqueness of Man (pp. 128–33)

p. 128, ¶ 2. It has probably been some time since pupils had occasion to refer to the Appendix of *DLT*. This is a good point at which to recall some ideas concerning classification. The order Primates is shown in some detail on p. 125 of that book.

p. 129, ¶ 3. Volumes of the brain cases of fossil hominids are subject to a considerable amount of dispute because they are mostly based on fragmentary fossils.

p. 130, Fig. 4.2. In addition to the points made in paragraph 3, compare the relative sizes of brain case and face, the brow ridges, the sizes of the mouths, and the positions of the ears. In both species the size of the external ear varies greatly.

p. 131, Fig. 4.3. For an animation of the cheetah gait, see the BSCS Laboratory Block by A. G. Richards, *The Complementarity of Form and Function* (Boston: D. C. Heath & Co., 1963).

p. 131, ¶ 2. The figures given here are extremes —records of the best athletes. Your pupils may want to contribute other athletic records or perhaps some that supersede these.

p. 132, ¶ 3. The menstrual cycle of the female (pp. 38–39) is sometimes emphasized as a human peculiarity, but it seems to be only a slight extension of a primate variation on the mammalian pattern. To what degree it has contributed to the evolution of human societal organization is an interesting question.

p. 133, ¶ 3, last sentence. Yet sociality is definitely a primate trend. A zoologist used to characterize primate behaviour for his pupils as 'social, dirty, noisy, and sexy'. The last three characteristics can probably be derived from the first.

p. 133, ¶ 4. Some of your more thoughtful pupils may not like the conservative implications of this paragraph. You might emphasize the fourth sentence: the discoveries of young inventors need not be lost; if useful, they become a part of that which is conserved—a kind of cultural natural selection.

INVESTIGATION 4.1

THE SKELETAL BASIS OF UPRIGHT POSTURE (pp. 134–6)

Materials

Cat skeletons are available from suppliers such as Gerrard and Haig, etc. Skeletons of other quadruped mammals may be used in place of the cat skeleton. (Rat skeletons, however, are too small for good observation by groups.) Human skeletons are expensive, but full-size replicas difficult to distinguish from the real thing are now available. The small plastic models are not satisfactory for this investigation.

Procedure (p. 134)

If only one skeleton of each species is available—as is usually the case—the observational part of the investigation can be done in shifts. The skeletons are large enough to be seen easily, so an observing group may be rather large. At least two groups can make all the observations in a single period.

The items under 'Studying the Data' and 'Summary' can be worked out by pupils individually at home; but if time and materials permit, it is preferable to have this done cooperatively within each group before class discussion is begun.

C. Without terms for individual bones, this may be a little difficult to state; but in general the changes must bring the eye sockets to a forward orientation, and nasal and jaw bones must be shifted so that they lie under the eye sockets.

G. This may be a little difficult for pupils to determine. Both the position of the forearm and the musculature around it result in the human skull being balanced rather than braced on the top of the vertebral column. (This makes us singularly liable to whiplash neck injuries in motor car accidents.)

L. Pupils sometimes have difficulty interpreting this question, but even in Fig. 4.1 the double curve of the human vertebral column is evident in comparison with the single curve of the gorilla's.

Q. The rather firm pectoral girdles of hominids (as compared with those of cats) are usually associated with presumed former locomotion by brachiation. If this presumption is correct, the shoulder architecture can be considered an example of *pre-adaptation*; that is, it was suited to the extensive use of the arms in wielding heavy tools and weapons when locomotion was shifted entirely to the posterior appendages.

Studying the Data (p. 135)

A. A greater cranium volume implies a greater brain. This is the kind of inference that is the basis for much paleontological interpretation.

B. Binocular vision.

C. Many answers are possible. All should concern the kinds of skeletal adaptations that have been stressed in the preceding questions.

E. A circus dog (which has a pectoral architecture similar to that of a cat—Fig. 1.4, *DLT*) may be taught to walk short distances on its forelegs. Hand walking in man is not easy either, but the problems are more of equilibrium and cranial circulation than of weakness of the supporting limbs.

F. See note to **Q** above.

H. It provides a springiness suited to leaping or a rapid getaway.

I. The tiring is merely a symptom of difficulty in balancing; it is remarkable enough that a man can balance himself on the full length of his feet.

Discussion

Because much interpretation is left to the teacher, the investigation should be concluded with a general class discussion. You may wish to refer to Investigations 1.1, *DLT*, and 4.1, *LO* in which man and other animals were compared with respect to various characteristics.

Any study of skeletons can become quite loaded with terminology. The terms used in the investigation are those that will be found useful for carrying on a discussion. You may add to them as you see fit.

Becoming Human (pp. 136–42)

p. 136, ¶ 2. Actually, as stated on p. 139, paragraph 2, the first remains of 'Neanderthal man' had been discovered in 1856, but the meaning of this find was then much disputed. You might ask pupils why human fossils are expected to be rare. Some points: populations of early hominids were apparently rather sparse (how far we have come in this respect!); most of the bones of hominids are rather fragile; in wet tropical conditions bone deteriorates rapidly unless buried under anaerobic conditions. It should be pointed out that the hominids, as contrasted with the early Hominoidea from which the family presumably evolved, are thought by many paleontologists to have belonged to savanna or grassland rather than forest ecosystems.

p. 137, ¶ 3, last sentence. Be sure pupils connect this with item 7 of the Procedure in Investigation 4.1.

p. 138, ¶ 2, first sentence. This is probably an overstatement. Perhaps the only thing clear in human paleontology is that nothing is clear. You should emphasize the state of flux in this field, but try to keep as up to date as possible. Pp. 136–41 were written in a restrained way that, it was hoped, would be not too misleading on the day they were printed; nothing could be done about the days after.

p. 138, ¶ 3. Notice that the discussion is tracing the chronology of evolving man. Unfortunately, the chronology of the discoveries is almost the reverse. To use the latter chronology would improve the investigative character of the discussion, but it would require a length of description unavailable in this short, penultimate chapter.

p. 139, Fig. 4.6. Concerning the caption question see p. 96, paragraph 1.

p. 139, Fig. 4.7. The questions are primarily intended to elicit observations from pupils, not to establish technical criteria. Perhaps the most important point is the ratio of brain case size to face size. The discussion on pp. 118–21 does not explicitly state that trinomials are used in nomenclature to designate subspecies, but this figure provides an example.

p. 140, Fig. 4.8. Much of the fossil evidence for *Australopithecus* and *Homo erectus* is fragmentary, and the skulls in Fig. 4.7 are themselves to some extent reconstructions; but most of the musculature is based directly on skeletal evidence. Details of fleshy structures, such as external ear and nose, are less sure; and hairiness is quite uncertain. Even in portraying modern man the artist has had to make compromises in the last respect, since different varieties of modern man (to say nothing of ages and sexes) vary markedly in the amount and character of hair.

p. 140, ¶ 2. A question which might be asked is what happened to 'Neanderthal man' when modern man appeared. Some anthropologists think that a good deal of hybridization occurred.

p. 142, ¶ 1. Refer pupils to Plate IV.

INVESTIGATION 4.2

BIOLOGICAL DISTANCE
(pp. 143–5)

In some classes this investigation can be assigned as homework. Not all pupils, however, are able to work their way successfully through it; therefore some class discussion must follow the work at home. In classes where slow pupils predominate, the investigation is best done entirely in class, under step-by-step teacher direction.

Some of your pupils may want to read further on the subject of genetic criteria for distinguishing human populations. A useful and very readable book on this topic is W. C. Boyd and I. Asimov, *Races and People* (New York: Abelard-Schuman, 1955).

Procedure (p. 143)

The data for British Columbia are derived from various tribal groups; those for New Mexico are confined to the Navahos. The data are obtained as blood types—A, B, AB, or O—and the allele frequencies are calculated from the phenotype frequencies.

A–C. After working on Investigations 2.2 and 3.2, some pupils may argue that the slight differences between the British Columbia and Navaho populations represent nothing more than sampling errors. Without knowing the number of persons in the sample, it is not possible to test for the significance of the difference. But, in general, blood-type data are abundant, and in the source from which the data were derived, the difference was considered significant.

D, E. Answers to these questions are likely to be as various as are the definitions of human races. Some pupils may react to these questions with: Why bother to distinguish racially between Eskimos and Indians? Or, phrased in the most likely form: Why bother? There is, of course, no rational answer to such questions. They are in the same category as: Why climb a mountain? With the course coming to an end, it is worthwhile to point out (again) that science is not *necessarily* concerned with 'practical' questions. Nevertheless, it brings about improvements in the circumstances of human life more frequently than any other enterprise.

I. Later $\mathbf{I^b}$-bearing populations would be most numerous nearest the point of origin; that is, closest to Asia.

J. The Eskimo population. This is supported by much other evidence.

K. On the basis of the $\mathbf{I^b}$ gene *only*, the biological distance between Central Asians and Basques equals that between Central Asians and Navahos.

L. This ought to be an obvious absurdity, but in matters of race there seems to be no criterion for absurdity. No racial, subspecific, or varietal status can be distinguished on the basis of one genetic locus. Moreover, all these terms denote a continuous interbreeding sub-population within a species population. Clearly Basques and Navahos seldom interbreed, though they are undoubtedly capable of doing so.

M. It is 0.57. If you wish to go into the practical consequences of $\mathbf{Rh^+}$ and $\mathbf{Rh^-}$ incompatibility, here is a good opportunity to do so. Often pupils bring the matter up, since it is now rather well known. But such a discussion is not necessary for the continued argument of the investigation.

N. It is 0.16. $(b)/(a)\times 100 = 28$ per cent. This means that the amount of mixing is 28 per cent of that required to make the difference between American and African Negroes with respect to $\mathbf{Rh^0}$ indistinguishable from the difference between Europeans and Africans with respect to $\mathbf{Rh^0}$. This method of calculation may seem faulty, since mixing must involve *both* populations. But under American conditions, most persons of mixed Negroid and Caucasoid ancestry are classified as Negroid, so that introgression of the $\mathbf{Rh^0}$ gene into the 'Caucasoid' population is undeterminable.

O. Thirteen generations.

P. About 2.2 per cent. The answer to this question is a biological deduction from verifiable observations. If the class climate is suitable, you may wish to project the situation into the future—a suitable prelude to Chapter 5, which is entirely forward-looking. Such a projection is still a matter of science, but it is so obscured by sociological implications that you may not wish to pursue it. However, if the matter is gone into, it should be kept in mind that the answer to item P assumes a constant rate of mixture, and this is highly unlikely. It is much more likely that the rate curve would show some of the characteristics of a growth curve. The **Rh⁰** gene would never disappear from a large population (in the absence of natural selection), but its frequency would become asymptotic to a level somewhat above that of the frequency in the European Caucasoid population. It should also be kept in mind that the point in time at which the Negroid and Caucasoid populations become indistinguishable with respect to frequency of the **Rh⁰** gene would not necessarily be a point at which the populations would be indistinguishable with respect to visible characteristics.

INVESTIGATION 4.3

HUMAN BLOOD GROUPS
(pp. 145–7)

This investigation is interesting to pupils and forms a fitting termination to the year's laboratory work. But it must be planned with particular care: First, arrangements must be made to take care of cases of fainting that occasionally occur; second, only sterile, disposable lancets should be used, and each lancet must be discarded after *one* use. The Department of Education and Science recommends that only staff should obtain blood from pupils.

The 'Background Information' should suffice for establishing the rationale of the procedure.

Materials

Serums and supplies of propan-2-ol (isopropanol) can be shared by two or three groups.

If serums are old or have not been kept under refrigeration, many errors may occur. These are likely to be systematic rather than random errors, and of course they will affect the percentages.

Procedure (p. 146)

Be sure that pupils understand the difference between *clumping* of red cells and *clotting* of blood.

Summary (p. 147)

A, B. When checking answers, you may be assisted by the following tables:

GROUP	PLASMA	RED CELLS
O	anti-a and anti-b	neither
A	anti-b	A
B	anti-a	B
AB	neither	A and B

Fig. T-21

	RECIPIENT			
	O	A	B	AB
O	−	−	−	−
A	+	−	+	−
B	+	+	−	−
AB	+	+	+	−

Positive reaction +
Negative reaction −

Fig. T-22

D. This depends upon circumstances, of which you will have to be the judge.

PROBLEMS (pp. 148–9)

1. See note to p. 136, paragraph 2 (p. T199).

4. Briefly, the Lamarckian argument might be that when canine teeth no longer were much used, they gradually became smaller; the Darwinian, that individuals with shorter canine teeth, no longer being at a disadvantage with respect to fighting and perhaps being better able to chew without the rather clumsy big canines, had first no disadvantage and then a positive advantage in survival over individuals with large canines.

7. This is a highly speculative question. But the pupil needs to know something about the social organization of wolf packs as well as the organization of human tribes. A considerable demand—and he had better plan to make the study a summer project. Advise a beginning with N. Tinbergen, *Social Behaviour in Animals* (reference: T160).

AUDIOVISUAL MATERIALS

Filmstrip

Man Inherits the Earth. Part 1A of 'Epic of Man'. *Life* Filmstrips. Colour. (Rather glib but colourful portrayal of early man.)

Films

Blood Groups, Skin Colour, and Gene Pools. AIBS Film Series. McGraw-Hill Text-Film Division. 16 mm, colour, 28 min. (Will serve as a review of some genetic principles important in human racial differentiation and as background to Investigation 4.2.)

Evolution of Man. AIBS Film Series. McGraw-Hill Text-Film Division. 16 mm, colour, 28 min. (Discusses the evidence for man's existence during the Pleistocene.)

TEACHER'S REFERENCE BOOKS

BEALS, R. L., H. HOIJER. *An Introduction to Anthropology.* Collier-Macmillan, 1971. (A simple clear and comprehensive presentation of the basic fundamentals of anthropology.)

BRACE, C. L., M. F. ASHLEY MONTAGU. *Man's Evolution.* Macmillan, 1965. (An interpretation of the data of the human fossil record.)

CARRINGTON, R. *A Million Years of Man.* Weidenfeld & Nicolson, 1963. (An account of the possible history of man related to its universal setting.)

COON, C. S. *The Origin of Races.* Cape, 1962.

COON, C. S. *The History of Man.* Cape, 1962. (Narrative account of human origins and prehistory.)

HARRISON, G. A., U. S. WEINER, U. M. TANNER, N. A. BARNICOT. *Human Biology.* Oxford University Press, 1964. (A survey of physical anthropology.)

HOWELLS, W. *Mankind in the Making.* Pelican, 1967. (A discussion of the evolution of man.)

5
MAN IN THE WEB OF LIFE

MAJOR IDEAS

1. Gradually, as his technology has improved, man's position in the biosphere has shifted. Primitive man was a member of a biotic community, important but not dominant; modern man overrides all natural community boundaries, fabricates new communities, and overbalances meterological and geological forces in reshaping the Earth.
2. Throughout his history, improvement of man's understanding has resulted in improvement in his technology, and improvement in his technology has resulted in improvement of his understanding. The problems of modern man lie as much in the proper application of understanding and technology as they do in the further improvement of either. And most of these problems are basically biological.
3. The roots of most major biological problems —from malnutrition to floods, from traffic to smog—lie in the rapidly increasing world density of human population. As in all other populations, this density involves the four determiners: natality, mortality, immigration, and emigration, of which the last two have (at present) zero values. It is clear that solutions to human overpopulation must be through a reduction of natality.
4. For man the living world is not only an entity that demands study, yielding understanding; it is also an experience that demands appreciation, yielding aesthetic pleasure. But scientific study is itself fraught with aesthetic considerations. Thus biology is doubly a humanistic enterprise.

PLANNING AHEAD

If you expect to use this course again, you undoubtedly have ideas for modifying it. These ideas should be worked out. As in all public performances, spontaneity in teaching is most effective when it springs from solid preparation.

The emphasis throughout this year on planning for laboratory work is justified—but it is not sufficient. Planning should involve more than logistics; it also should involve tactics and strategy—and, further, a philosophical consideration that (to extend the image) might be called polity. It is clear that no science teacher can ever rest from this kind of planning. Discover by courses or reading what is currently happening on the research frontiers of some branch of biology with which you have had little acquaintance. Read and ponder some work of biological humanism or biological philosophy.

GUIDELINES

Field and laboratory equipment has been inventoried and stored. But, hopefully, pupils' impressions of a year's contact with living things remain. These impressions are the materials for work on Chapter 5.

The first thing to do is to return to pp. 31–33, *WL*. The problems posed at the beginning of the pupil's biology course constitute the foundation for discussion in Chapter 5 of this book. They are not taken up sequentially, nor is explicit reference to each to be found in Chapter 5. The present plight of man in the biosphere is much too complex for such a simplistic approach to be valid. But the problems of Chapter 1, *WL*, are all implicit in Chapter 5. Having been exposed to eighteen intervening chapters, most pupils should be in a position to bring some further biological maturity to a few periods of reading, thought and discussion.

The form that such discussion takes will depend upon your shrewdness and the pupils' insight. The areas considered in the chapter should by no means limit the discussion. Wherever possible, the information and insights gained during the study of biology should be woven into the prospect of the future. Above all, endeavour to make each pupil feel his personal responsibility for applying whatever knowledge of science he possesses to the decisions that will need to be made by his generation.

TEACHING NOTES

Man and the Biological Community (pp. 151–5)

p. 152, Fig. 5.1. Just as species that have retained primitive characteristics help us to visualize organisms of the past, so primitive human cultures surviving in remote parts of the modern world suggest to us the conditions of life in ancient human cultures. But study of such surviving primitive cultures must proceed rapidly, for modern culture is penetrating everywhere.

p. 154, Fig. 5.3. The contrast between this figure and Fig. 5.2 is self evident, but the implications of the contrast require some thinking. For example, Fig. 5.3 represents an industrial country of today better than a panorama of steel mills does. Concentrations of industrial population are possible only because such wholesale production of food is possible. On the other hand, Fig. 5.3 implies a great industrial complex, for only that can produce and keep running such a fleet of machines.

p. 156, ¶ 3. Sharks remain a rather serious danger, as man has turned in recent years towards increasing exploitation of the seas. An active programme of research on protection from sharks has been carried on in recent years in America.

p. 157, ¶ 3. DDT is dichlorodiphenyltrichloroethane, or more precisely 1,1,1,-trichloro-2,2,-di(*p*-chlorophenyl)-ethane. Biology is not the only science that makes use of long terms! (See further, p. 173.)

p. 158, Fig. 5.6. The graph bears a striking resemblance to the one drawn in Investigation 2.1, *WL*, but pupils should be cautioned concerning the dangers of extrapolation.

p. 160, ¶ 2. But birth rate tells only part of the story; an important factor in population increase is the great number of people in the fertile age group (ages fifteen to forty-five). Population increases in proportion to (*a*) the rate of increase (similar to the interest rate in financial matters), and (*b*) the number of people doing the reproducing (equivalent to the amount of money drawing interest). Therefore, although the *rate of increase* is declining, there is an ever-expanding population doing the reproducing. Even if our rate of increase were reduced to 0.1 per cent, the population would continue to grow.

p. 160, ¶ 5. Concerning the possibility of emigration from the world, there are three major problems: (*a*) no planet is yet known that would support man, (*b*) the energy cost to lift any sizable part of our population to another planet is beyond all present consideration, to say nothing of the energy required to deliver support systems to the population in space, and (*c*) the by-product of missile blast in the world atmospheric pollution would be great and probably would result in major changes in climate. Even present measurements of the increase of atmospheric aerosols and their effects in increasing the reflection of solar radiation indicate a decrease in world temperature. (See p. 171.)

p. 162, ¶ 2. There are important programmes to supplement the protein-deficient diets of peoples who live in countries where they cannot afford the luxury of feeding plants to animals in order to get high-quality proteins. One programme involves adding to the carbohydrate diet low-cost proteins, for example fish meal, egg powder, yeast, and soybeans. Another high-protein food (used in Latin America) is called Incaparina; it includes 1 part of oilcake (residue after oils have been extracted from cottonseeds or soybeans) and 2 parts of grain, with additions of yeast and vitamin A. Look back at Fig. 1.10 on p. 21, *WL*, to recall why it takes so many joules (or calories) to produce animal protein.

p. 164, ¶ 2. Early concepts of forest conservation included the idea that water is conserved by the presence of forests in a watershed. Many now believe that if water supply is the *major* concern, a watershed with grass or low-shrub cover is superior to one with trees. This is true because much rainfall is intercepted by the trees and evaporates before joining the groundwater. Secondly, the trees remove much of the groundwater through transpiration. As always in resource management, all purposes must be considered before a management programme for a particular watershed is planned.

p. 167, ¶ 3. Wildlife and woods seem to have

an intrinsic appeal for fourteen- to sixteen-year-old males. This is as pronounced in urban as in rural areas.

p. 169, ¶ 2. 'For one species to mourn the death of another is a new thing under the sun. The Cro-Magnon who slew the last mammoth thought only of steaks. The sportsman who shot the last pigeon thought only of his prowess. The sailor who shot the last auk thought of nothing at all. But we, who have lost our pigeons, mourn the loss. Had the funeral been ours, the pigeons would hardly have mourned us. In this fact, rather than in . . . nylons or . . . bombs, lies objective evidence of our superiority over the beasts.'

Aldo Leopold, *Sand County Almanac and Sketches Here and There* (New York: Oxford University Press, Copyright © 1948).

Values (pp. 174–6)

p. 177. A summary for Chapter 5—a chapter that looks both backwards and forwards—seems inappropriate. Instead, the authors feel that a fitting close to the book is found in the words with which Charles Darwin grandly concluded *The Origin of Species*.

GUIDE QUESTIONS (pp. 177–8)

4. Complexity reduces population fluctuations because it provides more competing, predatory, or parasitic species to control surges in populations of any one species. Moreover, complexity makes it more difficult for a species to find a readily available or abundant food supply. For example, contrast the condition of an insect species that feeds only on pine needles in a community where only one out of every twenty-five trees is pine and in a plantation where every tree is a pine.

PROBLEMS (pp. 178–9)

1. There is no single good definition for 'life', but this should stimulate discussion of some of its major attributes.

2. Biological control involves the manipulation of either organism or environment, or both, to utilize inherent limiting factors for control purposes. For example, if in one developmental stage an insect injurious to a field crop is particularly vulnerable to flooding, fields might be flooded at the proper time. Biological control requires detailed knowledge of life cycles, behaviour, and ecological relationships. Biological control usually has minimal effects upon other species in a community.

3. This is primarily a thought problem for your more sophisticated pupils. Facts are difficult to come by, but there is evidence that food-import programmes may have actually decreased the food production of some importing countries while stimulating an increased rate of population growth.

4. The quotation simply shows that the Pygmies, like the American Indians before invasion by Europeans, recognized that they were a *part* of the ecosystem rather than *apart* from it.

6. (*a*) Ceylon has higher birth and death rates; therefore there is a great preponderance of young, and the life expectancy of an individual is low. (*b*) The young and the old age groups have increased relative to the 'middle-aged' group in the United Kingdom as a result of better control of childhood infectious diseases and of the diseases of old age. The increase in the young also results from a high natality rate. This distribution of age groups in the population places an increasing economic burden on the middle-aged class, which is the principal supporting one. (*c*) The birth rate for the population as a whole will decline, since the number of reproducing individuals declines. (*d*) The rate of population increase will be greater in Nation A, since the earlier age of delivery of the first child reduces the length of a generation. Therefore, more progeny will be produced per unit time—an increased rate. If you know the proportions of the population that were of child-bearing age in both nations, the hypothesis would be on firmer ground.

7. If wastes are quickly picked up and utilized by organisms so that they do not accumulate sufficiently to harm the ecosystem, they are not pollutants.

AUDIOVISUAL MATERIALS

Filmstrips

Competitive Land Use. McGraw-Hill Filmstrip. Colour. (A good visual presentation of some problems of planning for multiple land use.)

Environment: Nature Conservation in the British Isles: Air Pollution; *Water Pollution*. Diana Wyllie Ltd.

Films

Biology in Today's World. Coronet Films, available from Gateway Educational Films.

16 mm, colour, 11 min. (May be used to stimulate some thinking about the role of biology in maintaining our present civilization.)

Poisons, Pests and People. National Film Board of Canada. 16 mm, black and white, 30 min. (An old film, but it presents the pesticide problem in an unusually rational manner.)

Conservation and the Balance of Nature. Concorde Films Council Ltd. Colour, 18 min. (Stresses man's relationship and responsibility to the environment.)

Teesside? We only live here. Concorde Films Council Ltd. Colour, 27 min. (Deals with air and water pollution.)

After the Torrey Canyon. Royal Society for the Protection of Birds. Colour, 13 min. (Fits in well with the reference in the text.)

TEACHER'S REFERENCE BOOKS

ARTHUR, D. *Survival. Man and His Environment.* The English Universities Press, 1969. (Discusses man and his relationship to his environment.)

ARVILL, R. *Man and Environment.* Pelican, 1967. (Problems of managing the environment taking the British Isles as a model.)

COMMONER, B. *The Closing Circle.* Cape, 1971. (Discusses what the environmental crisis really means.)

DEPARTMENT OF SCIENTIFIC AND INDUSTRIAL RESEARCH. *Effects of polluting discharges on the Thames Estuary.* HMSO, 1965.

DORST, J. *Before Nature Dies.* Collins, 1970. (Deals with the dangers threatening man and nature and proposes a rational management of the Earth. Profusely illustrated with pictures from all over the world.)

DUMONT, R., B. ROSIER. *The Hungry Future.* Methuen, 1969. (Discusses the stark reasons for famines and suggests possible solutions.)

LOWRY, J. H. *World Population and Food Supply.* Arnold, 1970.

NICHOLSON, M. *The Environmental Revolution.* Pelican, 1972. (Reviews the causes of pollution to date and, using this as a base line, explores the future possibilities.)

TAYLOR, R. B. *The Doomsday Book.* Panther, 1970. (A powerful treatise on the perils of pollution.)

APPENDICES

Appendix A Summary of Materials and Equipment

The lists of materials and equipment on the following pages may prove helpful for teachers who are planning to use *Biology: An Environmental Approach* for the first time. Keep the points listed below in mind when using the lists.

1. A few items are omitted. These items are easily obtained from pupils (e.g. cardboard boxes, food samples).
2. Substitutes are feasible for some of the listed items. For suggestions, check the notes on each investigation.
3. Quantities for one class are specified on the basis of thirty pupils, usually working in groups of five.
4. If a school has more than one biology teacher, the quantities needed are not necessarily obtained by multiplying the quantities required for one teacher by the number of teachers.
5. For the most part, estimates are close to minimal, with only small allowances for loss and breakage.

Key to books, Appendix A:

1. The World of Life: the Biosphere
2. Diversity Among Living Things
3. Patterns in the Living World
4. Looking into Organisms
5. Man and his Environment

MATERIALS

BOOK 1	2	3	4	5	Item	1 Class
			√		Aceto-orcein solution	100 cm^3
	√	√	√	√	Agar	400 g
					Alcohols:	
√	√	√	√	√	ethanol	4000 cm^3
√			√	√	propan-2-ol	4000 cm^3
√					Aluminium foil	2 rolls
√					Ammonium hydroxide	
					Bacteria:	
		√			*Agrobacterium tumefaciens*	1 culture
	√				*Sarcina lutea*	1 culture
	√				*Serratia marcescens*	1 culture
	√	√			Beef extract	250 g

BOOK						
1	**2**	**3**	**4**	**5**	**Item**	**1 Class**
			√		Benedict's solution (or Fehling's solution, A and B)	500 cm^3
			√		Petrol	500 cm^3
	√				Bouillon cubes	1 packet
√					Bromothymol blue solution, aqueous	25 cm^3 or 5 g
			√		Calcium chloride	
			√		Carnoy's fluid, with chloroform	250 cm^3
					Cellophane sheets, 2 colours:	
			√		red	5 sheets
			√		blue	5 sheets
			√		Cellulose tubing	15 m
			√		Chalk, 3 colours	2 sticks per colour
			√	√	Cheesecloth/muslin	1 m^2
			√		Chromatography paper	1 roll
√	√	√	√	√	Cleansing tissue	1 box
			√		Clinitest tablets (or Clinistix)	1 bottle
			√		Copper(II) sulphate(VI), anhydrous	500 g
					Cotton wool:	
			√	√	absorbent	500 g
	√				non-absorbent	500 g
					Crayfish:	
		√			living	12
	√				preserved	12
√	√	√	√	√	Crayons, glass-marking	36
	√	√			Crystal-violet solution	100 g
	√		√		*Daphnia*, living	1 culture
√		√	√	√	Dextrose, CP (glucose)	2 kg
			√		Drinking straws	1 box
				√	*Drosophila* (*see* Fruit flies)	
	√				Earthworms, living	12
				√	Eggs, chicken, fertilized, unincubated	36
√			√		*Elodea* (*Anacharis*)	2 bunches
			√		Eosin	100 g
			√		Ethanoic (acetic) acid, glacial	500 cm^3
					Ethers:	
√				√	ethoxyethane (diethyl ether)	500 cm^3
			√		petroleum ether	500 cm^3
			√		Fehling's solution (*see* Benedict's solution)	
√	√	√	√	√	Filter paper, 11 cm diameter	5 packets
√			√		Formalin	2500 cm^3
	√	√	√	√	Frogs, living	12
					Fruit flies, living:	
				√	mutant (e.g. 'ebony')	1 culture
				√	wild type	1 culture
√		√	√		Fungicide—sodium chlorate(I) (hypochlorite) or commercial bleach	1000 cm^3
	√				Gauze bandage	1 box
√		√	√	√	Glucose (*see* Dextrose)	
				√	Glucose–1–phosphate	
					Graph paper:	
√					semilogarithmic	60 sheets
√		√	√		square coordinate	600 sheets
√	√	√	√	√	Gravel, small aquarium	5 kg
	√				*Hydra*, living	18

BOOK 1	2	3	4	5	Item	1 Class
		√	√		Hydrochloric acid	500 cm^3
			√		Hydrogen peroxide	500 cm^3
√			√	√	Iodine–potassium iodide solution	100 cm^3
√	√	√	√	√	Lens paper	4 books
			√		Limewater	2000 cm^3
				√	Maize meal	500 g
				√	Maize starch	500 g
			√		Manganese(IV) oxide (manganese dioxide), powder	500 g
√	√	√	√	√	Markers, felt-tip	6
			√		Methylene blue solution	25 g
		√			Methyl orange solution	25 g
				√	Oats, rolled	1 box
			√		Onion bulbs	6
√	√	√	√	√	Paper, heavy blotting	3 sheets
			√		Paper clips	1 box
			√		Paper cups, small	200
√	√	√	√	√	Paper tissues	12 boxes
√	√		√		Paraffin wax	500 g
√	√	√	√	√	Pencils, coloured	12
			√		Peppercorns	25 g
√	√	√			Peptone	125 g
			√		Petroleum jelly	1 jar
		√			Phenolphthalein solution	100 cm^3
			√		pH test paper, universal	1 roll
√	√	√	√	√	Pins, 'bank' or 'florist'	2 boxes
			√	√	Pipe cleaners	48
	√				Planarians, living	24
		√	√	√	Plants, potted (bean, tomato, sunflower, *Coleus*, *Zebrina*, etc.)	10
√	√	√	√	√	Plastic bags, sandwich-size	250
			√		Potassium chloride	
√					Potassium dihydrogenphosphate(V) (dihydrogen-orthophosphate)	500 g
			√		Potassium sodium tartrate	500 g
			√		Potatoes	6
				√	Pot labels, wooden	100
	√		√		Propane-1,2,3-triol (glycerine, glycerol)	250 cm^3
			√		Propanone (acetone)	1000 cm^3
				√	Propionic acid	500 cm^3
			√		Quinine sulphate	25 g
√	√	√	√	√	Rubber bands, assorted	1 box
					Rubber tubing:	
√	√	√	√	√	5 mm inside diameter, to fit 6 or 7 mm glass tubing	9 m
√	√	√	√	√	rubber gas tubing, for bunsen burner	6 m
					Sand:	
		√		√	coarse	1 kg
		√	√		fine, washed	1 kg
					Seeds:	
			√		bean	500 g
		√	√		maize, field-type	500 g
		√			lettuce	1 packet
					pea, genetic strains:	
			√	√	'round'	500 g
			√	√	'wrinkled'	500 g
		√			sunflower	2 packets

BOOK 1	2	3	4	5	Item	1 Class
		√		√	tobacco	1 packet
		√			radish	1 packet
		√			tomato	2 packet
		√			vetch	250 g
				√	Serums, blood-testing, anti-A and anti-B	1 set
√					Snails, living, small	12
			√		Sodium hydrogen carbonate (bicarbonate)	500 g
			√		Sodium carbonate	500 g
		√	√		Sodium chloride	500 g
			√		Sodium citrate	250 g
		√	√		Sodium hydroxide	500 g
		√	√		Sodium chlorate(I) (hypochlorite) (*see* Fungicide)	
			√		Spinach	500 g
			√		Sponges, synthetic	3
√					Stakes, wooden	48
			√		Starch, soluble	250 g
		√	√		String, cotton cord	1 ball
			√	√	Sucrose	500 g
			√		Tape, pressure-sensitive: masking; transparent	1 roll each
			√	√	Toothpicks	
			√		Treacle	250 g
√					Trichloromethane (chloroform)	250 cm^3
			√	√	Vermiculite	50 kg
			√		Wood-lice	60
√	√	√	√	√	Wrapping paper	1 roll
					Yeast:	
√			√	√	brewer's	100 g
√				√	dried	3 packets

EQUIPMENT

BOOK 1	2	3	4	5	Item	1 Class
√	√		√		Aquaria	3
√	√				Autoclave (or pressure cooker)	1
√	√	√	√	√	Balance, 0.1-g sensitivity	2
	√		√		Battery jars, large	4
			√		Beads, glass, pea-size	1 kg
				√	Beads, Poppit, 2 colours	360 per colour
					Beakers, Griffin low-form:	
√	√	√	√	√	50 cm^3	36
√	√	√	√	√	100 cm^3	36
√	√	√	√	√	250 cm^3	36
√	√	√	√	√	600 cm^3	12
√	√	√	√	√	1000 cm^3	10
√					Berlese apparatus	5
		√		√	Bottles, dropping, 10 cm^3	12
				√	Bottles, small, with plastic lids	48
					Bottles, wide-mouth, screw-cap:	
√			√		60 cm^3	30

BOOK 1	2	3	4	5	Item	1 Class
√			√		100 cm^3	24
√			√		125 cm^3	30
√			√		250 cm^3	18
					Brushes:	
	√				camel-hair	18
√			√	√	small, water-colour type	12
√	√	√	√		Bunsen burners	6
					Clamps:	
√	√	√	√	√	burette	12
√	√	√	√	√	pinch or spring	3
√					Clothesline, plastic (15 m)	6
√	√	√	√	√	Corks, assorted sizes	1 packet
√	√	√	√	√	Cover slips, circular diam. 16 mm (No. 1)	3 boxes (100 ea.)
√	√				Crystallizing dishes, glass or clear plastic, 250 cm^3	60
√					Culture bottles (specimen tubes) 25 mm × 100 mm	48
√					Culture tubes, screw-cap, test tubes, 19 mm × 150 mm	36
√	√	√	√	√	Dissecting needles, straight	60
			√		Dissecting dishes	10
				√	Etherizers	6
	√				Files, triangular	6
					Flasks, Erlenmeyer (conical):	
√	√	√	√	√	250 cm^3	24
√	√	√	√	√	500 cm^3	6
					Flowerpots:	
		√	√	√	10 cm diameter	24
		√	√	√	15 cm diameter	24
		√	√	√	15 cm diameter (shallow form)	24
		√	√	√	Flowerpot saucers, to fit 15 cm pots	24
					Forceps:	
√	√	√	√	√	fine-pointed	30
√	√	√	√	√	pupil grade	30
			√		Funnels, glass, long-stem	6
	√		√		Funnels, glass, 75 mm diameter	6
			√		Funnel supports	6
		√			Glass covers, 12.5 cm squares	48
	√		√		Glass rods	500 g
					Glass slides:	
			√		'micro-culture' (with centre depression)	12
√	√	√	√	√	ordinary microscope	100
	√		√		Glass tubing, 6 to 7 mm diameter	10 kg
					Graduated cylinders:	
√	√	√	√	√	10 cm^3	10
√	√	√	√	√	25 cm^3	10
√	√	√	√	√	100 cm^3	12
√	√	√	√	√	500 cm^3	2
√	√	√	√	√	Hammers (or small mallets)	3
	√	√		√	Hand lenses	36
√	√		√	√	Hotplate, electric 2-plate	1
		√		√	Incubator, 50-egg capacity	1
	√	√			Inoculating loops	12
					Jars, glass:	
√	√	√	√		1 litre	15
√	√	√	√		4 litre	6

BOOK 1	2	3	4	5	Item	1 Class
√			√		Jars, waste, crockery	6
√	√	√	√	√	Lamps, flexible (with 60 to 150 W bulbs)	6
				√	Lancets, sterile, disposable	1 packet
√	√	√	√	√	Medicine droppers	48
√			√	√	Metre sticks	9
					Microscopes:	
√	√	√	√	√	monocular	15–20
	√	√		√	stereo	8–12
		√	√	√	Mortar and pestle	6
			√		Nails	500 g
√	√	√	√	√	Petri dishes, Pyrex, 100 mm × 15 mm	50
					Pipettes:	
	√		√		1 cm^3; 10 cm^3	6 each
			√		Razor blades	12
		√			Refrigerator, 9 to 11 ft^3	1
	√		√		Ring stands	6
√	√	√	√	√	Rulers, metric, transparent	18
	√	√	√		Scalpels	12
					Scissors:	
√	√	√	√	√	dissecting	12
√	√	√	√	√	fine-pointed	12
	√		√		Seed boxes	6
					Skeletons, mounted:	
				√	cat	1
	√				frog	2
				√	human	1
					Slides, prepared:	
	√				*Hydra*, longitudinal sections	10
					planarians—	
	√				cross-sections	10
	√				whole mounts	10
	√				earthworms, cross sections	10
			√		onion-root tips, longitudinal sections	10
					Spatulas:	
			√		porcelain	6
√	√				stainless-steel, 10 cm blade	6
√	√	√	√	√	Stoppers, rubber, various, solid, 1 and 2 hole	80
					Test-tubes:	
√	√	√	√	√	12 mm × 100 mm	60
√	√	√	√	√	16 mm × 125 mm	120
√	√	√	√	√	19 mm × 150 mm	60
√	√	√	√	√	25 mm × 200 mm	24
√	√	√	√	√	Test-tube holders	12
√	√	√	√	√	Test-tube racks	12
√	√	√	√	√	Thermometers, −10 °C to +110 °C	18
				√	Tiles, plastic, white	10 squares
	√				Trowels, garden	6
			√		Vacuum bottles	6
			√		Volumeter	6
√			√	√	Watches, with second hands	10
√			√	√	Watch glasses, solid type	24
√	√				Wire basket, 50-tube capacity	1

Appendix B Index to Formulations

Appendix C Some Suppliers

General biological equipment

Stains, chemicals and reagents. Microscopical mounting media. Microscopical preparations. Microprojectors. Entomological collections and life histories. Skeletons and dissections—biological models. Specimens for dissection and demonstration living and preserved. Colour slides—biological, zoological and botanical. Collecting apparatus—nets, etc. Chemicals. Dissecting microscopes; lens stands and all microscopical accessories, glassware, etc:

Gerrard & Haig Ltd.,
Gerrard House,
Littlehampton,
E. Preston,
Sussex.

Harris Biological Supplies Ltd.,
Oldmixon,
Weston-super-Mare.

Griffin Biological Laboratories Ltd.,
Lavender Hill,
Tonbridge,
Kent.

Arnold R. Horwell Ltd.,
2 Grangeway,
Kilburn High Road,
London NW6.

Baird & Tatlock (London) Ltd.,
PO Box 1,
Freshwater Road,
Chadwell Heath,
Romford RM1 1HA.

Gallenkamp,
Technico House,
Christopher Street,
London EC2.

Laboratory chemicals and testing outfits

Hopkin & Williams Ltd.,
Freshwater Road,
Chadwell Heath, Essex.

The British Drug Houses Ltd.,
BDH Laboratory Chemicals Division,
Poole, Dorset.

May & Baker Ltd.,
Dagenham,
Essex.

W. B. Nicholson Ltd.,
Thornliebank,
Glasgow S3.

Live insects—especially silkworms, caterpillars, moths, butterflies

Hugh Newman,
Butterfly Farm,
Bexley,
Kent.

Microscopes and accessories

R. & J. Beck Ltd.,
Greycaine Road,
Bushey Mill Lane,
Watford,
Herts.

Nikon microscopes:
The Projectina Co. Ltd.,
8 Montgomerie Terrace,
Skelmorlie,
Ayrshire,
Scotland.

Russian Technical & Optical Equipment,
263/4 High Holborn,
London WC1

C. Z. Scientific Instruments,
Zeiss England House,
93/97 New Cavendish Street,
London WC1

Vickers Instruments Ltd.,
226 Purley Way,
Croydon.

Bausch & Lomb Optical Co. Ltd.,
Aldwych House,
Aldwych,
London WC2

Gillett & Sibert,
417 Battersea Park Road,
London SW11.

E. Leitz (Instruments) Ltd.,
30, Mortimer Street,
London W1.

W. R. Prior & Co. Ltd.,
London Road,
Bishop's Stortford,
Herts.

Bacterial cultures

National Collection of Industrial Bacteria,
Department of Trade & Industry,
Torry Research Station,
135 Abbey Road,
Aberdeen.

National Collection of Plant Pathological Bacteria,
Plant Pathology Laboratories,
Hatching Green,
Harpenden,
Herts.

Audiovisual materials

Boulton-Hawker Films,
Hadleigh,
Ipswich,
Suffolk IP7 5BG.

British Transport Film Library,
Melbury House,
Melbury Terrace,
London NW1.

Canada House Film Library,
Canada House,
Trafalgar Square,
London SW1.

Concord Films Council Ltd.,
Nacton,
Ipswich,
Suffolk IP10 0JZ.

Contemporary Films Ltd.,
55 Greek Street,
London W1.

Ealing Scientific Ltd.,
Greycaine Road,
Watford WD2 4PW.

Encyclopaedia Britannica International Ltd.,
Multi Media Division,
Dolcis House,
87/91 New Bond Street,
London W1.
(Films hired from National Audio-Visual Aids Library or the Scottish Central Film Library.)

Gateway Educational Films Ltd.,
St Lawrence House,
29/31 Broad Street,
Bristol BS1 2HF.

Guild Sound & Vision Ltd.,
Woodston House, Oundle Road,
Peterborough PE2 9PZ.

John Murray (Publishers) Ltd.,
50 Albemarle Street,
London W1X 4BD.

McGraw-Hill Publishing Co. Ltd.,
Shoppenhangers Road,
Maidenhead.
(McGraw-Hill films hired from Contemporary Films Ltd.)

National Audio-Visual Aids Library,
Paxton Place,
Gipsy Road,
London SE27.

National Committee for Audio-Visual Aids in Education,
Educational Foundation for Visual Aids,
33 Queen Anne Street,
London W1M 0AL.

The Petroleum Films Bureau,
4 Brook Street,
Hanover Square,
London W1Y 2AY.

Rank Film Library,
Rank Audio Visual Ltd.,
PO Box 70,
Great West Road,
Brentford,
Middlesex.

The Royal Society for the Protection of Birds,
The Lodge,
Sandy,
Bedfordshire.

Scottish Central Film Library,
16/17 Woodside Terrace,
Charing Cross,
Glasgow C3.

Shell International Petroleum Co. Ltd.,
Trade Relations Division,
Shell Centre,
London SE1 7NA.

Time–Life Education,
Time Building,
New York,
New York 10020,
USA.

Visual Information Service Ltd.,
12 Bridge Street,
Hungerford,
Berks.

John Wiley & Sons Ltd.,
Baffins Lane,
Chichester,
Sussex.

Diana Wyllie Ltd.,
3 Park Road,
Baker Street,
London NW1.

Some useful addresses

Publication—*School Science Review:*
The Association for Science Education,
College Lane,
Hatfield,
Herts.

BBC School Broadcasting Council,
The Langham,
Portland Place,
London W1A 1AA.

Postcards and booklets:
British Museum (Natural History),
Cromwell Road,
London SW7.

Publication—*Biology and Human Affairs*:
British Social Biology Council,
69 Eccleston Square,
London SW1.

Council for Nature,
Zoological Gardens,
Regents Park,
London NW1.

Posters and leaflets on how to keep pets:
Royal Society for the Prevention of Cruelty to Animals,
105 Jermyn Street,
London SW1.

The Scout Association,
25 Buckingham Palace Road,
London SW1.

Universities Federation for Animal Welfare,
230 High Street,
Potters Bar,
Herts.

School Natural Science Society,
(Hon. General Secretary, Miss M. J. Sellers),
2 Bramley Mansions,
Berrylands Road,
Surbiton,
Surrey.

Appendix D Concerning Measurements

Scientists are continually exchanging information so it is imperative that their units should be unambiguous and must be understood with minimum explanation. Units have to be few in number and internationally acceptable. To satisfy this aim an International System of Units (SI units) was adopted by the *Conférence Générale des Poids et Mesures* in 1960.

The *metre* is the basic SI unit of length and is defined as the length equal to 1 650 763.73 wavelengths in a vacuum of a specified radiation from an atom of krypton-86.

The *kilogram* is the basic SI unit of mass and is the mass of the international platinum–iridium prototype.

The International System of units is based on the *metric* system which was the system of measurement introduced by the French government in 1790 to replace the many systems in use in France at that time. British scientists have been using the metric system for a long time and some of its units, e.g. the *litre*, although not SI units, will be acceptable for some time to come.

British literature and magazines has a confusing mixture of units and the following equivalents may be helpful for reference:

1 Calorie = 1000 calories = 4200 joules
1 metre = 39.37 inches
1 kilometre = 0.62 miles
1 litre = 1.06 quarts
1 kilogram = 2.2 pounds
1 hectare = 100 × *are* = 10 000 square metres = 2.47 acres

Prefixes can be attached to SI units and have the following meanings:

Fraction	*Prefix*	*Symbol*	*Multiple*	*Prefix*	*Symbol*
10^{-3}	milli	m	10^{3}	kilo	k
10^{-6}	micro	μ	10^{6}	mega	M
10^{-9}	nano	n	10^{9}	giga	G
10^{-12}	pico	p	10^{12}	tero	T
10^{-15}	femto	f			
10^{-18}	atto	a			

Temperature is measured in degrees Kelvin in SI units but in this series of books we have used degrees Celsius. The degree Kelvin is equal to the degree Celsius.